An Introduction to
Modern Optics

He beholds the light and
               when it flows
He sees it in his joy

— Robert Blake

Ajoy K. Ghatak
Department of Physics
Indian Institute of Technology
New Delhi

An Introduction to
Modern Optics

McGRAW-HILL BOOK COMPANY
New York   St. Louis   San Francisco   Montreal   Toronto

Library of Congress Cataloging in Publication Data

Ghatak, Ajoy K   1939-
    An introduction to modern optics.

    Includes bibliographical references.
    1. Optics.   I.   Title.
QC357.2.G48   535   71-39600
ISBN 0-07-023156-7

Published 1971 by Tata McGraw-Hill Publishing
Company Limited.

Published in the United States by McGraw-Hill
Book Company 1972. Printed and bound by
Kingsport Press, Inc.

1 2 3 4 5 6 7 8 9 0 K P K P 7 9 8 7 6 5 4 3 2

to my wife, Gopa

Preface

Most undergraduate textbooks in optics suffer from an undue amount of emphasis on details of geometrical optics and a superficial treatment of physical optics, with little attention being paid to the nature of light. Many students have remarked that we are teaching them optics as it was probably taught in the 19th Century, ignoring the recent exciting developments that they have only heard about.

In this book an effort has been made to deal with the principles of optics in the context of the relevance of these principles to other branches of physics and incorporating some of the recent developments in the field. For example, in the first chapter, the phenomenon of waves, in all its generality, has been discussed. Thus, beginning with waves on strings and going on to waves on the surface of water and sound waves (which are easier to visualize), the student is led to the generation and propagation of electromagnetic waves.

In the next chapter, the Huygens' principle and the Huygens-Fresnel principle have been discussed. That the Huygens' principle is applicable to all wave phenomena is emphasized. Geometrical optics is introduced as an approximation to wave optics and the concept of 'ray' is brought out by using Fermat principle. The validity of the geometrical optics approximation has also been discussed. Only limited space has been devoted to discussing lens systems. As pointed out by Feynman, '. . . . in dealing with a system of lenses we simply chase the ray through the succession of lenses. That is all there to it. It involves nothing new in principle, so we shall not go into it.' Hence a discussion of topics like nodal planes, principal planes etc., has been avoided.

The third chapter opens with a discussion of some basic inferences from electromagnetic theory. This is followed by a discussion of the results of various experiments which show that light is an electromagnetic wave. The 'wire-grid polarizer' is then introduced to explain the polarization of light rather than the tourmaline crystal whose mode of operation is more difficult to understand. It is then quite easy to discuss polaroids as a simple extension of the wire-grid polarizer. Polarization by reflection and double refraction has been explained from first principles. The theory of double refraction has not been dealt with in detail.

In the fourth chapter, while discussing interference, it is emphasized that the phenomenon can be explained only in terms of wave theory. The accurate measurement of small distances, the flatness of surfaces and the analysis of the structure of a spectral line have been discussed as applications of the phenomenon of interference.

The fifth chapter is on coherence — an important concept in modern optics. Temporal and spatial coherence along with some laser experiments are discussed in some detail. The phenomenon of optical beats obtained by using two different laser beams has also been discussed.

In the sixth chapter, the traditional approach to the theory of Fraunhofer diffraction has been adopted, except for the inclusion of a section on diffraction-limited optics and some laser experiments.

The next chapter, on holography, appears in an undergraduate course probably for the first time. It gives a clean application of Huygens' theory and the superposition principle.

In the last two chapters some momentous developments in physics in this century are discussed by reopening and then resolving the wave — particle controversy. An account of the experiments which led to Einstein's photon hypothesis is followed by a discussion on why radio waves are so 'wave like' and gamma rays so 'particle like'. The splitting of photons and the explanation of the interference experiment on the photon hypothesis is discussed in detail. Emphasis is laid on the fact that the photon is neither a wave nor a particle and that this is true for all atomic objects. This is followed by discussions of the probabilistic interpretation of matter-waves, interference of two different photons, ending with uncertainty relations.

A chapter on the velocity of light has been deliberately omitted as it does not fit in with the existing chapters. A chapter on the velocity of light must include consideration of the Doppler effect, phase and group velocities, the Michelson-Morley experiment, etc., which should properly be covered in a course on the special theory of relativity.

A large number of problems are appended to each chapter and only few are of the simple substitution type. Many problems involve extension of the text material and derivations. Each chapter also contains quite a few worked out examples, with the simple substitution variety used only to emphasize a numerical magnitude.

Throughout the book, an endeavour has been made to avoid giving the impression that the subject is a closed body of knowledge. This is why many references to popular scientific articles as well as to books have been included. Selected references have been listed at the end of each chapter for elaboration of the material discussed or for treatment of points not touched on. A brief evaluation of each reference is provided to give the student some guidance as to the level of the material.

This book has been developed from an earlier version which was used in teaching this part of the course to the undergraduate students of the Indian Institute of Technology, Delhi. The preliminary version was distributed to the students in the form of cyclostyled notes and the present book incorporates numerous suggestions received from the faculty members associated with the course and the students. The presentation in this book is open for further revisions, and for this reason any remarks from interested readers are welcome.

AJOY K. GHATAK

Acknowledgements

By far the most pleasant task in writing a book is the compilation of acknowledgements.

I am grateful to Professor R. N. Dogra, Director, Indian Institute of Technology, Delhi, for his encouragement and support of this work. Special thanks are due to Professor M. S. Sodha for his valuable suggestions, critical comments and numerous illuminating discussions.

As mentioned in the Preface, this book has been developed from an earlier version which was used in teaching this part of the course to our undergraduate students. The course was taught by P. K. Srivastava, D. P. Tewari, I. C. Goyal, B. K. Sawhney and myself. The tutorials were taken by A. K. Chakravarty, H. C. Gupta and P. K. Dubey. I take this opportunity to thank my colleagues, as well as the students in the course, for the many comments and suggestions which they have made. Thanks are also due to B. B. Tripathi for his help in writing the portion on X-ray diffraction (without getting into Miller indices), and to R. S. Sirohi for several illuminating discussions on Holography. I am grateful to H. K. Sehgal and R. N. Singh for drawing my attention to some useful references. I also wish to thank my students D. P. S. Malik, Om Pal Singh, N. K. Bansal, Shashi Bala, Bimal Prasad, and S. Shobhana and my sister, Rita Ghatak, for their untiring help in checking the text and correcting the proofs.

Thanks are also due to V. N. Sharma, O. P. Virmani, T. R. Sharma, Jarnail Singh and T. N. Gupta for their help in the preparation of this manuscript as well as the earlier version and Balbir Singh for drawing the figures.

Thanks are also due to the various publishers for granting permission to reproduce some of the plates included in this volume.

Finally, I am indebted to my wife for reading through the manuscript and improving so many of my more tortured sentences and for her encouragement throughout the endless hours that have gone into the writing of this book.

AKG

Contents

Preface   vii

Introduction   xvii

Chapter 1 Waves   *1*

  1-1 Introduction   1
  1-2 The Definition of a Simple Wave in One Dimension   2
  1-3 Sinusoidal Waves: Concept of Frequency and Wavelength   7
  1-4 Phase   12
  1-5 Types of Waves   13
  1-6 General Wave Motion   13
  1-7 Energy Transport in Wave Motion   15
  1-8 Electromagnetic Waves   17
  1-9 Propagation of Electromagnetic Waves   20
  1-10 Generation of Electromagnetic Waves   21
  1-11 Composition of Waves: The Superposition Principle   23
  1-12 Reflection of Waves   24
  1-13 Standing Waves   27
  1-14 Stationary Light Waves—Wiener's Experiments   31
  1-15 Superposition of Two Sinusoidal Waves   32
  1-16 The Graphical Method of Adding Disturbances of the Same Frequency   34
  1-17 Superposition of Many Simple Harmonic Vibrations along the Same Line 35
  1-18 Nearly Equal Frequencies in the Same Straight Line   38

      Suggested Reading   40
      Problems   40

Chapter 2 Huygens' Principle and Geometrical Optics   *44*

  2-1 Introduction   44
  2-2 The Wave Model   45
  2-3 Huygens' Theory   46
  2-4 Huygens' Principle   47
  2-5 Rectilinear Propagation   50
  2-6 Effect of a Plane Wave on an Exterior Point—Fresnel's Half-period Zones
      and the Phenomenon of Diffraction   51
  2-7 Zone Plate   55

2-8   Huygens' Principle and the Law of Reflection   57
2-9   Huygens' Principle and the Law of Refraction   58
2-10 Reflection of Light from a Point near a Mirror   62
2-11 Refraction of a Spherical Wave by a Spherical Surface   63
2-12 Ray Treatment of Light: Geometrical Optics   64
2-13 Fermat's Principle   65
2-14 Applications of Fermat's Principle   67
2-15 Applications of Geometrical Optics   69
2-16 Refraction at a Spherical Surface   70
2-17 Thin Lenses   72
2-18 Compound Lenses   74
2-19 Aberrations   75
2-20 Optical Instruments   78
2-21 The Magnifying Glass   79
2-22 Compound Microscope   80
2-23 Astronomical Telescopes   81
2-24 Dispersion   89

      Suggested Reading   90
      Problems   91

Chapter 3   Electromagnetic Character of Light and Polarization   95

3-1   Introduction   95
3-2   Some Basic Consequences of Electromagnetic Theory   97
3-3   A Representation of an Electromagnetic Wave   97
3-4   Hertz's Experiments   98
3-5   Standing Electromagnetic Waves   100
3-6   Reflection and Refraction of Electromagnetic Waves   103
3-7   Energy and Intensity of an Electromagnetic Wave   105
3-8   Momentum of an Electromagnetic Wave   108
3-9   The Electromagnetic Spectrum   110
3-10 Describing an Electromagnetic Disturbance   112
3-11 Superposition of two Disturbances   113
3-12 Unpolarized Light   118
3-13 Production of Polarized Light   119
3-14 The Wire Grid Polarizer   119
3-15 H-Sheet: World's Most Popular Polarizer   122
3-16 Percentage Polarization and Malus' Law   122
3-17 Polarization by Reflection   124
3-18 Interpretation of Brewster's Law   126
3-19 Polarization of Scattered Light   127
3-20 Birefringence   128
3-21 Applications of Birefringence   131
3-22 Anomalous Refraction   135
3-23 Analysis of Polarized Light   140

      Suggested Reading   141
      Problems   142

Chapter 4   Interference   145

4-1   Introduction   145

4-2   Interference of Waves Originating from two Point Sources   147
4-3   Nodal Lines   149
4-4   Determination of Wavelength and Fringe Width   150
4-5   Phase   152
4-6   Interference in Light   155
4-7   Intensity Distribution   160
4-8   Determination of Wavelength   163
4-9   Interference with White Light   164
4-10   Displacement of Fringes   166
4-11   Other Apparatus for the Production of Interference Pattern   170
4-12   Phase Change on Reflection   172
4-13   Interference Involving Multiple Reflections   174
4-14   Reflection from a Plane Parallel Film   174
4-15   Non-reflecting Films   180
4-16   Newton's Rings   182
4-17   Application   183
4-18   Michelson Interferometer   184
4-19   Uses of the Michelson Interferometer   186

Suggested Reading   188
Problems   188

Chapter 5   Coherence   **193**

5-1   Introduction   193
5-2   Temporal Coherence: Concept of Path Length   194
5-3   The Purity of a Spectral Line   196
5-4   Spatial Coherence   199
5-5   Michelson's Stellar Interferometer   202
5-6   An Experiment with a Laser Beam   204
5-7   Optical Beats   205

Suggested Reading   209
Problems   209

Chapter 6   Diffraction   **211**

6-1   Introduction   211
6-2   Calculation of Single Slit Diffraction Pattern   213
6-3   The Diffraction Pattern   216
6-4   A Possible Home Experiment   220
6-5   Angular Width of a Diffraction Limited Beam   222
6-6   Application: Laser Beam versus Flashlight Beam   223
6-7   Calculation of the Half Width of the Principal Maximum in the Single Slit Diffraction Pattern   223
6-8   Some General Observations   224
6-9   Diffraction by a Rectangular Aperture   225
6-10   Diffraction by a Circular Aperture   226
6-11   Diffraction Limited Optics   229
6-12   Limit of Resolution   231
6-13   Limit of Resolution for Optical Instruments   232
6-14   Angular Resolution of the Human Eye   235
6-15   Diffraction Pattern produced by a Double Slit   235
6-16   Derivation of the Equation for Intensity   236

6-17 Position of the Maxima and Minima — 239
6-18 The Intensity Distribution from N Parallel Slits — 243
6-19 Principal Maxima — 244
6-20 Minima and Secondary Maxima — 245
6-21 The Diffraction Grating — 247
6-22 Dispersion — 248
6-23 Widths of the Principal Maxima — 251
6-24 Resolving Power of a Grating — 251
6-25 X-ray Diffraction — 253

Suggested Reading — 260
Problems — 260

Chapter 7 Photography by Coherent Light— Holography 264

7-1 Introduction — 264
7-2 Principle — 266
7-3 Theory — 268
7-4 Reconstruction — 270
7-5 Holograms — 271
7-6 Requirements — 272
7-7 Holography with Sound — 275

Suggested Reading — 276

Chapter 8 Particle Nature of Radiations: Photons 277

8-1 Introduction — 277
8-2 The Photoelectric Effect — 277
8-3 Einstein's Photon Theory — 282
8-4 Photon Momentum and Compton Effect — 283
8-5 Rest Mass of the Photon — 289
8-6 Splitting of Photon — 289
8-7 Photons and Electromagnetic Waves — 291
8-8 Matter Waves: Wave Particle Duality for Other Atomic Objects — 293

Suggested Reading — 296
Problems — 297

Chapter 9 Quantum Behaviour of Photons and Uncertainty Principle 298

9-1 Introduction — 298
9-2 Quantum Behaviour of Photons — 299
9-3 Experiment No. 1: Interference Experiment with Bullets — 299
9-4 Experiment No. 2: Interference Experiment with Water Waves — 301
9-5 Experiment No. 3: Interference Experiment with Photons — 303
9-6 Watching the Photons — 307
9-7 The Probabilistic Interpretation of Matter Waves — 310
9-8 Explanation of the Interference Pattern produced by an Apparatus like Michelson's Interferometer — 311
9-9 Interference of Two Different Photons — 313
9-10 Looking at Pictures — 314

9-11 The Uncertainty Principle   315

Suggested Reading   321
Problems   321

Appendix A *Some Mathematical Considerations of Wave motion*   325

      B *Simple Harmonic Motion*   337

      C *Electromagnetic Oscillators*   348

      D *Lasers*   351

      E *Optical Activity*   355

      F *Resolving Power of a Microscope*   357

      G *Miller Indices*   360

        Author Index   363

        Subject Index   365

Plates

1 : Diffraction through an aperture
2 : Interference pattern from two point sources
2 : Young's interference fringes
4 : Newton's rings
5 : Interference pattern produced by a thin non-uniform film
6 : Fringe pattern in a Michelson's interferometer
7 : Diffraction by a square aperture
8 : Diffraction by a circular aperture
9 : Images of two point sources
10 : Fraunhofer diffraction pattern
11 : Hologram and its reproduction
12 : Randomness in a picture due to arrival of individual photons

Introduction

And God said, "Let there be light"; and there was light. And God saw
that the light was good; and God separated the light from darkness.

— *Genesis*

The answer to the question 'What is light?' has changed several times in
the last three centuries. Each time the answer has assumed more fundamental importance in the physicist's picture of the universe.

Isaac Newton described light as a stream of corpuscules. To Newton,
the lack of evidence of diffraction was proof that light cannot be a wave.
He explained his experiments associated with color phenomena in glass
plates (the so-called Newton's Rings) by assuming 'fits of easy reflection
and easy transmission' of the light beam. So compelling was his authority,
that his successors were more convinced to the corpuscular theory of
light than Newton himself.

The opponents of Newton's corpuscular hypothesis led by Christiaan
Huygens, impressed by the lack of evidence for interaction between two
light beams which traverse the same portion of a medium simultaneously,
put forward the wave model of light. However, the phenomena of reflection and refraction could be equally well explained by either the
corpuscular theory or wave hypothesis.

An important experiment performed in 1803 by Thomas Young seemed
to settle the controversy. Young showed that when a monochromatic
beam of light passed through two pinholes, an interference pattern is
set up which resembled the patterns observed for 'water waves' and

'sound waves.' From this experiment the wavelength of the light waves was determined and its value was found to be very small compared to the sizes of typical macroscopic objects*. Thus, rectilinear propagation was explained and colour proved to be related to the wavelength.

At about the same time Augustin Jean Fresnel and Dominique Francois Arago showed that the light transmitted by Huygens' blocks of calcite crystal is polarized and that light waves cannot be longitudinal compression waves but must be transverse waves oscillating at right angles to their direction of propagation.

The nature of the wave phenomenon and of the medium which supports it remained a mystery until, later in the 19th Century, James Clerk Maxwell propounded the electromagnetic theory of light. Hertz confirmed Maxwell's theory by demonstrating the reality of much longer electromagnetic waves radiated from electrical circuits and showing their qualitative similarity to light. At about the same time the experiment of Michelson and Morley disproved the concept of an all-pervading solid medium in which the elastic vibrations of Fresnel's theory could be supported.

Maxwell's recognition of light being electromagnetic waves was one of the greatest syntheses in physics. Maxwell's theory unified the description of a great diversity of phenomena and enabled physicists to predict with confidence the existence and properties of previously unknown forms of electromagnetic radiation. By 1890, the tremendous success of Maxwell's electromagnetic theory of light appeared to many to be a sign that the task of fundamental science was nearing completion.

However, the momentous developments in physics in the 20th Century showed that such wave properties as interference and diffraction, so well demonstrated by light, are also exhibited under suitable circumstances by the atomic objects such as electrons. Further, it was also shown that light, in its interaction with matter, behaves as though it is composed of many individual corpuscles called photons. Thus, Planck and Einstein explained the spectrum of radiation from incandescent substances, photoelectric effect and the Compton effect, and the combination principle of atomic spectra by reviving the corpuscular hypothesis; now, however, this had to be done in a way which could be reconciled with the firmly established wave theory of light.

Modern quantum theory is the result of this synthesis. As long as one is interested in the propagation of an electromagnetic wave, it is permissible to use Maxwell's theory in which intensities are continuous functions of space and time. However, in any interaction of light with matter, the experiments show that the exchanges of energy and momentum are

---

* It is for this reason that while designing simple optical instruments the propagation of light can usually be described, to sufficient accuracy, by geometrical optics, the approximation in which diffraction is unimportant and only the direction of energy transport is of concern.

not continuous; they occur only in discrete steps (or multiples thereof) of magnitude proportional to the frequency of the wavefield. However, even for the violet end of the spectrum, the frequencies are of the order of 10^{15} sec^{-1} and each quantum step is so minute that a vast number of individual quantum exchanges will be involved in the transfer of even small amounts of energy. One is therefore led to believe that electromagnetic wave is truly corpuscular, and the field intensities in the wave merely expresses probabilities for the presence of energy quanta or 'photon.' In the first half of the present century the work of Louis de Broglie, Erwin Schrodinger, Werner Heisenberg, Paul Dirac and many others has developed quantum theory to a highly sophisticated mathematical structure to interpret quantitatively the interaction of light with matter. However, quantum theory principally consists of rules of computations, the justification for which rests mainly on their success, rather than on an underlying physical content.

We conclude by noting that among the newly observed effects* which call for new refinements in optical theory are the amplification of light, the mixing of light waves to create new frequencies, the generation of sound waves by light, the ionization of atoms by visual or even infrared light, etc. These effects come under the study of 'non-linear optics.'

* These topics are not discussed in this book.

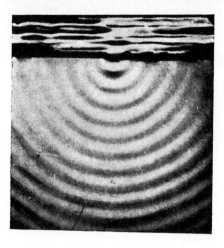

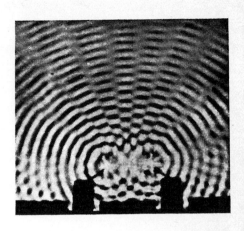

PLATE 1: Plane waves exhibit a large amount of diffraction in passing through an aperture of about the same size as their wavelength. The picture corresponds to waves in a ripple tank.

PLATE 2: A photograph of the interference pattern from two point sources in phase. Notice the nodal lines radiating outward. (From PSSC PHYSICS, Second Edition, D. C. Heath and Company, Lexington, 1965.)

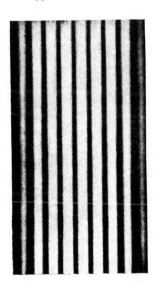

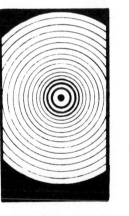

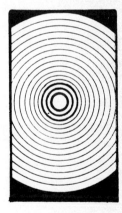

(a)

(b)

PLATE 4: Schematic diagram of Newton's rings. If the convex surface of a lens is placed in contact with a plane glass plate (Fig. 4-28) a thin film of air is formed between the two surfaces; and, since the loci of points of equal thickness are circles (concentric with the point of contact) a system of rings is seen around the point of contact. These are Newton's rings shown in (a). The same phenomenon is also seen with transmitted light but with a much smaller contrast (b). (After *Fundamentals of Optics*, by F. A. Jenkins and H. E. White, III Ed. McGraw-Hill Book Company, 1957.)

PLATE 3: Interference fringes produced by Young's slits using monochromatic light.

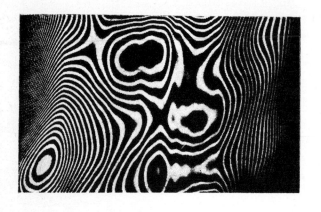

PLATE 5: If two surfaces, placed in contact with one another, are accurately plane, the entire area of contact will be dark. When the film surfaces are not flat, a system of fringes is seen as shown. Each fringe describes a locus of equal thickness of the film. Between two fringes the change in thickness is $\lambda/2\mu$, where μ is the refractive index and λ the wavelength of the light used. (From *Atlas of Optical Phenomena*, by M. Cagnet *et al.*, Berlin-Göttingen-Heidelberg: Springer, 1962.)

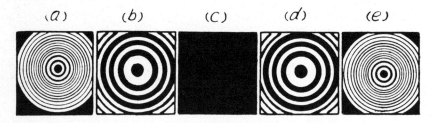

PLATE 6: Schematic diagrams of the various types of fringes observed in Michelson's interferometer. (After *Fundamentals of Optics*, by F. A. Jenkins and H. E. White, III Edition. McGraw-Hill Book Company, 1957.)

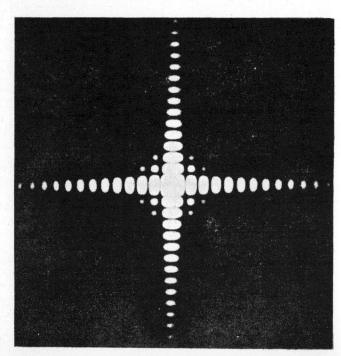

PLATE 7. Diffraction pattern produced by a square aperture. (From *Atlas of Optical Phenomena* by M. Cagnet *et al.*, Berlin - Göttingen - Heidelberg: Springer, 1962.)

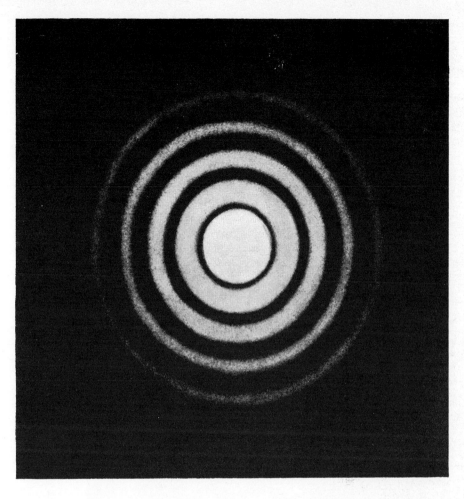

PLATE 8: Diffraction rings produced when plane waves are diffracted by a circular aperture. (From *Atlas of Optical Phenomena*, M. Cagnet *et al.*, Berlin-Göttingen-Heidelberg: Springer, 1962.)

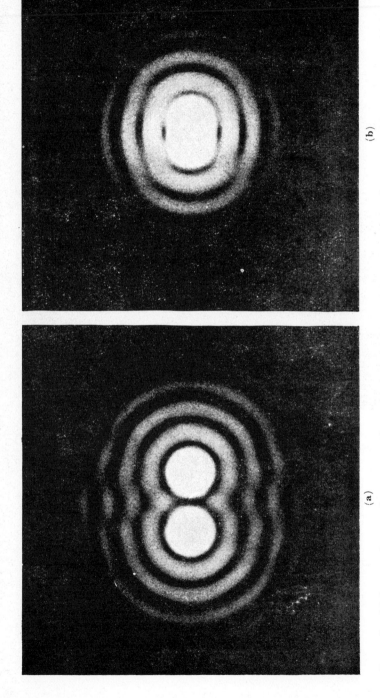

(a)

(b)

PLATE 9: (a) Images of two point sources with first dark ring tangent, (b) images of two point sources which are just resolved according to Rayleigh's criterion. (From *Atlas of Optical Phenomena*, M. Cagnet *et al.*, Berlin-Göttingen-Heidelberg: Springer, 1962.)

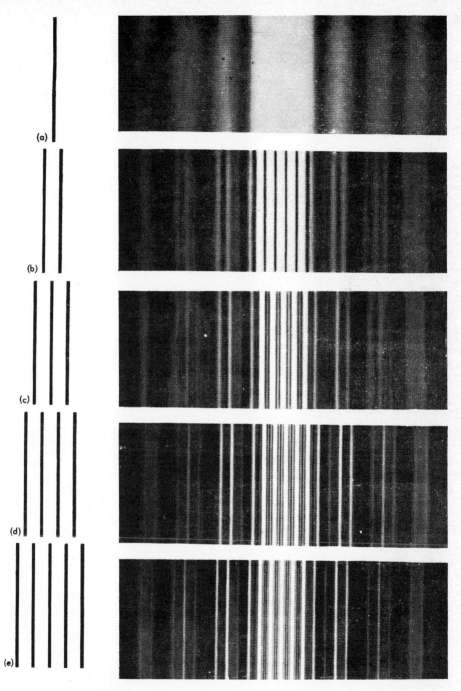

PLATE 10: (a) Fraunhofer diffraction pattern of a single slit; (b), (c), (d), (e) represent diffraction patterns of 2, 3, 4 and 5 slits respectively. (From F. W. Sears, *Optics*, Third Edition, 1949, Addison-Wesley, Reading, Mass., USA.)

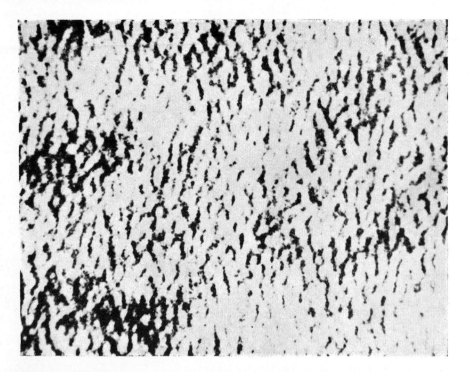

(b)

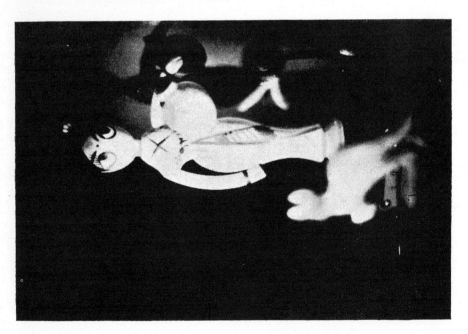

(a)

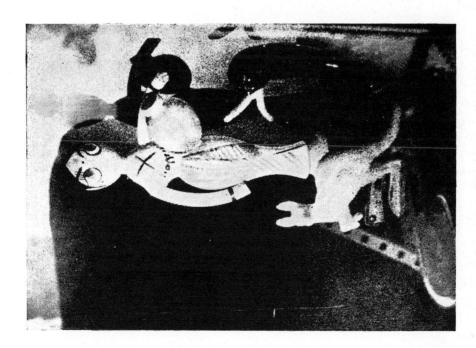

PLATE 11: (a) Ordinary photograph is made by illuminating the object (here an Indian doll) with normally incoherent light and recording a two dimensional image of the scene on a photographic film. Reflected light is focussed on a film by a camera lens. (b) Hologram recording of the scene shown in (a) using the process of wave-front construction photography. The visible structure of the hologram bears no resemblance to the original scene than would be contained in an ordinary photograph. (c) Reconstructed image was made by directing a laser beam through the hologram. The reconstructed wave were then passed through a lens and brought to a focus, thereby forming an image of the original scene, even though the object had long since been removed. (From 'Holography: Theory and Applications', by D. Venkateswarulu, *Journal of Scientific & Industrial Research*, Vol. 29, November 1970, © Council of Scientific & Industrial Research. Used by permission.)

(c)

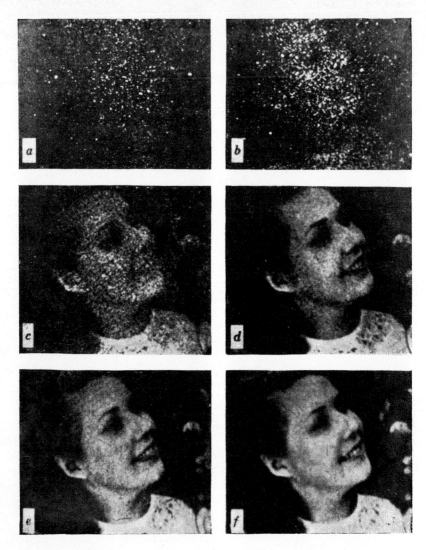

PLATE 12: Series of photographs showing the quality of picture obtainable from various number of photons. (a), (b), (c), (d) and (e) correspond to 3×10^3 photons, 1.2×10^4 photons, 9.3×10^4 photons, 7.6×10^5 photons, 3.6×10^6 photons and 2.8×10^7 photons respectively (From A. Rose *Quantum effects in Human Vision, Advances in Biological and Medical Physics*, Vol. V, Academic Press, 1957, © Academic Press.)

Waves

> 'If you are dropping pebbles into a pond and do not watch the spreading rings, your occupation should be considered as useless,' said the fictional Russian philosopher, Kuzma Prutkoff. And, indeed we can learn much by observing these graceful circles spreading out from the punctured surface of calm water.
>
> —G. Gamow and J. M. Cleveland, *Physics: Foundations and Frontiers*, p. 115

1-1 Introduction

In this chapter we shall discuss the phenomenon of waves. This phenomenon appears in almost every branch of physics; and, in a sense, everybody knows what a wave is. We all know that if a small stone is dropped into a quiet pool of water, a circular pattern spreads out from the point of impact. Such a disturbance is called a *wave*, and if we watch the water closely we will find that as the wave moves across the surface, the water does not m ove forward with the wave. This can easily be shown if we watch a small piece of wood or a patch of oil that may be floating on the pool. The wood (or the patch of oil) moves up and down as the wave passes; it does not travel along with the wave. In other words, a wave can travel for long distances but, once the disturbance has passed, every drop of water is left where it was before.

Si milarly we are all familiar with the fact that, when a vertical string under tension is set oscillating back and forth at one end, a disturbance moves along the string with the string particles vibrating at right angles to the direction of propagation of the disturbance. We also know that sound is conveyed to our ears by means of waves and that our radios

1

detect radio waves. However, we cannot see radio waves and sound waves in the same way as we can see water waves and waves on strings. We also learn that light consists of waves but there is no obvious connection between waves and the sensation of vision.

Now, when we observe water waves we see the crests and troughs, regularly spaced, and the distance between successive crests, which we call the wavelength, appears as a fundamental property of the wave motion, as does also the speed with which the crests and troughs travel over the surface of water. Radio waves are not so evident to our senses, but we can detect such properties as wavelength and velocity of these by means of special techniques. Similarly, with the help of suitable apparatus, we can measure the wavelength and speed of light waves and of sound waves. The theory of wave motion, therefore, forms a descriptive system appropriate, with minor modifications, to sound and light, as well as to waves propagated along the surface of a liquid. In the general account of wave theory given in this chapter, and in most of the applications discussed in later chapters, it is not necessary to specify the detailed physical properties of the disturbance which represents the wave. Often it is even a matter of indifference whether the disturbance considered is a scalar quantity, like the pressure of a gas, or a vector quantity, like electric or magnetic field vectors.

The theory of wave motion can be said to involve three important concepts:

(*a*)   There is some physical property which, at any given instant, has a defined and measurable value at every point (this property may be displacement of a particle, or an electric or magnetic field vector).
(*b*)   The value of this property at any given point undergoes a periodic fluctuation.
(*c*)   A disturbance at one point at a given time produces a similar disturbance at a neighbouring point at a slightly later time so that the pattern of the disturbance is continuously transferred from one place to another.

1-2   The Definition of a Simple Wave in One Dimension

Restricting our attention initially to waves in one dimension, we begin by considering the propagation of a wave on a rope.

Imagine yourself holding one end of a long rope and your friend holding the other end without letting the rope sag. If you move your end of the rope up and down a few times, you will generate a short train of waves (Fig. 1-1). These travel down the rope toward your friend, maintaining roughly the shape of the train as it first appeared. This ability to travel along the string without changing shape is a crucial characteristic of what we call a wave. It is apparent that in this experiment, the rope itself

))) 1

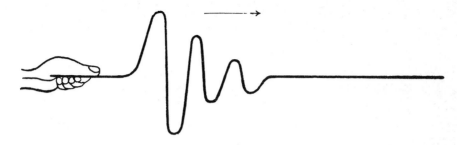

Fig. 1-1: Wave on a rope.

does not move forward along with the wave; as is obvious from the fact that however long you may vibrate the rope, it will never get away from your hand. Thus, *a wave is the propagation of a disturbance through a medium without change in form and without any translation of the medium in the direction of propagation of the waves.*

Further, it takes a certain amount of your energy to get the wave started. When the wave reaches the other end of the rope, your friend feels the rope exerting forces on his hand. If his hand moves, work is done on it. This indicates that the energy you put into generating the wave has reached him. *Thus, the wave carries energy.*

Let us follow the propagation of the disturbance as it travels down the rope: of course in most real ropes the wave gets smaller as it progresses; it also changes shape somewhat. However, we will suppose that we have an ideal rope in which such tendencies are negligibly small. Such a rope can in fact be made, so the idealization is not entirely artificial.

Then, how do we define a wave? In our example, it is a change of position of the rope so that it lies along a new curve. However, this new configuration of the rope changes with time in a very special way, and we will now try to describe this change in configuration with time.

We have already stated that the wave travels along the rope without change of shape; therefore, if we take a photograph of the rope at a particular time (say at $t = t_0$) and another one later (say at $t = t_0 + \Delta t$), then the two photographs will show similar geometrical structure except for the fact that one will show displacement of this structure in relation to the other (Fig. 1-2).

Now, let us assume the rope to be extremely long and let there be two persons; the first person always standing at position A; and the second person, starting at B, carrying a metre stick and running along the rope (i.e. in the $+ x$ direction) at such a speed that the displacement of the rope at every point indicated on the metre stick remains the same (Fig. 1-2). (This speed is called the *wave speed* and is usually denoted by v). Thus, the displacement of every point on the rope can be described *as a function which depends only on the x-coordinate as measured by the running man.*

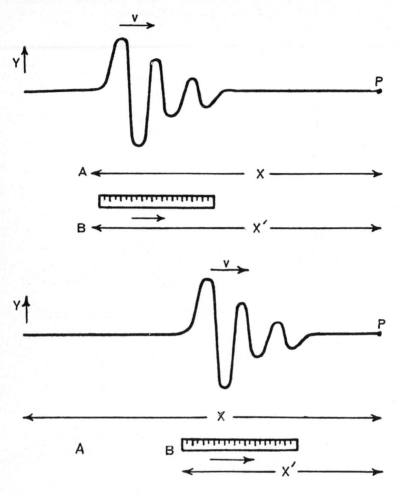

FIG. 1-2: Successive 'snapshots' of a disturbance propagating in the direction of the arrow.

If x and x' denote the coordinates of any point P as measured by A and B respectively, then, as time goes on, the value of x remains the same whereas x' goes on decreasing (see Fig. 1-2). In fact

$$x' = x - vt \qquad (1\text{-}1)$$

Now at $t = 0$, let $y(x')$ represent the displacement of each point on the rope as measured by B. We have already shown, that with respect to the moving person, the displacement of each point remains the same. Thus, the shape of the displacement remains $y(x')$. Obviously, for a person who is not running, the displacement will be of the type

$$y\,(x - vt) \qquad (1\text{-}2)$$

where y can be any arbitrary function of its argument. Eq. 1-2 describes the most general form of a wave propagating in the $+x$ direction in which x and t do not appear independently but only as $x-vt$. Examples of such functions are $\mathrm{Sin}\ k\ (x-vt)$, $\sqrt{R^2-(x-vt)^2}$, $\exp\left[-\alpha\ (x-vt)^2\right]$, $\exp\left[-\beta\ (x-vt+\gamma)^2\right]$, $1/[(x-vt)^2+x_0^2]$ etc.*, where k, R, α, β, γ and x_0 are constants and independent of x and t. It should be noted that in these functions also x and t do not appear independently but only as $x-vt$. That the above functions do represent a disturbance propagating in the $+x$ direction with speed v can easily be shown by noting that an arbitrary function of $x-vt$ can be written in the form

$$y\ (x-vt) = y\ (x_0 + \Delta x - \Delta x - vt_0)$$
$$= y\left[x_0 + \Delta x - v\ (t_0 + \frac{\Delta x}{v})\right] \tag{1-3}$$

Eq. 1-3 shows that the value of the function y for $x=x_0+\Delta x$ at time $t=t_0+\Delta x/v$ is identical to the value of the function for $x=x_0$ at time $t=t_0$. This indicates that the wave has moved through a distance Δx in time $\Delta x/v$. In other words, the disturbance is said to be propagating in the $+x$ direction with speed $\Delta x/(\Delta x/v) = v$.

In an exactly similar manner, the wave $y\ (x+vt)$ represents† a disturbance propagating in the $-x$ direction with speed v. This can easily be shown by noting that an arbitrary function of $x+vt$ can be written in the form

$$y\ (x_0+vt_0) = y\ (x_0-\Delta x + \Delta x + vt_0)$$
$$= y\left[x_0-\Delta x + v\ (t_0 + \frac{\Delta x}{v})\right] \tag{1-4}$$

Eq. 1-4 shows that the value of the function y for $x=x_0-\Delta x$ at time $t=t_0+\dfrac{\Delta x}{v}$ is identical to the value of the function for $x=x_0$ at time $t=t_0$. Thus the wave has moved through a distance $-\Delta x$ in time $\dfrac{\Delta x}{v}$ indicating that the wave is propagating in the $-x$ direction with speed v.

---

* $\exp[y] \equiv e^y$. Those coming across the term e for the first time may just consider this as a pure number equal to 2·7183. However, this number, which is actually defined as the sum of an infinite series

$$e \equiv 1 + \frac{1}{\underline{|1}} + \frac{1}{\underline{|2}} + \frac{1}{\underline{|3}} + \dots$$

has many interesting properties. Two of them are

(i) $\exp(x) \equiv e^x = 1 + x + \dfrac{x^2}{\underline{|2}} + \dfrac{x^3}{\underline{|3}} + \dots$

(ii) $e^{i\theta} = \mathrm{Cos}\ \theta + i\ \mathrm{Sin}\ \theta$, $i = \sqrt{-1}$

† See Appendix A for a discussion on wave equation [The general solution of the one-dimensional wave equation is of the form $y\ (x \pm vt)$].

Example

Study the propagation of a semicircular pulse (propagating along a string) for which the instantaneous displacement $y(x, t)$ is given by

$$\left.\begin{array}{ll} y(x, t) = +\sqrt{\{R^2 - (x - vt)^2\}}, & \text{when } x - vt < R \\ \quad\quad = 0 & \text{when } x - vt > R \end{array}\right\} \quad (1\text{-}5)$$

Solution

We first note that the positive sign in front of the radical indicates the fact that the displacements lie on the *upper* half of the semicircle. Now at $t = 0$, the centre of the semicircle is at the origin (i.e. $x = 0$); and at a later time, say at $t = t_0$, the displacement will be given by

$$y(x, t_0) = \sqrt{R^2 - (x - vt_0)^2} \quad (1\text{-}6)$$

indicating that the centre of the semicircle has shifted by a distance vt_0 in time t_0 (see Fig. 1-3).

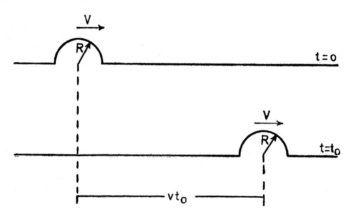

FIG. 1-3: Propagation of a semicircular pulse along a string in the $+x$ direction.

Alternatively, if it is given that a semicircular pulse is propagating in the $+x$ direction with speed v and at $t = 0$, the displacement has the form

$$y(x, t = 0) = +\sqrt{R^2 - (x - x_0)^2} \quad (1\text{-}7)$$

then, at a later time t, the displacement would be given by

$$y(x, t) = \sqrt{R^2 - (x - x_0 - vt)^2} \quad (1\text{-}8)$$

This immediately follows from the fact that $y(x, t)$ has to be of the form $y(x - vt)$ and at $t = 0$, $y(x, t)$ must be given by Eq. 1-7.

Example

A pulse is propagating in the $-x$ direction with speed v. Its shape at $t = t_0$ is of the form:

$$y (x, t = t_0) = a \exp [-\alpha (x - x_0)^2] \qquad (1\text{-}9)$$

(Such a pulse is known as a Gaussian pulse.)
What will be its shape at an arbitrary time t and at $t = 4t_0$?

Solution

Since the pulse is propagating in the $-x$ direction, its shape at an arbitrary time t would be given by

$$y (x, t) = a \exp [-\alpha (x + vt - vt_0 - x_0)^2] \qquad (1\text{-}10)$$

(Obviously at $t = t_0$, Eq. 1-10 becomes Eq. 1-9).

At $t = 4t_0$

$$y (x, t = 4t_0) = a \exp [-\alpha (x + 3vt_0 - x_0)^2] \qquad (1\text{-}11)$$

Eqs. 1-9 and 1-11 are plotted in Fig. 1-4. The figure shows the propagation of the disturbance in the $-x$ direction with speed v.

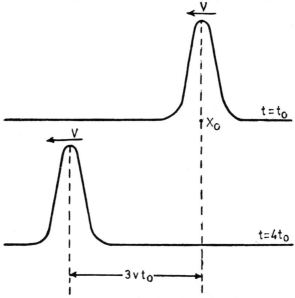

Fig. 1-4: Propagation of a Gaussian pulse along a string in the $-x$ direction.

1-3 Sinusoidal Waves: Concept of Frequency and Wavelength*

So far we have confined our discussion to waves which are produced by a disturbance which lasts only for a short period of time. Such a wave of short duration is called a pulse.

---

* This section assumes a knowledge of simple harmonic motion, which happens to be the simplest and most fundamental of the many types of vibratory motion which may give rise to waves. Students who are not familiar with simple harmonic motion must study Appendix B before reading this section.

Some waves are periodic in nature, i.e. the disturbance repeats itself periodically. Regular periodic wavetrains, as shown in Figs. 1-5 and 1-6,

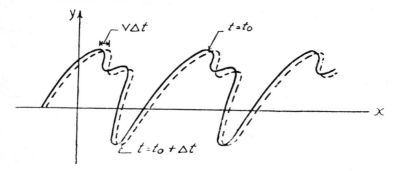

FIG. 1-5: The propagation of a general periodic wavetrain in the $+x$ direction. The solid and the dotted curves give the displacement as a function of x at $t = t_0$ and $t = t_0 + \Delta t$ respectively.

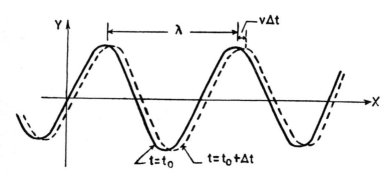

FIG. 1-6: The propagation of a sinusoidal wavetrain in the $+x$ direction. The solid and the dotted curves give the displacement as a function of x at $t = t_0$ and $t = t_0 + \Delta t$ respectively.

are excited by sources having a regular, oscillatory motion of some definite frequency such as a vibrating tuning fork or a vibrating rod producing ripples at a water surface. For a periodic wave, if the displacement y has the form

$$y = a \, \text{Sin} \, [k \, (x + vt)] \tag{1-12}$$

or, to be more general, if

$$y = a \, \text{Sin} \, [k \, (x + vt) + \phi] \tag{1-13}$$

then such a wave is known as a *sinusoidal wave*. The wave form produced by a tuning fork vibrating in air is very nearly sinusoidal, i.e. we can to a very good approximation represent the experimentally observed waveforms by a sine or cosine function. Another example is a stretched string, one end of which is made to vibrate in simple harmonic motion with amplitude a and frequency ν. A wave travels along the string with a certain speed v. Each particle of the string vibrates in a simple harmonic motion* with a common frequency ν, but each successive particle has a later phase of vibration. Now, if the source makes ν complete oscillations in one second, it will emit ν complete waves in one second. If we take up a point of observation in the medium, we will observe ν complete waves (i.e. ν crests or ν troughs) pass by in each second. Thus the frequency ν associated with the periodic wavetrain is identical with that of the source. The period T of the wave is the time taken by one complete wave to pass a given point in space (obviously $T = 1/\nu$). The wavelength, λ, is the distance from one crest to the next or the distance between two consecutive points which have identical displacements at a particular instant (Fig. 1-6). Further, if we watch a crest that passes a particular point A at $t = 0$ (Fig. 1-7), then one second later the crest will have advanced to position B, a distance v away from A. Since ν complete waves pass by each point in one second, the space AB will contain ν wavelengths.

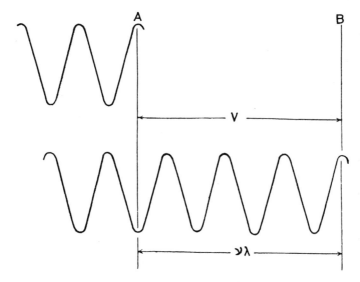

FIG. 1-7: Wavecrest, initially at point A, advances through a distance v in one second. Length $AB = v$ contains ν wavelengths.

Therefore

$$v = \nu\lambda \qquad (1\text{-}14)$$

We shall now study in some detail the passage of a sinusoidal wave

---

* It is important to point out that as the wave progresses any point on the string is executing simple harmonic motion because the motion imparted to the end is simple harmonic. This illustrates the general principle that a wave of any type can be generated by exciting, at any point in the wavefield, just the motion which is proper to the wave.

through a string in which the wave motion is maintained by the vibration of the particle at O (Fig. 1-8). Let us consider the motion of a particular particle of the string, say at point P. At $t = 0$, it is in the position

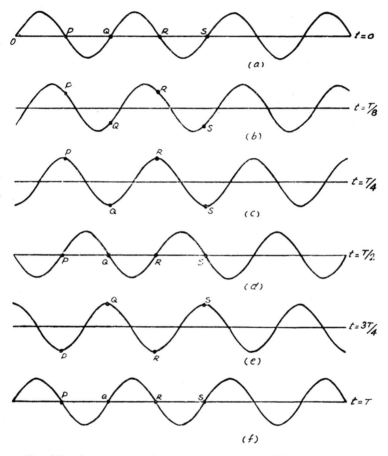

FIG. 1-8: Successive snapshots of a string through which a transverse sinusoidal wave is propagating.

of zero displacement and is moving upwards; at $t = T/4$, it has reached the position of maximum displacement in the upward direction; at $t = T/2$, it has reached the position of zero displacement once more, but this time it is moving in the downward direction; at $t = 3T/4$, it has reached the position of maximum displacement in the downward direction, and at $t = T$, it is once more in the state in which it started, i.e. at the position of zero displacement moving upwards. So while one complete wave passes the point in space normally occupied by the particle P, that particle executes one complete vibration; and, in fact, the passage of the wave merely involves the vibration of the particle without any translation in the direction of the wave. If we now consider another particle, say at Q, we see that the same thing is true in this case also. It just executes vibrations as the wave passes along, one complete vibration occupying the same time as one complete wave takes to pass a given point in space. But the particles P and Q do not vibrate in phase, i.e. they do not reach the point of maximum upward displacement simultaneously. On the other hand, particles P and R do vibrate in phase and similarly particles R and S also vibrate in phase.

We can derive a mathematical relation for the wave motion (i.e. Eq. 1-13) from a consideration of the waves of Fig. 1-8. As shown in Fig. 1-8, we take the time $t = 0$ at the instant the vibrating particle at $x = 0$ is at zero displacement and moving downward. At this instant the shape of the string will be obtained by expressing the displacement of each particle of the string as a function of x:

$$y = a \, \mathrm{Sin} \, \frac{2\pi x}{\lambda} \tag{1-15}$$

At a time one eighth of a period later, i.e. at $t = T/8$, the displacement is given by

$$y = a \, \mathrm{Sin} \left(\frac{2\pi x}{\lambda} - \frac{\pi}{4} \right) \tag{1-16}$$

and at $t = T/4$,

$$y = a \, \mathrm{Sin} \left(\frac{2\pi x}{\lambda} - \frac{\pi}{2} \right) \tag{1-17}$$

In general, the displacement at any point x and at any time t is given by the equation

$$y = a \, \mathrm{Sin} \left(\frac{2\pi}{\lambda} x - \frac{2\pi t}{T} \right) \tag{1-18}$$

which is precisely of the same form as Eq. 1-12. The above equation represents a harmonic wave of amplitude a travelling in the *positive x* direction.

If we recall that $\nu = 1/T$ and $v = \nu\lambda$, we can write the equation of the wave in several different forms:

$$y = a \, \mathrm{Sin} \left(\frac{2\pi}{\lambda} x - 2\pi\nu t \right) \tag{1-19}$$

or

$$y = a \, \mathrm{Sin} \, \frac{2\pi}{\lambda} (x - vt) \tag{1-20}$$

or

$$y = a \, \mathrm{Sin} \, k \, (x - vt) \tag{1-21}$$

where

$$k = \frac{2\pi}{\lambda} \tag{1-22}$$

For a wave travelling in the negative x direction we have to merely reverse the direction of the velocity. Therefore, for such a wave the displacement is given by

$$y = a \, \text{Sin} \, k \, (x + vt) \qquad (1\text{-}23)$$

The symmetry of waves with respect to the two variables x and t can be easily seen if we make two plots: the first one showing the variation of y with respect to x at a particular value of t (Fig. 1-9); the second one showing the variation of y with respect to t at a particular value of x

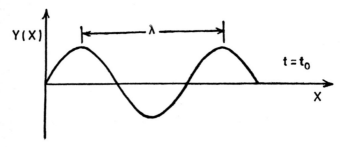

FIG. 1-9: Snapshot of the function y (x, t). The graph shows the variation of y (x, t) with x for $t = t_0$.

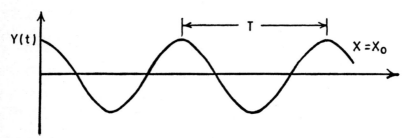

FIG. 1-10: 'Local' chart of the values of y (x, t) measured by a stationary observer (at $x = x_0$) as the wave passes him.

(Fig. 1-10). It is immediately seen that the wavelength λ in Fig. 1-9 plays the same role as the time period T in Fig. 1-10. This symmetry of λ and T is best expressed by rewriting the function as

$$y = a \, \text{Sin} \, 2\pi \left(\frac{x}{\lambda} - \frac{t}{T} \right) \qquad (1\text{-}24)$$

Another commonly used notation is

$$y = a \, \text{Sin} \, (kx - \omega t), \quad \omega = 2\pi\nu \qquad (1\text{-}25)$$

1-4 Phase

The wave described above requires that $y = 0$ at $x = 0$ when $t = 0$. This restriction may be removed by the inclusion of a *phase factor* ϕ:

$$y = a \operatorname{Sin}\left[\frac{2\pi}{\lambda}(x + vt) + \phi\right] \qquad (1\text{-}26)$$

Thus the phase factor may be thought of as specifying the choice of the zero of time and the origin of coordinates. For any particular wave, we can always choose the zero of time or choose appropriately the origin of coordinates such that ϕ becomes zero. However, if we have more than one wave propagating in a medium and if both the disturbances are to be considered simultaneously, the phase factors for *both* the waves cannot be set arbitrarily as equal to zero (see Appendix B).

1-5   Types of Waves

Waves can be classified according to their broad physical properties. For example, we have water waves, sound waves, light waves, etc. We can also distinguish different kinds of waves by considering how the motion of the vibrating particles of matter is related to the direction of the waves themselves. If the motions of the particles conveying the wave are perpendicular to the direction of propagation of the wave itself, we have a *transverse* wave.* For example, when a string under tension is set oscillating back and forth at one end a transverse wave travels down the string; the disturbance moves along the string but the string particles vibrate at right angles to the direction of propagation of the disturbance (see Fig. 1-8).

If, however, the motion of the particles conveying a wave is back and forth along the direction of propagation, we have a *longitudinal* wave. For example, when a vertical spring under tension is set oscillating up and down at one end, a longitudinal wave travels along the spring, the coils vibrate back and forth in the direction in which the disturbance travels along the spring (Fig. 1-11). Sound waves in a gas are also longitudinal waves.

Some waves are neither purely longitudinal nor purely transverse. For example, associated with the waves on the surface of water, the particles of water move both up and down and back and forth, tracing out roughly circular orbits about their original positions with an average velocity of zero, and so never undergo any permanent displacement (Fig. 1-12).

1-6   General Wave Motion

We have developed the sinusoidal wave form by considering the displacement of a string through which a transverse wave is propagating. This

---

* It must be pointed out that light waves can propagate through vacuum. They do not require the presence of any medium. However, as we shall later discuss, light waves are transverse in character and almost all mathematical descriptions developed in this chapter are applicable to light waves.

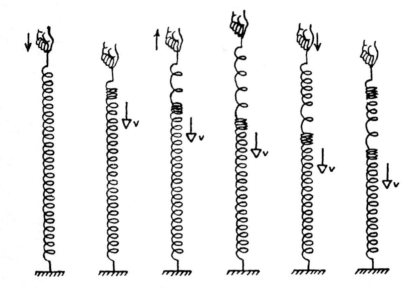

FIG. 1-11: Generation of a longitudinal wave through a spring. (In a longitudinal wave the particles of the medium, here of the spring, vibrate in the same direction as that in which the wave is propagating.)

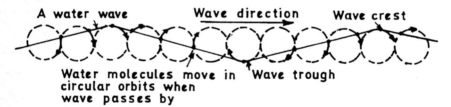

FIG. 1-12: Water molecules move in approximately circular orbits about their original positions when a wave passes by. At the crest of a wave the molecules are moving in the direction the wave is travelling; in the trough the molecules are moving in the opposite direction. There is no net movement of water involved in the motion of a wave. The amplitude of the water wave is the height of its crest above normal sea level, which is equal to the depth its trough lies below sea level (After S. Borowitz & A. Beiser, *Essentials of Physics*, p. 418, Addision-Wesley, 1966).

wave could also be described by specifying the velocity of each particle as the wave progresses. In fact any property of a medium that varies harmonically as the wave passes may be used to describe a wave. Such properties include displacement, velocity, pressure, density, electric and magnetic fields, etc. So instead of using the function $y\,(x \pm vt)$, which describes the wave on a string, we will use the general form $f\,(x \pm vt)$, where the function f may represent the displacement or even time-varying electric (or magnetic) fields. For a sinusoidal wave in one dimension, if f represents the property that varies in the medium then

$$f = a \operatorname{Sin} \frac{2\pi}{\lambda} (x - vt)$$

$$= a \operatorname{Sin} 2\pi \left(\frac{x}{\lambda} - \frac{t}{T} \right) \tag{1-27}$$

When the wave is confined to a single direction and there is no dissipation of energy, the amplitude remains constant. However, if the waves travel in all directions from the source, the amplitude would decrease as the waves travel further from the source. This decrease in amplitude occurs because the energy is spread over successively larger areas.

1-7 Energy Transport in Wave Motion

As mentioned at the beginning of this chapter, wave motion represents a mode of transporting energy. Solar light waves transport energy from the sun to the earth. This energy can be absorbed by a substance and manifested as thermal energy. Sound waves transport acoustical energy from the source to the receiver. This energy can be absorbed by the tympanic membrane of the ear. It should be noted that not a single particle moves from the source to the receiver; it is the disturbance itself that transports the energy. Further, the energy travels through the medium in the direction in which the wave travels.

Now, if the wave is a periodic sinusoidal wave, each particle executes simple harmonic motion and, if there is no damping, the energy of the vibrating particle changes from kinetic energy to potential energy and back, with the total energy remaining constant. We may calculate this constant energy E from the maximum kinetic energy. Thus

$$E = \tfrac{1}{2}m \, (v^2)_{max} \tag{1-28}$$

where m is the mass of the vibrating particle and v its speed. For a particle executing simple harmonic motion, the velocity varies with time as $a\omega \operatorname{Cos} (\omega t + \phi)$ (see Eq. B-6 in Appendix B). Thus

$$(v^2)_{max} = a^2\omega^2 \, [\operatorname{Cos}^2 (\omega t + \phi)]_{max}$$
$$= 4\pi^2 a^2 v^2$$

(because the maximum value of $\operatorname{Cos}^2\theta$ is 1)

Substituting the above expression for $(v^2)_{max}$ in Eq. 1-28 we obtain

$$E = \tfrac{1}{2}m \, (4\pi^2 v^2 a^2)$$

or

$$E = 2\pi^2 m v^2 a^2 \tag{1-29}$$

As a wave passes through the medium, the energy per unit volume, E, in the medium is the energy per particle times the number of particles per unit volume, i.e.

$$E = n.2\pi^2 m v^2 a^2 \tag{1-30}$$

But nm is the density ρ of the medium; hence

$$E = 2\pi^2 \rho v^2 a^2 \tag{1-31}$$

For waves in more than one dimension one introduces the concept of the *intensity* of the waves which is defined as the rate of flow of energy across unit area perpendicular to the direction of propagation. This is also known as the *energy flux*. Since the speed of propagation is v, a wave train of length v crosses any area perpendicular to the direction of propagation in one second, and if the area is unity the volume containing the waves which cross the area in one second is also v. Thus the intensity, I, at the point considered is v times the energy per unit volume, i.e.

$$I = 2\pi^2 v \rho \ v^2 a^2 \tag{1-32}$$

We see that the intensity is directly proportional to the square of the amplitude and to the square of the frequency.

Let us next consider a wave travelling out in a uniform isotropic* medium from a point source. If the source is emitting W ergs of energy per second uniformly in all directions and if there is no absorption in the surrounding medium, then at a distance r from the source W ergs per second pass through a spherical surface of radius r. The intensity, or energy flux, at this distance is therefore $\dfrac{W}{\text{area of the sphere}}$, i.e.

$$I = \frac{W}{4\pi r^2} \tag{1-33}$$

Thus under these conditions the intensity obeys the *inverse square law*. Further, the amplitude must then vary as $1/r$, and one may write the displacement for a spherical wave† as

$$f = \frac{a}{r} \operatorname{Sin} \frac{2\pi}{\lambda} (r - vt) \tag{1-34}$$

Here a represents the amplitude at unit distance from the source.

It should be noted that as a wave passes through any medium, if any of the energy is transformed into heat, i.e. if there is *absorption*, the loss of intensity of spherical waves will be more rapid than is required by the inverse square law.

Example
The speed of sound waves in air at standard temperature and pressure is is 330 metres/sec. A point source of frequency 500 vibrations per second

---

* Isotropic substances are ones in which physical properties (like the velocity of propagation of a particular kind of wave) do not vary with direction. Examples of anisotropic substances, in which the physical properties vary with direction, will be given in Chapter 3.

† When waves emanate from a point source in a uniform isotropic medium, the locus of equal displacement occurs on the surface of a sphere and, therefore, the waves are known as spherical waves. At large distances from the source, over a small area, the spherical waves will essentially be plane waves.

radiates energy uniformly in all directions at the rate of 10 watts. What is the intensity of the wave at a distance of 10 metres from the source? What is the amplitude of the wave there?

Solution

The intensity I is given by

$$I = \frac{W}{4\pi r^2}$$

For $r = 10$ metres

$$I = \frac{10}{4\pi \times 10 \times 10} = 7 \cdot 97 \times 10^{-3} \text{ watts/sq. m}$$

Now for air

$$\rho = 1 \cdot 29 \text{ gm/litre}$$
$$= 1 \cdot 29 \text{ kg/m}^3$$

Thus from Eq. 1-32

$$a^2 = \frac{I}{2\pi^2 v \rho \nu^2}$$
$$= \frac{7 \cdot 97 \times 10^{-3}}{2 \times 3 \cdot 14 \times 3 \cdot 14 \times 330 \times 1 \cdot 29 \times 500 \times 500}$$
$$= 3 \cdot 79 \times 10^{-12} \text{ m}^2$$

or

$$a \approx 1 \cdot 95 \times 10^{-6} \text{ m} = 1 \cdot 95 \times 10^{-4} \text{ cm}$$

1-8 Electromagnetic Waves

An important group of waves which we have not considered so far are the electromagnetic waves. Here the quantities which appear in the equations of wave motion are not displacements (of position) of particles but components of electric or magnetic fields. To understand the propagation of electromagnetic waves, it is necessary to outline Maxwell's* formulation of the principles of electromagnetism. We first note the following two basic principles of electromagnetism established by the work of Oersted, Ampere, Henry and Faraday:

---

* JAMES CLERK MAXWELL (1831-1879) contributed a paper to the Royal Society of Edinburgh when he was only fourteen years old. By the time he was seventeen he had published three papers on the results of his original research. In 1856 he became professor of Physics at the University of Aberdeen in Scotland. His investigations in many fields of physics bear the stamp of genius. He was one of the main contributors to the kinetic theory of gases and to two other important branches of physics: statistical mechanics and thermodynamics. His greatest achievement was his electromagnetic theory. Because of his tremendous contributions, Maxwell is generally regarded as the greatest physicist between the time of Isaac Newton and that of Albert Einstein.

1. An electric current in a conductor produces a magnetic field. Fig. 1-13 illustrates the principle.

2. An electromotive force is induced when there is any change of magnetic flux linked by the conductor. Fig. 1-14 illustrates the principle.

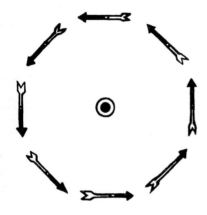

FIG. 1-13: An array of compass needles near a wire carrying a strong current. The black ends of the compass needles are their north poles. The circle denotes the cross-section of the wire which is placed at right angles to the plane of the page. The dot (inside the circle) shows that the current is emerging from the page.

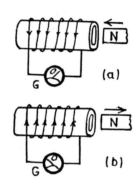

FIG. 1-14: A to and fro motion of the magnet induces a current in the coil.

According to Maxwell's theory, the above principles* of electromagnetism must be modified to include the following:

1. *A changing electric field produces a magnetic field.* It is not only currents in conductors that produce magnetic fields. Changing electric fields in insulators (such as glass or air or vacuum) also produce magnetic fields. The induced magnetic field vector is at right angles to the electric field vector. The magnitude of the magnetic field vector depends on the rate at which the electric field is changing and also on the position at which the magnetic field is being measured.

2. *A changing magnetic field produces an electric field.* The induced electric field vector is at right angles to the magnetic field vector and the magnitude of the electric vector depends on the rate at which the magnetic field vector is changing and also on the position at which the field is being measured.

* The mathematical formulations of these principles, the Maxwell equations, are given in Appendix A.

The first principle can be understood by considering a pair of conducting plates connected to a source of current (Fig. 1-15). As charges move to the plates through the conductors, the electric field in the space between the plates changes with time. This changing electric field produces a magnetic field that varies with distance from the region between the plates. This effect was predicted by Maxwell. Previously it was thought

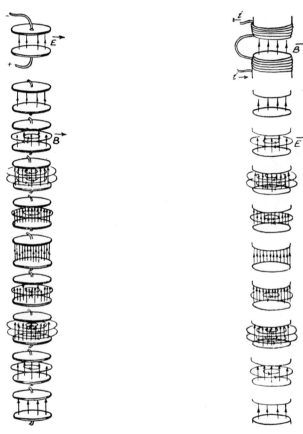

FIG. 1-15: When the electric field **E** between a pair of charged plates starts to increase in intensity, a magnetic field **B** is induced. The faster **E** changes, the more intense **B** is. When **E** diminishes, field **B** is again induced, but in the opposite direction.

FIG. 1-16: When the magnetic field **B** between the poles of an electromagnet starts to increase, an electric field **E** is induced. The faster **B** changes, the more intense **E** is. When **B** diminishes, field **E** is again induced, but in the opposite direction.

that only a current in a conductor produced a magnetic field. The additional magnetic field that Maxwell predicted would arise from a changing electric field is so small compared with the magnetic field produced by the current in the conductors that it was not possible to measure it directly. But Maxwell's theory nevertheless predicted consequences that could be experimentally verified.

As an illustration of the second principle, which was known before Maxwell's work, consider the changing magnetic field produced by, say, increasing the current in an electromagnet. This changing magnetic field (Fig. 1-16) induces an electric field in the region around the magnet. If a conductor is aligned in the direction of the induced electric field, the free charges in the conductor will move under its influence, producing a current in the direction of the induced field. This electromagnetic induction was discovered experimentally and explained by Faraday.

1-9 Propagation of Electromagnetic Waves

Let us now consider the propagation of electromagnetic waves. Suppose we create, in a certain region of space, electric and magnetic fields that change with time. According to Maxwell's theory, when we create an electric field E that is different at different times, this field will induce a magnetic field B that varies with time and with distance from the region where we created the changing electric field. In addition, the magnetic field that is changing with time induces an electric field that changes with time and with distance from the region where we created the magnetic field.

This reciprocal induction of time- and space-changing electric and magnetic fields makes possible the following unending sequence of events: A time-varying electric field in one region produces a time- and space-varying magnetic field at points near this region. This *magnetic* field produces a time- and space-varying *electric* field in the surrounding space. *This* electric field, in turn, produces time- and space-varying magnetic fields in its neighbourhood, and so on. An electromagnetic disturbance initiated at one location by oscillating charges can travel to distant points through the mutual generation of the electric and magnetic fields. The electric and magnetic fields 'join hands', so to speak, and 'march off' through space in the form of an electromagnetic wave.

We know that waves occur when a disturbance created in one region produces at a later time a disturbance in adjacent regions. Oscillation of one end of a rope produces, through the action of one part of the rope on the other, a displacement at later times at points farther along the rope. Dropping a small stone into a pond of water produces a disturbance that moves away from the source as a result of the action of one part of water on the neighbouring parts. Time-varying electric and magnetic fields produce disturbances that move *away from the source** as the varying fields in one region create varying fields in neighbouring regions.

---

* Why the disturbance moves *away* from the source can only be shown from the solution of the Maxwell equations.

1-10   Generation of Electromagnetic Waves

Electromagnetic waves always originate from accelerated charges. A simple example is an oscillating dipole. Fig. 1-17 shows such a dipole,

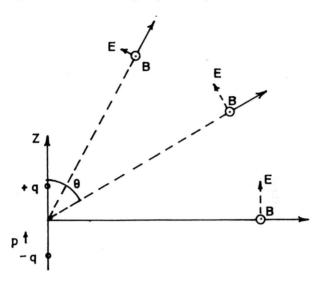

Fig. 1-17: Electromagnetic radiation from a simple oscillating dipole. The dotted arrow denotes the electric vector and the solid arrow denotes the direction of propagation of the wave. The magnetic vector is perpendicular to the electric vector and to the direction of propagation and is emerging from the page. The amplitude of the electric (and the magnetic) field is maximum in a direction perpendicular to the direction of the oscillating dipole and zero along the line of the dipole.

represented by two equal and opposite charges. Its dipole moment, represented by the arrow **p** in the figure is assumed to oscillate sinusoidally (along the z-axis) as the charges oscillate. Thus, we may write

$$p_x = p_y = 0, \qquad\qquad p_z = p_o \, \mathrm{Sin}\, \omega t$$

It can be shown that for such an oscillating dipole, the electromagnetic waves propagate radially outward and, at distances far away* from the oscillator, the direction of the electric and magnetic fields are as shown in Fig. 1-17. The direction of the electric field is perpendicular to the radius vector in a plane containing the z-axis and the radius vector. The magnetic field is perpendicular to this plane. The electric and the

---

* To be precise, at distances large compared with the size of the antenna and the wavelength. The exact expressions are discussed in Appendix C.

magnetic fields are at right angles to each other and also at right angles to the direction of propagation of the wave.* The magnitudes of the electric and magnetic field are of the form

$$E = E_o \frac{\text{Sin}\theta}{r} \quad \text{Cos } (kr - \omega t) \tag{1-35}$$

and

$$B = B_o \frac{\text{Sin}\theta}{r} \quad \text{Cos } (kr - \omega t) \tag{1-36}$$

Thus, we find a strong field in a direction perpendicular to the direction of the oscillating dipole (i.e. when $\theta = \pi/2$); and the fields are zero along the line of the oscillating dipole. Since the electric and magnetic fields are at right angles to the direction of propagation, the wave is transverse. Also the electric and magnetic fields oscillate 'in phase'. It should be noted that Eqs. 1-35 and 1-36 represent spherical waves (cf. Eq. 1-34).

Another simple example is a wire which is oriented vertically as shown in Fig. 1-18. If it carries a steady current I, it will be surrounded by a

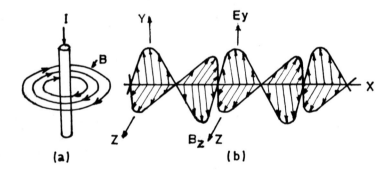

FIG. 1-18: Sinusoidal variation of current in the wire causes emission of a sinusoidal wave train as sketched in (*b*). The sinusoidal variations in *E* and *B* are shown along a single radial line. The electromagnetic wave is plane polarized with all *E*-vectors parallel to the wire and all *B*-vectors lying in planes perpendicular to the wire. A radio transmitting antenna emits radio waves in essentially this fashion.

magnetic field that decreases in intensity with increasing radial distance from the wire but does not vary with time. Under these circumstances, no pulses or waves propagate. In other words, when electric charges move with uniform velocity no electromagnetic waves are generated. However, if we start with zero current and zero field and abruptly increase the current to the value I an electromagnetic pulse will be propagated outward. If we vary the current in the wire sinusoidally with time

* In fact, the direction of propagation is perpendicular to the plane of the electric and magnetic vectors in the sense of advance of a right-handed screw rotated from the electric to the magnetic vector through the smaller angle between their positive directions.

at frequency ν, a sinusoidal wavetrain of the same frequency will be emitted.

The electromagnetic wavetrain illustrated in Fig. 1-18 is said to be plane polarized because the electric vector is confined to variation in the *x-y* plane whereas the magnetic vector is confined to variation in the *x-z* plane.* It should be noted that the electric vector varies sinusoidally in the *y* direction while the magnetic vector varies sinusoidally in the *z* direction.

Thus we find that a source of electromagnetic waves, like the vibrating tuning fork, must be an oscillating current—a system in which electrical charge is accelerated, either in a transient fashion to produce a pulse or periodically to produce a continuous wavetrain.† Further, Maxwell's theory suggests that electromagnetic waves will be emitted whenever electrical charge is accelerated, regardless of whether or not it is confined within conducting wires.

Maxwell also showed that one can think of energy as being stored in any region of space in which there exist electric and magnetic fields and that electromagnetic waves transport energy outward from a source just as do water ripples or sound waves.‡ Hence we may speak of electromagnetic radiation in the same sense as we speak of acoustic radiation. If such radiation is absorbed in matter, it will cause an increase in temperature.

1-11   Composition of Waves: The Superposition Principle

In the wave phenomena discussed thus far, the wave motion originates in one region of the medium and propagates in one direction. But it may happen that two wave motions originate in different regions and propagate toward each other, such as the two wave pulses shown at the top of Fig. 1-19. When the two pulses are separated, no problem arises, but what happens when they reach the same region of the string and overlap? Observation shows that, after they pass each other each pulse has the same original shape. Furthermore, during the time of overlap the total displacement of each point on the string is just the sum of the displacements corresponding to the separate pulses. That is, each pulse moves exactly as though the other were not present, and the total displacement of any point at a given time is obtained by adding the corresponding displacements of the individual pulses. Similar experiments can be carried

---

* We have previously discussed that when one end of a long stretched string is set oscillating in the up and down direction, a transverse wave propagates along the string (see Fig. 1-8). The vibration of such a string is also confined to a single plane and is said to be plane polarized.

† We will show in Chapter 3 that light waves are also electromagnetic. Therefore we must assume that atoms emitting light are also oscillating dipoles. Since atoms contain electrons in their outer layers, this raises no difficulties.

‡ The energy associated with an electromagnetic wave is discussed in Chapter 3.

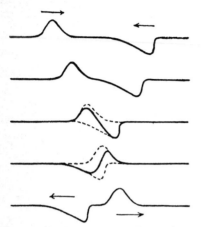

FIG. 1-19: Two pulses of different shapes, moving toward each other, are produced in two separate sections of a string. The positions of the individual pulses during the time of overlap are shown as broken lines. The actual displacement, shown as a solid line, is the sum of displacements corresponding to the separate pulses. After the two pulses pass, they proceed with their original shapes. (After H. D. Young, *Fundamentals of Optics & Modern Physics*, p. 15, McGraw-Hill, 1968.)

out for other wave situations (such as for waves created by dropping two particles simultaneously in a quiet pool of water), and one can generalize the results by stating that: *two or more waves can traverse the same space independently of one another and the resultant displacement of any particle at a given time is simply the vector addition of the displacement caused by each wave independently.* This is the *superposition principle* and was first clearly stated by Young* in 1802. Although introduced here with reference to mechanical waves, it has important applications to other types of waves as well. The fact that two beams of light may cross and continue unchanged is an example of the principle of superposition for light waves, or, more precisely, for electromagnetic waves.

1-12   Reflection of Waves

We can use the superposition principle to simplify the description of a wave undergoing reflection. First, let us look at the physical details of such a reflection.

Consider a wave pulse travelling along a string. We can describe it as $f(x - vt)$ only as long as the string is infinitely long. But, if the string has an end (and most strings do), we must take this into account. For instance, we may fix the string to a wall so that point B cannot move (Fig. 1-20). At a time well before the pulse encounters the wall, its leading edge accelerates the string (upward in Fig. 1-20). This happens because, as Newton's second law tells us, unbalanced forces on a section of the string result in its being accelerated. But the wall is too massive to be moved, so it must exert a force back on the string according to Newton's third law. Since this force is equal in magnitude and opposite in direction to the one which the string exerts on the wall, it is just the force which would be exerted by the leading edge of an inverted pulse coming

---

* Thomas Young (1773-1829). English physician and physicist, usually called the founder of the wave theory of light. His work on intereference, which we shall describe in a later chapter, constituted the most important contribution on light since Newton. His early work proved the wave nature of light but was not taken seriously by others until it was corroborated by Fresnel.

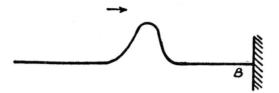

Fɪɢ. 1-20: Pulse propagating on a string with one
end fixed.

from an (imaginary) string extending past the wall. This result can be
best understood if we consider a specific problem.

Suppose there is a semicircular disturbance propagating through a
string in the $+x$ direction with speed v. Let, at $t = 0$, the shape of the
pulse be of the form (Fig. 1-21):

$$y(x, 0) = + \sqrt{R^2 - (x - x_0)^2} \qquad (1\text{-}37)$$

which represents the upper half of a semicircle of radius R whose centre

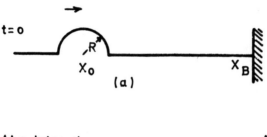

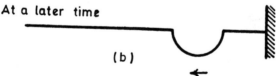

Fɪɢ. 1-21: A semicircular pulse propagating on a string and
getting reflected.

is at x_0. Since the disturbance is propagating in the $+x$ direction with
speed v, the shape of the pulse at time t will be

$$y(x, t) = \sqrt{R^2 - (x - x_0 - vt)^2} \qquad (1\text{-}38)$$

The above expression can be easily understood by noting that, at a
later time, say at $t = 15x_0/v$, the shape of the pulse would be
$\sqrt{R^2 - (x - 16x_0)^2}$, indicating the centre has moved by a distance
$15x_0$ in the $+x$ direction. In other words, the disturbance has propagated
through a distance $15x_0$ in time $15x_0/v$ (obviously!). Further, the displacement at $x = x_B$ (as a function of time) produced by this wave
would be

$$y\,(x_{\rm B},t) = \sqrt{R^2 - (x_{\rm B} - x_o - vt)^2}$$

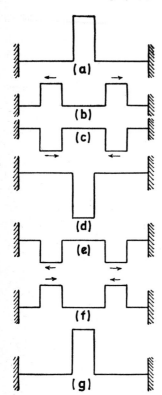

Now, the reflected wave should be such that the displacement it produces at $x = x_{\rm B}$ must be equal to

$$y_{refl}\,(x_{\rm B},t) = -\sqrt{R^2 - (x_{\rm B} - x_o - vt)^2}$$

This will ensure that the resultant displacement at the point B is zero for all values of t. The displacement produced by the reflected wave will be given by

$$y_{refl}\,(x, t) = -\sqrt{R^2 - (-x + 2x_{\rm B} - x_o - vt)^2}$$

The total disturbance, $y_{\rm T}$, will be

$$
\begin{aligned}
y_{\rm T} &= y\,(x, t) + y_{refl}\,(x, t) \\
&= \sqrt{R^2 - (x - x_o - vt)^2} - \\
&\quad \sqrt{R^2 - (-x + 2x_{\rm B} - x_o - vt)^2}
\end{aligned}
$$

Clearly, for all times, the value of $y_{\rm T}$ at $x = x_{\rm B}$ is zero. The above result can easily be generalized for an arbitrary shape of the disturbance. If the incident wave is $f\,(x - vt)$ then the reflected wave is $-f\,(-x + 2x_{\rm B} - vt)$ and the superposed disturbance is

$$f\,(x - vt) - f\,(-x + 2x_{\rm B} - vt) \quad (1\text{-}39)$$

On the other hand, if both ends of a spring are clamped and if the spring is initially disturbed from its equilibrium position and then released as in Fig. 1-22(a), two pulses are generated;* the resultant shape of the spring shortly after release is shown in Fig. 1-22(b). Next, the pulse gets reflected by the wall [Fig. 1-22(c)] and the two pulses superpose to form the negative of the initial disturbance [Fig. 1-22(d)]. In Fig. 1-22(e), they are approaching the wall again and in Fig. 1-22(f), they are reflected as positive pulses. In Fig. 1-22(g) they combine again to form the initial disturbance. If there are no energy losses to the surroundings, this process can continue indefinitely.

FIG. 1-22: (a) At $t = 0$, a spring is distorted from its equilibrium position and then released.

(b) Two pulses are generated; one moves to the right and one to the left.

(c) Both pulses have undergone a reversal of phase when reflected from the clamped ends.

(d) The pulses combine to form the negative of the initial disturbance.

(e) They approach the wall again.

(f) The pulses are reflected with a reversal of phase.

(g) They again combine to form the initial disturbance.

(We have assumed that there are no energy losses to the surroundings.)

* The generation and propagation of the two pulses are discussed in Problems 12 and 13.

1-13 Standing Waves

Another interesting and practical application of the superposition principle is the analysis of the possible sinusoidal waves (called standing waves) that can exist on a string clamped at both ends. The waves propagating in the string are reflected from the clamps. Each such reflection gives rise to a wave travelling in the string in the opposite direction. The reflected waves add to the incident waves according to the principle of superposition.

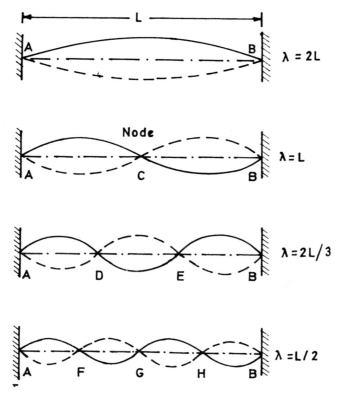

Fig. 1-23: Standing waves on a string. The points *A, B, C, D, E, F, G* and *H* are nodes and always remain at rest.

Now in Fig. 1-23, if end *B* is assumed to correspond to $x = 0$, and if the wave incident at *B* is of the form

$$f_{inc} = a \operatorname{Sin} \frac{2\pi}{\lambda} (x - vt) \qquad (1\text{-}40)$$

then the displacement at $x = 0$ produced by the incident wave will vary with time as $-a \operatorname{Sin} \frac{2\pi}{\lambda} (vt)$. For zero resultant displacement at $x = 0$, the displacement produced by the reflected wave must be $+ a \, \text{Sin} \, \frac{2\pi}{\lambda} \, (vt)$.

Thus, the reflected wave will be of the form

$$f_{refl} = a \, \text{Sin} \, \frac{2\pi}{\lambda} \, (x + vt) \tag{1-41}$$

The superposed waveform will be given by

$$f_T = a \, [\text{Sin} \, \frac{2\pi}{\lambda} \, (x - vt) + \text{Sin} \, \frac{2\pi}{\lambda} \, (x + vt)]$$

$$= 2a \, \text{Sin} \, \frac{2\pi}{\lambda} \, x \, \text{Cos} \, 2\pi vt \tag{1-42}$$

where we have made use of the relation $v = v/\lambda$. If the coordinate of the end A is given by $x = - L$, then the displacement at A has got to be zero for all times and, therefore

$$\text{Sin} \, \frac{2\pi}{\lambda} \, L = 0$$

or

$$\frac{2\pi}{\lambda} \, L = n\pi, \quad n = 1, \, 2, \, 3, \ldots$$

or

$$\lambda = 2L, \, L, \, \frac{2L}{3}, \, \frac{L}{2} \ldots \tag{1-43}$$

Eq. 1-43 tells us that if the string is clamped at both ends, then there can exist on the string only those vibrations which are compatible with the zero displacement at the ends. Fig. 1-23 shows some of the possible ways in which the string can vibrate. Waves of this type are known as *standing waves* or *stationary waves*. When $\lambda = 2L$, the string is said to vibrate in its *fundamental mode*. In general, when $\lambda = 2L/n$, the string is said to vibrate in its $(n - 1)^{\text{th}}$ harmonic.,

In a stationary wave (see Fig. 1-23) there is no net transference of energy in either direction, though there is an energy density in the medium. This energy density is not uniformly distributed. It is maximum at points for which $\text{Sin}^2 \, \frac{2\pi}{\lambda} \, x = 1$, i.e. at $x = - \lambda/4$, $x = - 3\lambda/4 \ldots$

(we have chosen negative values of x because the string has been defined to lie between $x = - L$ and $x = 0$). The points where the energy is maximum are called *antinodes*; those where it is a minimum are called *nodes*. If a detector which measures the energy density is moved along the x-axis it gives a maximum reading at the antinodes and zero at nodes. The distance from a node to the nearest antinode is $\lambda/4$ and the distance between successive nodes (or successive antinodes) is $\lambda/2$.

Example

Two similar strings are attached to two different points on a wall at $x = 0$. A transverse sinusoidal wave is sent down through each of the strings with the displacements given by

$$y_1 = a \operatorname{Sin} 2\pi \left[\frac{x}{\lambda_1} + v_1 t \right]$$

and

$$y_2 = a \operatorname{Sin} 2\pi \left[\frac{x}{\lambda_2} + v_2 t \right]$$

where $v_1 = 0.7 v_2$. After the waves are reflected from the ends, standing waves are formed.
Find:

(a) The kinetic energy per unit length at a point $1.75\lambda_1$ from the wall on string 1.
(b) The kinetic energy per unit length at a point the same distance $(1.75\lambda_1)$ from the wall on string 2.
(c) The distances from the wall at which nodes on the two strings coincide.

Solution

We first note that since the strings are identical, the speed of the wave is same on each string.
Thus

$$\lambda_1 v_1 = \lambda_2 v_2 = v$$

Therefore

$$\lambda_2 = 0.7 \lambda_1$$

The two strings are shown in Fig. 1-24. Since there is a node at the wall, nodes will also occur at

$$x = \frac{n\lambda}{2}, \ n = 1, 2, \ldots$$

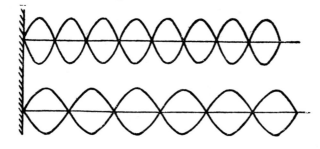

FIG. 1-24: Reflection of sinusoidal waves.

Antinodes will occur at $x = \dfrac{n\lambda}{2} - \dfrac{\lambda}{4}$.

(a)   The point $x = 1 \cdot 75\lambda_1$ is an antinode on string 1. So the kinetic energy density at that point is

$$\text{Kinetic energy/unit length} = \tfrac{1}{2}\rho(2a)^2 \, \omega^2 \mathrm{Sin}^2 \omega t$$
$$= 2\rho a^2 \, \omega^2 \, \mathrm{Sin}^2 \, \omega t$$

where $\omega = 2\pi\nu$ and ρ is the mass per unit length.

(b)   Here $x = 1 \cdot 75\lambda_1 = 2 \cdot 50\lambda_2$ is at a node on string 2. The string does not move at a node so kinetic energy/unit length $= 0$.

(c)   Nodes on string 1 occur at $x_1 = n_1\lambda_1/2$ and those on string 2 at $x_2 = n_2 \, \lambda_2/2$. For a coincidence

$$\frac{n_1\lambda_1}{2} = \frac{n_2\lambda_2}{2}$$

or

$$\frac{n_1}{n_2} = \frac{\lambda_2}{\lambda_1} = \frac{7}{10}$$

Since n_1 and n_2 are integers the possible values of $n_1 = 0, 7, 14, 21\ldots$ and $n_2 = 0, 10, 20, 30\ldots$, and positions of coincidence are given by

$$x_{\text{coinc}} = \frac{n_1\lambda_1}{2} = 0, \; 3 \cdot 5\lambda_1, \; 7\lambda_1, \; 10 \cdot 5\lambda_1, \ldots$$

or

$$x_{\text{coinc}} = \frac{n_2\lambda_2}{2} = 0, \; 5\lambda_2, \; 10\lambda_2, \; 15\lambda_2 \ldots$$

giving the same value of x.

Next, let us consider the reflection of waves which are being emitted by a source. The reflected waves superpose with the incident waves and give rise to stationary waves. In general, the reflection is not perfect and one introduces the coefficient of reflection ρ, which is defined as the ratio of the energy of the reflected beam to the energy of the incident beam. This implies that the ratio of the amplitudes is $\rho^{1/2}$. Eq. 1-41 must now be replaced by

$$f_{\text{refl}} = a\rho^{1/2} \, \mathrm{Sin} \, \frac{2\pi}{\lambda} \, (x + vt)$$

and Eq. 1-42 is replaced by

$$f_{\text{T}} = 2a\rho^{1/2} \, \mathrm{Sin} \, \frac{2\pi}{\lambda} \, x \, \mathrm{Cos} \, 2\pi \, vt + a \, (1 - \rho^{1/2}) \, \mathrm{Sin} \, \frac{2\pi}{\lambda} \, (x - vt) \qquad (1\text{-}44)$$

The wave is now partly stationary and partly progressive. A detector moved along the x-axis will still record maxima and minima. The readings at minima will not be zero since at points where the first term of Eq. 1-44 is zero the detector gives a reading proportional to the square of the amplitude of the second term.

A series of observations made on stationary electromagnetic waves* is shown in Fig. 1-25. These waves are produced by allowing radiation emitted by a small high-frequency oscillator to fall upon a large sheet of metal placed at $x = 0$ (Fig. 1-26). The waves are measured by means of a small detector. The metal has a high coefficient of reflection for these waves and the waves are mainly stationary waves with small progressive components. From the distance between the nodes the wavelength is found to be 11·6 cm.

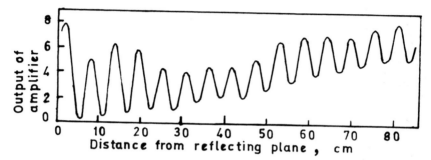

FIG. 1-25: Experimental observations on stationary electromagnetic waves (After R. W. Ditchburn, *Light*, p. 63, Blackie, 1963).

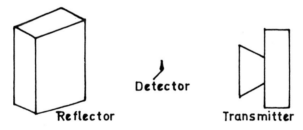

FIG. 1-26: Apparatus for producing stationary electromagnetic wave.

1-14 Stationary Light Waves† —Wiener's Experiments

The detection of stationary light waves is difficult because, as we shall see in Chapter 4, the wavelength of light waves (and therefore the distance between nodes and antinodes) is very small, being of the order of 10^{-5} centimetres. The experimental difficulties were first overcome by Wiener in 1890. Before discussing his work we shall describe the

---

* Hertz's experiment, which is based on similar principles is discussed in Chapter 3.

† So far we have not discussed light waves. Students may skip this for the time being and come back to it after reading Chapter 4.

experiments of Ives which, though more difficult to carry out, are easy to understand. The experimental arrangement is outlined in Fig. 1-27. Ives prepared a special photographic plate with a thin fine-grained emulsion. The emulsion side of the plate was placed in contact with a film of mercury and a parallel beam of monochromatic light (i.e. light of one particular wavelength) was directed normally upon the glass side of the plate as shown in Fig. 1-27. The light passed through the glass and the film, and was reflected at the mercury surface. The superposition of the incident and reflected waves gave rise to standing waves which were formed in the film. The nodes form a series of laminae which are $\lambda/2$ apart, with the antinodes half way between them. In Ives' experiment the waves are detected by cutting a section of the film in a plane normal to the surface of the glass and examining it under a high-power microscope. The nodes and antinodes can be clearly seen; 250 successive laminae were counted in one particular experiment.

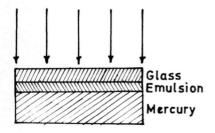

FIG. 1-27: Ives's experiment.

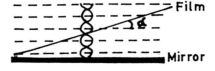

FIG. 1-28: Wiener's experiment.

Wiener used a thin transparent photographic film, about 2×10^{-6} cm thick, for the detection of stationary waves. Wiener arranged the film to lie at a very small angle with the system, so that it cut the successive antinodal lines as shown in Fig. 1-28. The photographic emulsion is blackened along a series of lines where the plane of the film cuts the antinodal laminae. The angle α (shown in Fig. 1-28) was about 10^{-3} radian and the distance between successive fringes was then about 1000 times greater than the distance between the laminae. From these experiments the wavelength of green light was estimated to be $5\cdot5 \times 10^{-5}$ cm, and it was shown that the wavelength of red light is nearly double that of blue light.

1-15 Superposition of Two Sinusoidal Waves

Frequently it is necessary to find the resultant disturbance at a point when a number of disturbances arrive there simultaneously. According to the superposition principle, the resultant disturbance is simply the sum of the separate disturbances. Consider, for example, the disturbance at a *point* where two disturbances of the *same frequency* arrive with different phases. The displacement due to two disturbances can be represented by

$$x_1 = a_1 \, \text{Cos} \, (\omega t + \phi_1) \Big\}$$
$$x_2 = a_2 \, \text{Cos} \, (\omega t + \phi_2) \Big\} \qquad (1\text{-}46)$$

where a_1, ϕ_1 and a_2, ϕ_2 represent the amplitudes and phases of the two vibrations. The displacements are here taken along the same line (along the x-axis). It should be noted that ω is the same for both waves, since we have assumed them to be of the same frequency. Since the two displacements are in the same direction, the resultant displacement, x, is obtained by merely adding the two:

$$x = x_1 + x_2 = a_1 \, \text{Cos} \, (\omega t + \phi_1) + a_2 \, \text{Cos} \, (\omega t + \phi_2)$$
$$= \text{Cos} \, \omega t \, [a_1 \, \text{Cos} \, \phi_1 + a_2 \, \text{Cos} \, \phi_2] - \text{Sin} \, \omega t \, [a_1 \, \text{Sin} \, \phi_1 + a_2 \, \text{Sin} \, \phi_2]$$

or, we may write

$$x = A \, \text{Cos} \, (\omega t + \theta) \qquad (1\text{-}45)$$

where

$$A \, \text{Cos} \, \theta = a_1 \, \text{Cos} \, \phi_1 + a_2 \, \text{Cos} \, \phi_2 \qquad [1\text{-}46(a)]$$
$$A \, \text{Sin} \, \theta = a_1 \, \text{Sin} \, \phi_1 + a_2 \, \text{Sin} \, \phi_2 \qquad [1\text{-}46(b)]$$

If we square Eqs. 1-46(a) and 1-46(b), add them and take the positive square root we would get:

$$A = \sqrt{a_1{}^2 + a_2{}^2 + 2a_1a_2 \, \text{Cos} \, (\phi_1 - \phi_2)} \qquad (1\text{-}47)$$

Further, if we divide Eq. 1-46(b) by Eq. 1-46(a), we would obtain

$$\tan \theta = \frac{a_1 \, \text{Sin} \, \phi_1 + a_2 \, \text{Sin} \, \phi_2}{a_1 \, \text{Cos} \, \phi_1 + a_2 \, \text{Cos} \, \phi_2} \qquad (1\text{-}48)*$$

Thus the resultant of two simple harmonic vibrations of equal frequencies in the same straight line is another simple harmonic vibration of the same frequency. The amplitude and the phase constant of the resultant motion can be calculated from those of the component vibrations by using Eqs. 1-47 and 1-48.

Since the resultant intensity at the point (which is proportional to A^2) is not simply the sum of the intensities due to the separate disturbances (which would be proportional to $a_1{}^2 + a_2{}^2$), the two disturbances are said to *interfere*. In particular, if $a_1 = a_2$ and $\phi_2 - \phi_1 = \pi$, they interfere destructively to give zero intensity. At first sight this would seem to violate the principle of conservation of energy, but closer inspection shows that whenever there are points where destructive interference

---

* It appears that if θ is a solution of Eq. 1-48, then $\theta + \pi$ is also a solution. However, this ambiguity is removed if we note that the resultant A has to be positive and that θ has to satisfy two equations (viz. Eqs. 1-46). When Sinθ and Cosθ are both known, the angle θ is determined unambiguously.

causes the resultant intensity to be less than the sum of the separate intensities, there are also other points where constructive interference causes the resultant intensity to be greater than the sum of the separate intensities. In Chapter 4 we shall study the superposition of disturbances that travel from a source to a point via two different routes, giving rise to an interference pattern.

It must be remembered that this discussion has been concerned with the superposition of disturbances, each of which is completely specified by a single scalar quantity. This corresponds to the superposition of simple harmonic oscillations along the same straight line. For disturbances that are not along the same straight line it would be necessary to specify the directions of the various oscillations.*

1-16 The Graphical Method of Adding Disturbances of the Same Frequency

A simple geometrical construction can also be used to find the resultant of two sine waves of the same frequency. (This method is particularly useful when we have to study the superposition of a large number of sine waves corresponding to the same frequency.)

The method consists in representing the amplitudes of the component vibrations, a_1 and a_2, by vectors making angles ϕ_1 and ϕ_2 with the x-axis (Fig. 1-29). The resultant amplitude, A, is then obtained by vectorially adding a_1 and a_2. The angle, θ, that this resultant vector makes with the x-axis is the phase constant of the resultant. To prove this proposition, we first note from Fig. 1-29 that in the parallelogram $PQOS$, the diagonal OP $(= A)$ is given by

$$A^2 = a_1{}^2 + a_2{}^2 + 2a_1a_2 \, \text{Cos} \, (\phi_1 - \phi_2)$$

or

$$A = \sqrt{a_1{}^2 + a_2{}^2 + 2a_1a_2 \, \text{Cos} \, (\phi_1 - \phi_2)} \qquad (1\text{-}49)$$

which is identical to Eq. 1-47. Eq. 1-48 can also be obtained from elementary geometrical considerations.

That the resultant motion is also simple harmonic can be easily seen from the fact that a simple harmonic motion may be represented as the projection on one of the diameters of a point moving with uniform circular motion (see Appendix B). Fig. 1-29 is drawn for $t = 0$ and, as time progresses, the vectors OS and OQ rotate (in the anticlockwise direction) with angular velocity ω. Obviously the vector OP also rotates with the same angular velocity and its projection on the x-axis executes simple harmonic motion.

---

* In Chapter 3 we will consider superposition of two vibrations which are at right angles to one another.

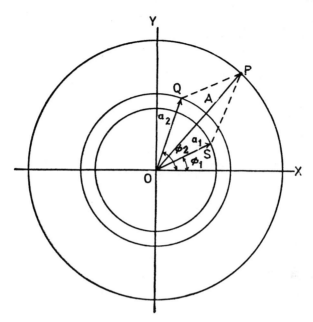

Fig. 1-29: Graphical composition of two simple harmonic motions of the same frequency.

1-17 Superposition of Many Simple Harmonic Vibrations along the Same Line

Let us next consider the superposition of n simple harmonic vibrations

$$
\left.\begin{aligned}
x_1 &= a_1 \, \text{Cos} \, (\omega t + \phi_1) \\
x_2 &= a_2 \, \text{Cos} \, (\omega t + \phi_2) \\
&\vdots \qquad \vdots \\
x_n &= a_n \, \text{Cos} \, (\omega t + \phi_n)
\end{aligned}\right\} \tag{1-50}
$$

The resultant of these vibrations is given by:

$$
\begin{aligned}
x &= x_1 + x_2 + \ldots + x_n \\
&= \text{Cos} \, \omega t \, [a_1 \, \text{Cos} \, \phi_1 + a_2 \, \text{Cos} \, \phi_2 + \ldots + a_n \, \text{Cos} \, \phi_n] \\
&\quad - \text{Sin} \, \omega t \, [a_1 \, \text{Sin} \, \phi_1 + a_2 \, \text{Sin} \, \phi_2 + \ldots + a_n \, \text{Sin} \, \phi_n] \\
&= a \, \text{Cos} \, (\omega t + \delta) \tag{1-51}
\end{aligned}
$$

where

$$
a = [(a_1 \, \text{Cos} \, \phi_1 + \ldots + a_n \, \text{Cos} \, \phi_n)^2 \\
+ (a_1 \, \text{Sin} \, \phi_1 + \ldots + a_n \, \text{Sin} \, \phi_n)^2]^{1/2} \qquad [\text{1-52}(a)]
$$

and

$$
\text{tan} \delta = \frac{a_1 \, \text{Sin} \, \phi_1 + \ldots + a_n \, \text{Sin} \, \phi_n}{a_1 \, \text{Cos} \, \phi_1 + \ldots + a_n \, \text{Cos} \, \phi_n} \qquad [\text{1-52}(b)]
$$

As a special case, let us assume that the amplitudes of the component vibrations are equal with phases increasing in arithmetical progression. The resultant will be given by

$$x = a\,[\text{Cos}\,\omega t + \text{Cos}(\,\omega t + \phi) + \text{Cos}\,(\omega t + 2\phi) + \dots$$
$$\dots + \text{Cos}\,\{\omega t + (n-1)\,\phi\}] \quad (1\text{-}53)$$

where ϕ is the phase difference between one source and the next one. (We have substituted $\phi_1 = 0$, without any loss of generality.) We will add all the terms by the graphical method (Fig. 1-30). The first one is of length a and it has zero phase. The next one is also of length a and it has phase equal to ϕ. The next one is again of length a and it has phase equal to 2ϕ and so on. So we are evidently going around an equiangular polygon with n sides (Fig. 1-30).

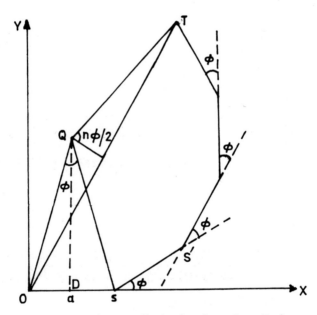

Fig. 1-30: The resultant amplitude of $n = 5$ equal amplitude sources with net successive phase difference ϕ.

The vertices, of course, all lie on a circle, and we can find the resultant amplitude most easily if we find the radius of that circle. Suppose that Q is the centre of the circle. Then, since

$$OQ = SQ = S'Q = R(\text{say})$$
$$\angle QOS = \angle QSO = \angle QSS' = \angle SS'Q$$

and
$$\angle OQS + \angle QOS = \angle QSS' + \phi$$
$$\therefore \quad \angle OQS = \phi$$

Further from $\triangle QOD$

$$R \ \text{Sin} \ \phi/2 = a/2$$

But the large angle OQT is equal to $n\phi$, and we thus find that the resultant amplitude $A \ (= OT)$ will be given by

$$\tfrac{1}{2}A = R \ \text{Sin} \ \frac{n\phi}{2}$$

or, eliminating R, we obtain

$$A = a \ \frac{\text{Sin} \ \dfrac{n\phi}{2}}{\text{Sin} \ \dfrac{\phi}{2}} \tag{1-54}$$

The resultant phase $(= \angle TOX)$ is obtained by considering the exterior angles in Fig. 1-30 and is equal to $\tfrac{1}{2}(n-1)\phi$. Thus the resultant vibration is given by:

$$x = a \ \frac{\text{Sin} \ \dfrac{n\phi}{2}}{\text{Sin} \ \dfrac{\phi}{2}} \ \text{Cos} \ [\omega t + \tfrac{1}{2}(n-1)\phi] \tag{1-55}$$

We would have obtained the same expression for the resultant vibration from trigonometric considerations. For, if we write

$$a \ [\text{Cos} \ \omega t + \text{Cos} \ (\omega t + \phi) + \ldots + \text{Cos} \ \{\omega t + (n-1)\phi\}]$$
$$= A \ \text{Cos} \ (\omega t + \delta)$$

then

$$A\text{Cos}\,\delta = a \ [1 + \text{Cos}\phi + \ldots + \text{Cos} \ (n-1) \ \phi] \tag{1-56}$$

and

$$A\text{Sin}\,\delta = a \ [\text{Sin}\phi + \ldots + \text{Sin} \ (n-1) \ \phi] \tag{1-57}$$

Multiplying the first series (i.e. Eq. 1-56) by $2 \ \text{Sin}\tfrac{1}{2}\phi$ and making use of the trigonometric relation

$$2\text{Cos}\alpha \ \text{Sin}\beta = \text{Sin} \ (\alpha + \beta) - \text{Sin} \ (\alpha - \beta)$$

we obtain

$$2A\text{Cos}\,\delta \ \text{Sin}\tfrac{1}{2}\phi = a \ [2 \ \text{Sin}\tfrac{1}{2}\phi + \{\text{Sin}\tfrac{3}{2}\phi - \text{Sin}\tfrac{1}{2}\phi\}$$
$$+ \{\text{Sin}\tfrac{5}{2}\phi - \text{Sin}\tfrac{3}{2}\phi\} + \ldots + \{\text{Sin} \ (n-\tfrac{1}{2}) \ \phi - \text{Sin} \ (n-\tfrac{3}{2}) \ \phi\}]$$
$$= a \ [\text{Sin}\tfrac{1}{2}\phi + \text{Sin} \ (n-\tfrac{1}{2}) \ \phi]$$

or

$$A\text{Cos}\,\delta \ \text{Sin}\tfrac{1}{2}\phi = a \ \text{Sin}\tfrac{1}{2}n\phi \ \text{Cos}\tfrac{1}{2} \ (n-1) \ \phi \tag{1-58}$$

Similarly, if we multiply Eq. 1-57 by $2 \ \text{Sin}\tfrac{1}{2}\phi$ and carry out similar manipulations we will obtain

$$A \operatorname{Sin} \delta \operatorname{Sin} \tfrac{1}{2} \phi = a \operatorname{Sin} \tfrac{1}{2} n\phi \operatorname{Sin} \tfrac{1}{2} (n-1) \phi \qquad (1\text{-}59)$$

If we square Eqs. (1-58) and (1-59), add them together and take the positive square root we get

$$A = a \frac{\operatorname{Sin} \tfrac{1}{2} n\phi}{\operatorname{Sin} \tfrac{1}{2} \phi} \qquad [1\text{-}60(a)]$$

On the other hand, if we divide Eq. (1-59) by Eq. (1-58) we obtain

$$\delta = \tfrac{1}{2} (n-1) \phi \qquad [1\text{-}60(b)]$$

Both the results are obviously identical to the ones derived from graphical methods.

Now in the limit of a very large number of vibrations differing continuously in phase, the vectors lie on the circumference of a circle. The resultant is represented by closing the chord of this arc. If the phase difference between the first and the last vibrations be π, the curve takes the form of a semicircle and the diameter represents the resultant. If the number of components is further increased, the resultant decreases, becoming zero when the circle is completed and the first and the last vibrations are in phase. For vibrations of this kind the resultant amplitude lies between zero and the diameter of the circle, however large may be the number of the component vibrations.

An interesting modification of this case arises when the amplitudes of successive vibrations decrease gradually. In this case, the resultant of the first few vibrations (for which the phase difference between the first and last vibration is π) will approximately lie on the circumference of a semicircle (Fig. 1-31). For the next few, the semicircle is slightly smaller than the first one and so on. It is seen (Fig. 1-31) that the resultant amplitude oscillates and in the limit it is approximately equal to half of the diameter of the first semicircle.

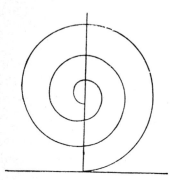

Fig. 1-31: Superposition of a large number of vibrations whose phases increase regularly and whose amplitudes go on decreasing gradually.

1-18 Nearly Equal Frequencies in the Same Straight Line

Consider two vibrations of angular velocity ω_1 and ω_2, amplitudes a_1 and a_2 and phases ϕ_1 and ϕ_2, such that ω_2 and ω_1 differ by a small amount only. Thus

$$x_1 = a_1 \operatorname{Cos} (\omega_1 t + \phi_1)$$
$$x_2 = a_2 \operatorname{Cos} (\omega_2 t + \phi_2)$$

The combined effect will be given by the equation

$$x = a_1 \text{ Cos } (\omega_1 t + \phi_1) + a_2 \text{ Cos } (\omega_2 t + \phi_2)$$

Since ω_2 is nearly equal to ω_1 we may write $\omega_2 = \omega_1 + m$ where m is small compared to ω_1 and ω_2. Therefore

$$x = a_1 \text{ Cos } (\omega_1 t + \phi_1) + a_2 \text{ Cos } (\omega_1 t + mt + \phi_2) \qquad (1\text{-}61)$$

Since m is small, the term mt can be associated with the phase ϕ_2 of the second vibration. We can, therefore, look upon the problem as one of composition of two simple harmonic vibrations of the same period but constantly differing phase. Since the phase is now a function of time, the amplitude and the resultant phase also will be functions of time. The resultant vibration can, thus, be expressed as

$$x = a \text{ Cos } (\omega t + \delta) \qquad (1\text{-}62)$$

where

$$a = \sqrt{a_1{}^2 + a_2{}^2 + 2a_1 a_2 \text{ Cos } (mt + \phi_2 - \phi_1)} \qquad (1\text{-}63)$$

and

$$\tan \delta = \frac{a_1 \text{ Sin } \phi_1 + a_2 \text{ Sin } (mt + \phi_2)}{a_1 \text{ Cos } \phi_1 + a_2 \text{ Cos } (mt + \phi_2)} \qquad (1\text{-}64)$$

This represents simple harmonic vibrations with amplitude a and phase δ. The time period of these vibrations is the same as that of the component vibrations but the amplitude does not remain constant as in other case. Its value depends upon the factor $\text{Cos } (mt + \phi_2 - \phi_1)$ and, therefore, a varies between $(a_1 + a_2)$ and $(a_1 - a_2)$. This variation depends upon the value of m and repeats itself after a time interval $2\pi/m$. The amplitude of the resultant vibrations increases and decreases $m/2\pi$ times every second. This phenomenon is known as *beats*. Such beats can be heard when two sources of sound of nearly the same frequency are sounded together. The effect is more pronounced when $a_1 = a_2$ so that the amplitude varies between $2a_1$ and 0. In this case for $\phi_1 = \phi_2 = 0$, we have

$$a = \sqrt{2a_1{}^2 (1 + \text{Cos } mt)}$$

$$= 2a_1 \text{ Cos } \frac{mt}{2}$$

It is worthwhile pointing out that beats have also been observed for light waves (see Chapter 5).

Before we conclude this chapter, it should be pointed out that there are many real situations in which the equations governing wave motion are not linear and the superposition principle fails. Physically, this happens when the wave disturbance is relatively large and the ordinary laws of mechanical action no longer hold. For example, beyond the elastic limit Hooke's law no longer holds and the linear relation that the restoring force is proportional to the displacement is no longer valid. As for sound, violent explosions create shock waves. Although shock waves are longitudinal waves in air, they do not superpose like ordinary sound waves.

SUGGESTED READING

COULSON, C. A., *Waves*, Oliver and Boyd (1955). Gives an excellent mathematical treatment of many different types of wave motion.

CRAWFORD, F.S., *Waves and Oscillations*, Berkeley Physics Course, Vol. III, McGraw-Hill (1968). Recommended for students who want a more thorough and advanced treatment on waves than is offered in our text.

HARVARD PROJECT PHYSICS, *An Introduction to Physics, 4*, 'Light and Electromagnetism', (Interim Version) (1968-69). Chapter 16 gives a good account of electromagnetic radiation without going into Maxwell equations. The mathematical level is elementary.

PHYSICAL SCIENCE STUDY COMMITTEE, *Physics*, D. C. Heath and Co., (1967). Chapters 15 and 16 give a good introduction to waves. The mathematical level is elementary.

WALDRON, R. A., *Waves and Oscillations*, Van Nostrand Momentum Book No. 4, Van Nostrand (1964). The author investigates specific wave properties and shows how phenomena such as resonance and interference are manifested in waves of many kinds. The mathematical details have been reduced to a minimum.

PROBLEMS

1. (*a*) A string which is clamped at both ends is vibrating in its fundamental mode (see Fig. 1-23). Show that all the points are vibrating in the same phase but with different amplitudes.

   (*b*) In the above problem, if the string is assumed to be vibrating in any one of its higher harmonics, then show that any two points are either vibrating in phase or completely out of phase.

2. A wave pulse on an infinitely long stretched string is described by the equation

$$y(x, t) = \frac{0 \cdot 05}{1 + (x - 5t)^2}$$

   where x and y are measured in centimetres and t in seconds.
   (i)   Plot the pulse at $t = 0$ and $t = 0 \cdot 05$ sec.
   (ii) In what direction does the pulse move?
   (iii) What would be the speed of the wave?
   (iv) What would be the equation describing a similar pulse moving in the opposite direction?

3. Consider a triangular pulse moving from right to left on a string with a speed of $\frac{1}{2}$ cm/sec with one end of the string, T, fixed. Fig. 1-32 shows the displacement of the string at $t = 0$ ($AB = BC = AC = 1$ cm).

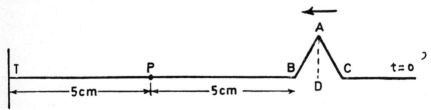

FIG. 1-32

(a) What will be the shape of the pulse at $t = 1$ sec, 20 sec, 21 sec, 22 sec and 25 sec?

(b) Plot the time variation of the displacement of point P.

4. (a) A transverse sinusoidal wave is generated at one end of a long horizontal string. The end moves up and down through a distance of 0·2 cm. The motion is continuous and is repeated regularly twice each second. If the speed of the wave is 20 cm/sec and if we assume that the wave moves from right to left and that at $t = 0$, the end of the string described by $x = 0$ is in its equilibrium position $y = 0$, show that the equation of the wave is given by

$$y = 0·1 \ \text{Sin} \ (0·2\pi x + 4\pi t)$$

where x and y are measured in centimetres and t in seconds.

(b) Find the velocity and acceleration of a particle 10 cm from the end.

(c) What would have been the equation of motion if the wave were moving from left to right?

5. Fig. 1-33 shows two triangular pulses travelling along a string in opposite directions. The speed of the wave is 40 cm/sec and at $t = 0$ the pulses are 120 cm apart. The distances $AB = BC = AC = A'B' = B'C' = A'C' = 1$ cm.

(a) Sketch the patterns after 0·25, 1·0, 1·5, 2·0 and 3 sec.

(b) Discuss the energy associated with the wave at $t = 1·5$ sec.

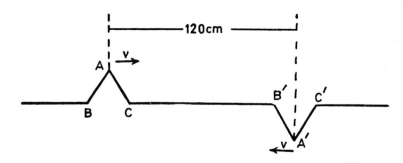

FIG. 1-33

6. A long string vibrates according to the equation

$$y = 0·2 \ \text{Sin} \ \frac{\pi x}{4} \ \text{Cos} \ 20\pi t$$

where x and y are measured in centimetres and t is in seconds.

(a) What are the amplitude, frequency, speed and direction of propagation of the component waves whose superposition can give rise to the above vibrations?

(b) What is the distance between two consecutive antinodes? Will this be the same as the distance between two consecutive nodes?

(c) What is the velocity and acceleration of a particle at the position $x = 2$ cm, when $t = 9/4$ sec?

7. The speed of transverse vibrations of a string under tension (like a sonometer) is given by

$$v = \sqrt{T/m}$$

where T is the tension on the string and m, the mass per unit length. The outline of an experimental arrangement is shown in Fig. 1-34. The length of the wire on which we would like to set up standing waves is 100 cm. A tuning fork (with a frequency of 256 vibrations/ sec) is placed on the end B so as to produce transverse vibrations in the string. What tension must be applied at the end C so that the string executes stationary vibrations with one loop and with four loops?

8. A string is clamped at $x = 100$ cm. A pulse is propagating in the positive x-direction with a speed of 10 cm/sec and the instantaneous displacement at $t = 10$ sec is given by

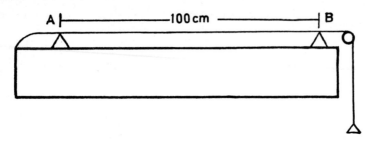

FIG. 1-34

$$y\,(x, t = 10\text{ sec}) = \frac{1}{(x - 10)^2 + 100}\ \text{cm, for } 0 < x < 20$$
$$= 0 \text{ elsewhere}$$

where x is in centimetres. Find the shape of the pulse at $t = 18$ sec, 19 sec, 20 sec and 30 sec.

9. A transverse sinusoidal wave is propagating through a string in the plus x-direction. Consider a point P on the string. The instantaneous displacement of this point is given by

$$y\,(t) = 0 \cdot 5\ \text{Sin}\ \frac{t}{\pi}\ \text{cm}$$

where t is measured in seconds. The wavelength of the wave is 5 cm.
(a) What is the instantaneous displacement at a distance of $1 \cdot 25$ cm on the right and on the left of the point P? Interpret this physically.
(b) Repeat the (a) part for a distance of 1 cm.

10. Two musical notes have frequencies 988 cps and 990 cps respectively. What should be the beat note?

11. In Fig. 1-25, the distance between two consecutive nodes is approximately $5 \cdot 8$ cm. Assuming the speed of electromagnetic waves to be 3×10^{10} cm/sec, calculate the frequency of the wave.

12. A string clamped at $x = 0$ and $x = 100$ cm is pulled at the centre by 1 cm in such a way that the string is on a sinusoidal curve, and at $t = 0$ it is released. Find the shape of the string at $t = 1$ sec from the time of release (velocity of waves $= 20$ cm/sec).
(*Hint:* The general displacement is given by

$$y = f\,(x - 20t) + g\,(x + 20t)$$
$$\therefore\quad \frac{dy}{dt} = -20 f'\,(x - 20t) + 20 g'\,(x + 20t)$$

where primes denote differentiation with respect to the argument. At $t = 0$ apply the condition that $y = \text{Sin}\ \dfrac{\pi x}{100}$ and $\dfrac{dy}{dt} = 0$. A trivial integration of the second equation will give $f\,(x) - g\,(x) = C$ and one may, without any loss of generality, assume $C = 0$. The general displacement is then given by

$$y = \text{Sin}\ \frac{\pi x}{100}\ \text{Cos}\ \frac{\pi}{5}\,t$$

Show that if one had not assumed $C = 0$ the general displacement would have remained the same. Discuss.)

13. Using the method developed in problem 12 show how the two pulses are generated in Fig. 1-22.

14. A string clamped at $x = 0$ and $x = 100$ cm is pulled at the center by 1 cm. The string is allowed to come to rest and at $t = 0$ the displacement is given by

$$y = \frac{1}{50} x \text{ for } 0 \leqslant x \leqslant 50$$

$$= \frac{1}{50} (100 - x) \text{ for } 50 \leqslant x \leqslant 100$$

The string is released at $t = 0$ sec. Obtain an expression for the shape of the string at an arbitrary time t and find the time for which the result displacement is everywhere zero (Velocity of the waves $= 20$ cm/sec).

15. What condition must be imposed on a general function of x and time so that it represents a vibrating string clamped at the point $x = L$. Show that the function becomes a periodic function (of period $2L$) if a similar condition is also imposed at the point $x = 2L$.
    [*Hint:* If the general displacement is written as $f(x - vt) + g(x + vt)$, then the condition that the string is clamped at $x = L$ will lead to $g(x + vt) = -f(2L - x - vt)$. The condition at $x = 2L$ would then show that the function f is periodic.]

16. Prove that for any transverse wave on a string, the slope at any point x is equal to the ratio of the instantaneous transverse speed of the point to the wave speed.

17. Two waves on a string are described by

$$y_1(x, t) = 3 \ exp \left[-\frac{(x + vt)^2}{25 \text{cm}^2} \right]$$

and

$$y_2(x, t) = 2 \ Sin \left[\frac{2\pi}{4 \text{cm}} (x - vt) \right]$$

between $x = (vt - 8)$ and $x = (vt + 8)$. Elsewhere $y_1(x, t) = y_2(x, t) = 0$. Graph the resultant disturbance at $t = -4$ sec, $t = 0$ sec and $t = +4$ sec ($v = 3$ cm/sec).

Huygens' Principle and Geometrical Optics

I never saw a moor,
I never saw the sea;
Yet know I how the heather looks,
And what a wave must be.

— Emily Dickinson

2-1   Introduction

Until about the middle of the seventeenth century, it was generally believed that light consisted of a stream of corpuscles or particles. These corpuscles were supposed to be emitted by sources of light, such as the sun or a candle flame, and to travel outward from the source in straight lines, penetrating transparent materials but not opaque materials. It was believed that when the corpuscles entered the eye, the sense of sight was stimulated.

If the test of the adequacy of a theory is its ability to account for known experimental facts with a minimum of hypotheses, we must admit that the corpuscular theory was an excellent one. This theory could explain why light appeared to travel in straight lines, why it could propagate through vacuum, why it was reflected from a smooth surface such as a mirror with the angle of reflection equal to the angle of incidence, and why and how it was refracted at a boundary surface, such as that between air and water or air and glass. For all of these phenomena, the corpuscular theory provides simple explanations.*

* A good description of Newton's particle model of light and an explanation of some optical phenomena on this basis is provided in Physical Science Study Committee, *Physics*, Chapter 14.

However, there are many aspects of the behaviour of light which are poorly described by the model and are sometimes in contradiction with it. For example, according to the particle theory, the speed of light in a refractive material must be greater than the speed c in vacuum (Problem 4). Indeed, if one uses Newton's particle model one obtains $v_{water} = \mu_{water}c$ where μ_{water} is the refractive index of water with respect to vacuum. Consequently, on this model, the speed in water should be about 4/3 of that in air, and in carbon disulfide ($\mu = 5/3$) the speed should be about 5/3 of that in air. The observed experimental values are almost 3/4 and 3/5 of the speed in air. Not only have the experiments failed to show that ratios that we expected; they indicate that the ratios are inverted.

Further, as we shall see later, light exhibits the phenomenon of 'diffraction', according to which when a beam of light passes through a narrow slit, it spreads out to a certain extent into the region of the geometrical shadow. There is no simple explanation of this phenomenon based on the corpuscular model.

Thus, quoting from the PSSC PHYSICS,

we are now in a situation in which physicists often find themselves. From time to time, a theory which successfully connects a whole group of experimental results fails to account for a new observation. Then we must modify the theory, or start over again and try to invent a new one. It is almost always true, however, that the efforts spent on the old model have not been wasted. . . . The particle nature of light, which here appears unsatisfactory, has played a very important role in the history of physics, and *it still contributes to our understanding of light.*

We shall have more to say about it in Chapter 8. For the moment we put it aside and try a new model.

2-2 The Wave Model

Historically, by the middle of the seventeenth century, while most of the workers in the field of optics accepted the corpuscular theory, the idea had begun to develop that light might be a wave motion of some sort. In 1678, the Dutch physicist, Christiaan Huygens,* showed that the laws of reflection and refraction could be explained on the basis of a wave theory and that such a theory furnished a simple explanation of the then

* Christiaan Huygens (1629-95), Dutch mathematician, astronomer and physicist, was one of the greatest scientists in a century that produced several scientific giants. His contributions were tremendous. To name a few, he was the first to apply spiral springs to watches; he made fundamental contributions to dynamics and he immortalized himself by formulating the wave theory of light. His book, *Treatise on Light,* enuciates the wave theory and presents the experimentation supporting it.

recently discovered phenomenon of double refraction. Huygens' theory simply assumes that light is a wave phenomenon rather than, say, a stream of particles. It says nothing about the nature of the wave and, in particular, gives no hint of the electromagnetic character of light. Huygens did not know whether light was a transverse wave or a longitudinal one; he did not know the wavelengths of visible light; he had only a little knowledge of the speed of light.* Thus, Huygens was able to explain phenomena which are purely related to the wave characer of light, i.e. phenomena which are related to phase, superposition principle, etc.

The wave theory was not immediately accepted and it was not until 1827 that the experiments of Young and Fresnel on interference, and the measurements of the velocity of light in liquids by Foucault at a somewhat later date, demonstrated *the existence of certain experimental results for whose explanation the corpuscular theory of Newton was inadequate.* In later chapters, we will show that the phenomena of interference and diffraction are to be expected if light is assumed to be a wave motion. Further, Young's experiments on interference enabled him to measure the wavelength of the waves, and he found that the wavelengths associated with light are very small (of the order of 10^{-7} metres).

2-3 Huygens' Theory

Huygens' theory is based on a geometrical construction called Huygens' principle. However, before we make this geometrical construction, we have to understand what a wavefront is. A *wavefront* is defined as the locus of points which are in the same phase. Thus, when we touch a quiet water surface with a pencil point, we see a circular wave pulse spread out from the origin of the disturbance. By vibrating the pencil point up and down we can send out a continuous rain of waves, successive crests and troughs forming concentric circles around the source. Any of these circles, where all points are in the same phase, is a wavefront; similarly, in the case of sound waves spreading out in all directions from a point source, any spherical surface concentric with the source is a wavefront. Some spherical surfaces are the loci of points at which the pressure is a maximum, others where it is a minimum, and so on, but the phase of the pressure waves is the same over any spherical surface. It is customary to draw only a few wavefronts which pass through the maxima and minima of the disturbance [Fig. 2-1(a)]. Such wavefronts are separated from one another by one-half of a wavelength. At a sufficiently great distance from the source, where the radii of the spheres have become very large, the spherical surfaces can be considered planes and

---

* Nevertheless, his theory was a useful guide to experiment for many years and remains useful today for quantitative explanation of many experimental phenomena and for other practical purposes. However, it cannot be expected to yield the same wealth of detailed information that the more sophisticated electromagnetic theory does.

we have a train of plane waves as in Fig. 2-1(*b*).

(If the wave is a light wave, the quantity which corresponds to the pressure in a sound wave is the electric or magnetic intensity.)

2-4 Huygens' Principle

Huygens' principle is a geometrical method for finding, from the known shape of a wavefront at some instant, what the shape will be at some later instant. The principle states

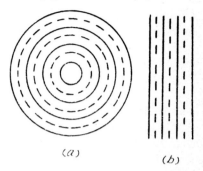

(a)

(b)

FIG. 2-1: Successive loci of points of crests and troughs produced by waves spreading out from a point source.

that every point of a wavefront may be considered as the source of 'secondary' wavelets, which spread out in all directions from their centres with a velocity equal to the velocity of propagation of the wave. The new wavefront is then found by constructing a surface tangent to the secondary wavelets or, as it is called, the *envelope* of the wavelets. If the velocity of propagation is not the same at all portions of the wavefront, the appropriate velocity must be used for the various wavelets.

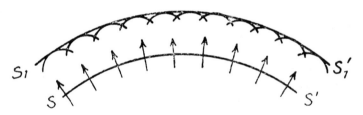

FIG. 2-2: The propagation of a spherical wave in free space as described by Huygens' constuction.

Huygens' principle is illustrated in Fig. 2-2. The original wave front, *SS'*, is travelling as indicated by the arrows. Let v represent the velocity of propagation which is assumed to be the same at all points in all directions. In order to find the shape of the wavefront after a time interval t, we construct a number of circles (traces of spherical wavelets) of radius $r = vt$, with centres along *SS'*. The envelope of these wavelets $S_1 S'_1$, is the new wavefront.

In the simple form given above, Huygens' principle is not fully satisfactory. For instance, the secondary wavelets, if they spread in *all* directions, should also combine to give a 'back' wave which is not observed. This result is avoided by assuming that the *intensity* of the spherical wavelets is not uniform in all directions but varies continuously from a maximum in the forward direction to a minimum of zero in the reverse direction, which, indeed, is what is obtained on the basis of the more rigorous electromagnetic theory. Huygens' method was put on a firm mathematical basis by Fresnel (1788-1827) and can be applied quantitatively to all wave phenomena. Let us consider a particular example.

Example

A vibrating source is moving through water at a speed v_s which is greater than the speed V of water waves. Applying Huygens' construction to the water waves, show that a conical wavefront is set up and that its half-angle α is given by

$$\text{Sin } \alpha = V/v_s$$

Solution

Fig. 2-3 shows wavecrests emitted at regular intervals by a vibrating source. From the figure it is easy to deduce the dependence of the bow-wave

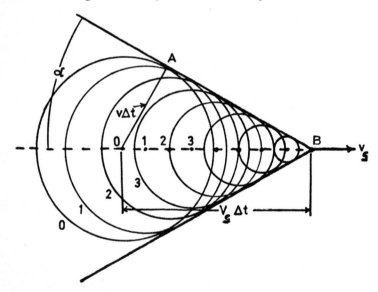

FIG. 2-3: Successive crests emitted by a vibrating source moving at velocity v_s, greater than the wave propagation velocity V. Crest O was emitted when source was at position O, crest 1 from position 1 etc. The 'envelope', or the line tangent to the various crests forms the so called bow waves.

angle α on the velocities v_s and V. During the time interval Δt, the source moves from O to B, a distance $v_s \Delta t$. During the same interval, the first wavecrest emitted at O moves a distance $OA = V\Delta t$. Then

$$\text{Sin } \alpha = \frac{V\Delta t}{v_s \Delta t} = \frac{V}{v_s} \qquad \text{Q.E.D.}$$

A similar phenomenon occurs when a source of sound is moving faster than the speed of sound. Suppose at a given moment a sound wave is generated from the source at point O_1 (Fig. 2-4), then, as the source moves to the point O_2, the wave from O_1 expands by a radius r_1 which is smaller than the distance that the source moves; and of course another wave starts from O_2. When the sound source has moved still farther, to O_3, and a wave is starting there, the wave from O_2 has now expanded to r_2, and the one from O_1 has expanded to r_3. Thus, instead of a source generating spherical waves, as it would if it were standing still, it generates a wavefront which forms a *cone* in three dimensions, or a pair of lines

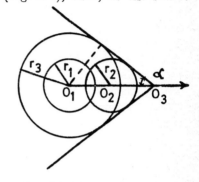

FIG. 2-4: The shock wavefront.

in two dimensions. The angle of the cone is again given by:

$$\text{Sin } \alpha = \frac{V}{v_s}$$

where V now represents the speed of the wave and v_s represents the speed of the source.

Incidentally, although we implied that it is necessary to have a *source* of sound, it turns out, very interestingly, that once the object is moving faster than the speed of sound, it will *make* sound, i.e. it is not necessary that it should have a certain tone vibrational character. Any object moving through a medium faster than the speed at which the medium carries waves will generate waves on each side automatically, just from the motion itself. This also occurs in the case of light. At first one might think nothing can move faster than the speed of light. However, as we shall soon discuss, the speed of light in glass (or in water) is less than the speed of light in vacuum, and it is possible to shoot a *charged* particle of very high energy through a block of glass such that the particle velocity is close to the speed of light in vacuum, while the speed of light in the glass may be only 2/3 the speed of light in vacuum. A charged particle moving faster than the speed of light in the medium will produce a conical wave of light with its apex at the source. By measuring the cone angle, we can determine the speed of the particle. This is used to determine the speeds of particles in high energy physics. The direction of the light is all that needs to be measured. This light is called Cerenkov radiation, because it was first observed by Cerenkov.*

* How intense this light will be was analysed theoretically by Frank and Tamn. The 1958 Nobel Prize for physics was awarded jointly to all three for this work. For further details on shock waves see *The Feynman Lectures on Physics* by R. P. Feynman, R. B. Leighton and M. Sands, Vol. 1, Chapter 51.

2-5   Rectilinear Propagation

One of the main objections which was first put forward against the wave theory was its failure to account for the rectilinear propagation of light and the formation of shadows. It was known that sound and water waves bend around the corners of obstacles (a phenomenon which is known as diffraction), and it was naturally argued that if light consisted of a wave motion, it should behave in a similar manner. The objection was partially answered by Huygens, though it remained for Fresnel to give the complete explanation.

Let us suppose we have a point source of light at O (Fig. 2-5) and a

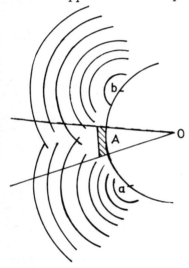

small obstacle (like a coin) at A. At the time of Huygens, it was believed that no light penetrates into the conical region behind the obstacle.* Now, the question put forward to Huygens was that, if all the points on the wavefront are acting as independent sources, why does not the entire area appear luminous to an eye behind the obstacle? Huygens' answer to this question was that these secondary waves produced no appreciable effect at a point unless they were *at that point* enveloped by a common tangent plane. This explanation of the rectilinear propagation of light amounted simply to an assumption-that only one point on the secondary wavelet was effective in producing light.

FIG. 2-5: Spherical wavelets emanating from points a and b on the wavefront.

Fresnel† was the first to give a really satisfactory explanation of the phenomenon by supplementing Huygens' construction with the postulate that the *secondary wavelets mutually interfere.* This combination of Huygens' construction with the principle of interference is called the *Huygens-Fresnel principle.* According to Fresnel's theory it was no longer necessary to assume that only a minute portion of the secondary wave was operative in producing light, which, as a matter of fact, is contrary to experimental evidence as can be shown by allowing a plane wave to fall on an opaque screen perforated with a *very small* aperture. The point

---

* Now we know that light *does* penetrate inside the conical region, but as the wavelength of light is extremely small ($\approx 10^{-5}$ cm) the diffraction effects are difficult to observe. This point is discussed later and also in Chapter 6.

† Augustin Fresnel (1788-1827). Notable French contributor to the theory of light. His mathematical investigations gave the wave theory a sound foundation.

on the wavefront not cut off by the screen acts as a centre of disturbance which spreads out into the space behind the obstacle (Fig. 2-6 and Plate 1). A screen placed behind the aperture, as shown in Fig. 2-6, will be illuminated over an area many times greater than that of aperture. Sound waves also behave in a similar manner and it is actually possible to photograph the secondary wavelets.*

Waves on the surface of water can also be made to exhibit diffraction in passing through an aperture whose size is comparable to the wavelength.†

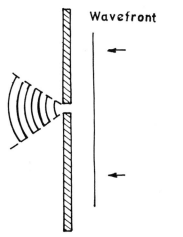

Wavefront

Fig. 2-6: Spreading out of waves after passing through a small aperture

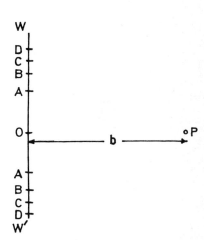

Fig. 2-7

2-6   Effect of a Plane wave on an Exterior Point—Fresnel's Half-period Zones and the Phenomenon of Diffraction

Let us consider a plane wave WW' moving towards a point P (Fig. 2-7). We would like to find out the effect at this point of all the secondary wavelets emanating from the wavefront. Now wavelets from different parts of the wavefront are, in general, at different distances from P and so will arrive at P with different phases. In order to calculate the resultant disturbance at P, Fresnel divided the wavefront into zones so that at P the wavelets from one zone are just π out of phase with those from the next. When a crest arrives at P from one zone, troughs are arriving from the next and it is easy to find the resultant disturbance. This division into half period zones is done in the following way. Draw a perpendicular

* See, for example, R. W. Wood, *Physical Optics*, Macmillan (1934). His original work appeared in an article entitled Photography of Sound Waves, *Philosophical Magazine*, August 1899.

† The diffraction of plane waves in water in a ripple tank arrangement is well described in PSSC, *Physics* (Chapter 16).

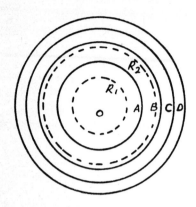

FIG. 2-8: Fresnel's half-period zones.

from the point P to the wavefront, intersecting it at O, the pole of the wave with respect to P. With the point P as centre draw spheres of radii $b + \lambda/2$, $b + 2\lambda/2$, $b + 3\lambda/2, \ldots$. These spheres will intersect the wavefront in circles* (Fig. 2-8) and will be such that the secondary disturbances coming from any circle will reach P half a wavelength ahead of those coming from the circle encircling it. If we consider the region on the circle R_1 in the first zone (Fig. 2-8) then there is always a corresponding circle R_2 in the second zone such that troughs from R_1 arrive at P at the same time as crests from R_2.

We regard the effect of the disturbances coming from each ring as proportional to its area and decreasing with increasing distance and obliquity.

Let us calculate the area of the rings. The radius of the nth ring, r_n, will be given by

$$r_n = \sqrt{\left(b + \frac{n\lambda}{2}\right)^2 - b^2}$$

or

$$r_n \cong \sqrt{bn\lambda} \tag{2-1}$$

where $OP = b$ and we have neglected quantities which depend on λ^2. (The wavelength of visible light ranges from 4×10^{-5} cm to 8×10^{-5} cm.) Thus the radii of the circles are (approximately) $\sqrt{b\lambda}$, $\sqrt{2b\lambda}$, $\sqrt{3b\lambda}$, etc., and the corresponding areas being $\pi b\lambda$, $2\pi b\lambda$, $3\pi b\lambda$, etc. Therefore, the areas of the central circle and each surrounding zone are the same, each equal to $\pi b\lambda$.

The effect due to the disturbances coming from one of the zones will be proportional to its area and inversely proportional to its distance from P. It can be easily shown (Problem 1) that the slight increase in the area of the zones (if we do not neglect the term involving λ^2 in Eq. 2-1) as we recede from the centre of the system is compensated by the increased distance, so that, other things remaining equal, we can regard the successive zones as producing equal and opposite effects at the point. The zones, however, become less and less effective as we recede from the centre owing to the increased obliquity. We can therefore, represent the

---

* The annular region between the circles are said to be Fresnel half-period zones. Fresnel's zone construction for a spherical wavefront can easily be carried out [see for example, M. Born and E. Wolf, *Principles of Optics*, Chapter VII, Pergamon Press (1969)].

resultant effect by a series of terms of alternate sign which would decrease slowly, eventually becoming zero. Thus we may write for the resultant amplitude:

$$S = K_1 - K_2 + K_3 - \ldots + (-1)^{n+1} K_n \qquad (2\text{-}2)$$

where K_n denotes the net amplitude of all the secondary wavelets from the nth zone. As discussed before, the zones are constructed in such a way that the net amplitude from the nth zone is π out of phase from the $(n-1)^{\text{th}}$ or $(n+1)^{\text{th}}$ zone. Hence the terms have alternately negative and positive signs.

The series in Eq. 2-2 can be approximately summed by a method due to Schuster. First we write the series in the form

$$S = \frac{K_1}{2} + \left(\frac{K_1}{2} - K_2 + \frac{K_3}{2}\right) + \left(\frac{K_3}{2} - K_4 + \frac{K_5}{2}\right) + \ldots \qquad (2\text{-}3)$$

the last term being $\frac{1}{2}K_n$ or $(\frac{1}{2}K_{n-1} - K_n)$ according to n being odd or even. Let us assume the obliquity factor to be such that

$$K_j > \frac{K_{j-1} + K_{j+1}}{2} \qquad (2\text{-}4)$$

Then each of the bracketed term in Eq. 2-3 is negative and it follows that

$$S < \frac{K_1}{2} + \frac{K_n}{2} \qquad \text{when } n \text{ is odd}$$

and

$$S < \frac{K_1}{2} + \frac{K_{n-1}}{2} - K_n \quad \text{when } n \text{ is even} \qquad (2\text{-}5)$$

We can also write Eq. 2-2 in the form

$$S = K_1 - \frac{K_2}{2} - \left(\frac{K_2}{2} - K_3 + \frac{K_4}{2}\right) - \left(\frac{K_4}{2} - K_5 + \frac{K_6}{2}\right) - \ldots \qquad (2\text{-}6)$$

the last term now being $-\frac{1}{2}K_{n-1} + K_n$ when n is odd and $-\frac{1}{2}K_n$ when n is even. We may now write

$$S > K_1 - \frac{K_2}{2} - \frac{K_{n-1}}{2} + K_n \qquad (n \text{ odd})$$

and

$$S > K_1 - \frac{K_2}{2} - \frac{K_n}{2} \qquad (n \text{ even}) \qquad (2\text{-}7)$$

From Eqs. 2-5 and 2-7 we can write approximately

$$S = \frac{K_1}{2} + \frac{K_n}{2} \qquad (n \text{ odd})$$

and

$$S = \frac{K_1}{2} - \frac{K_n}{2} \qquad (n \text{ even})$$

For large values of n, since $|K_n| \ll |K_1|$, the problem reduces to a determination of the effect due to one half of the central zone. The secondary wavelets from this zone unite into a disturbance, the phase of which is midway between those of the wavelets from the centre and rim; for, we may divide the zone into a series of concentric rings of equal area the effects of which at the point are equal in ampliude and of phases ranging over half of a complete period. These vibrations may be compounded as vectors by the method given in Figs. 1-30 and 1-31. The resultant amplitude will be very nearly the diameter of a circle, the semicircumference of which is made up of the vectors which represent the amplitudes contributed by the elementary zones into which we have divided the central circle (see Fig. 1-30). The direction of the diameter makes an angle of 90° with that of the first vector. *Consequently the phase of the resultants is $\pi/2$ behind that due to the element at the centre.*

The above theory can be put to direct experimental test by screening off all of the wavefront except the central circle of the zones. This can be accomplished by placing a screen provided with a small circular aperture *at such a distance from the point P that the area of the aperture is equal to the area of the central zone* (Fig. 2-9). This would double the resultant amplitude

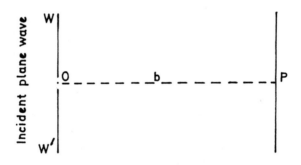

FIG. 2-9: Diffraction by a circular aperture.

and the illumination becomes four times that due to the unobstructed wave. Obviously, the actual sizes on the wavefront in a given plane depends on the distance of the point P. As this distance (i.e. b) increases, the zones widen out.

Next, if we increase the size of our aperture until it contains another zone, then disturbances coming from the second zone will be out of phase with those coming from the central circle and the two will almost destroy each other. *Thus, by increasing the size of the hole we can reduce the illumination to (almost) zero!*

If we substitute a small circular disc for the aperture we find that the illumination on the axis of the shadow is almost unaffected by the interposition of the circular disc. By increasing the size of the disc we cut off another zone, still without influencing the illumination, and this may be continued, not indefinitely, but until, owing to the increasing obliquity, the effect of the zone begins to diminish appreciably. Therefore, we find that the centre of the shadow of a circular body may, under certain conditions, be almost as brightly illuminated as the surrounding field.

Let us now work out the radius of the first zone for a typical experiment in which a plane wave falls on an obstacle which casts a shadow on the screen some 30 cm away. Thus $b = 30$ cm and taking λ as 6×10^{-5} cm, we get $r_1 = \sqrt{30 \times 6 \times 10^{-5}} \approx 0 \cdot 04$ cm, so that the diameter of the first zone is less than a millimetre and that of $10{,}000^{\text{th}}$ zone is less than 10 cm.

Now let a plane wave represented by WW' (Fig. 2-10) fall on an obstacle T such as a sheet of cardboard, the shadow being cast on a screen some 30 cm beyond the obstalce. The illumination at a point such as P_1, whose pole O_1 is so far from the edge of the obstacle that the first 100 or so zones pass it, will be almost the same as if the obstacle were removed, for we may assume that the zones after the hundreth add very little to the amplitude

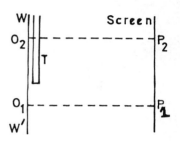

FIG. 2-10: Explanation of rectilinear propagation of light.

at P_1. Hence the presence of the obstacle will not be noticed when O_1 is farther from the edge of the cardbaord than r_{100} (≈ 4 mm) (which is also the distance of P_1 from the edge of the *geometrical shadow*). Similarly, if P_2 is so far inside the geometrical shadow that its pole O_2 is further from the edge of the obstacle than 4 mm, only zones of higher order than the hundreth can send wavelets to P_2. The shadow will then be quite dark, just as if light travelled in straight lines. It is only when O_2 is within 4 mm of the edge of the obstacle and P within 4 mm of the edge of the geometrical shadow that we would observe the effects due to diffraction of light. Thus, due to the very small wavelength of light, the deviations from rectilinear propagation will be observed only within a millimetre or two of the edge of the shadow.*

2-7 Zone Plate

An interesting verification of Fresnel's theory is furnished by the *zone plate*. If on a large sheet of white paper we describe circles, the radii of which are proportional to the square roots of the natural numbers, we shall have very nearly an exact drawing of the zone system (neglecting of

---

* It is interesting to point out that for sound waves the wavelength is roughly 100 cm. Therefore, for $b = 30$ cm, $r_1 = \sqrt{b\lambda + \lambda^2/4} \approx 70$ cm. For a case such as this the propagation cannot be regarded as rectilinear.

course, terms containing λ^2). If we now blacken the alternate rings with ink (Fig. 2-11) and take a greatly reduced photograph of the whole on glass, we shall obtain a device called zone plate, which will enable us to screen off the alternate zones on a wavefront. Suppose we intercept a plane wave with such a plate and consider the illumination at a point so situated behind the plate that the central circle of the plate corresponds in size and position to the first zone on the wavefront. The black rings stop all the secondary disturbances from the alternate or odd zones, which previously neutralised those coming from the even ones. Consequently, all the secondary disturbances coming from that portion of the wavefront covered by the plate reach the point in the same phase, and the illumination will be very intense. The whole surface of the zone plate will send light to the point, the action being very similar to that of a convex lens.

FIG. 2-11: The zone plate.

Example
Light of wavelength 5×10^{-5} cm is passed through a narrow circular aperture of diameter 2 mm. Find the positions of the brightest and darkest points on the axis.

Solution
For the brightest point on the axis, only one half-period zone should be covered by the aperture. If the distance of this point from the aperture is b cm, then

$$r_1{}^2 = (0 \cdot 1)^2 = b\lambda$$

or

$$b = \frac{0 \cdot 01}{5 \times 10^{-5}} = 200 \text{ cm}$$

Similarly for the darkest point on the axis, only two half-period zones should be covered by the aperture. If the distance of this point from the aperture is a cm, then

$$r_2{}^2 = (0 \cdot 1)^2 = 2a\lambda$$

or

$$a = \frac{0 \cdot 01}{2 \times 5 \times 10^{-5}} = 100 \text{ cm}$$

Example

In a zone plate, the radii of the circles are given by

$$r_n = 0 \cdot 05 \ \sqrt{n} \ \text{cm}$$

Thus the radii of the circles are proportional to the square roots of natural numbers. The 2nd, 4th, 6th....zones are darkened. Calculate the focal length of the zone plate. Will there be any other bright points on the axis? ($\lambda = 5 \times 10^{-5}$ cm)

Solution

For the brightest point on the axis, the first circle must contain only the first half-period zone. Thus the focal length, f, will be given by

$$r_1^2 = f \lambda \tag{2-9}$$

or

$$f = \frac{(0 \cdot 05)^2}{5 \times 10^{-5}} = 50 \ \text{cm}$$

There will be other focii on the axis for which the first circle will contain 3, 5, 7...half-period zones, but these points will not be as intense as the first focus. Their distances will be given by

$$r_1^2 = (2n + 1) \ f_n \lambda; \ n = 1, 2, 3, \ldots \tag{2-10}$$

or

$$f_n = 16 \cdot 7 \ \text{cm}, \ 10 \ \text{cm}, \ 7 \cdot 1 \ \text{cm etc.}$$

Between these focii there will be points of low intensity for which the first circle will contain an even number of half-period zones.

2-8 Huygens' Principle and the Law of Reflection

We now return to Huygens' principle and show how it can be used to determine the reflection of a plane wavefront from a surface. It will be assumed that the surfaces are optically smooth, which means the irregularities are small compared to the wavelength. Under these conditions the reflection is termed *specular*. Polished metals, liquid surfaces and mirrors (which reflect in such a way that the reflected pencils are sharply defined) are said to be *specular reflectors*. (Materials such as white paper give diffuse reflection.)

We will consider a plane wavefront ABC incident on a plane reflecting surface MM' (Fig. 2-12). At $t = 0$, the wave front just touches the reflecting surface and secondary wavelets start to spread out from A. The part of the wavefront at C will not reach the reflecting surface, and begin to be reflected, until it has reached C'. If the distance $CC' = a$, then the secondary wavelet starting from A will have, by this time, reached a position given by an arc of a circle whose centre is at the point

A and whose radius is equal to *a*. The part of the wavefront starting at *B*, midway between *A* and *C*, will travel a distance *a*/2 before reaching the reflecting surface at *D*, midway between *A* and *C'*. The secondary wavelet starting from *D* will therefore have only half the time to travel than the one from *A* has and, therefore, we draw an arc of a circle with centre *D* whose radius is equal to *a*/2. The envelope of all the spherical surfaces (shown as circles in the Fig. 2-12) will be found to be a plane (*A'B'C'* in the Fig. 2-12) and travelling in the direction *AA'*. The plane *A'B'C'* is the reflected wave front.

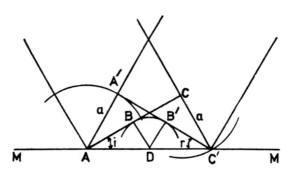

FIG. 2-12: Reflection of a plane wavefront by a plane mirror

The angle *i* between an approaching wavefront and the reflecting surface is called the *angle of incidence* and the angle *r* between a receding wave front and the reflecting surface is called the *angle of reflection*. Because

$$\text{Sin } i = \frac{CC'}{AC'}, \quad \text{Sin } r = \frac{AA'}{AC'}$$

and

$$AA' = CC'$$

we have

$$\text{Sin } i = \text{Sin } r,$$

i.e. the angles *i* and *r* are equal. Thus we have the basic law of reflection: *the angle of reflection of a plane wavefront with a plane mirror is equal to the angle of incidence.*

2-9 Huygens' Principle and the Law of Refraction

Another phenomenon which can be analyzed using Huygens' principle is the refraction of light—that is, the change in direction of a light beam when it passes from one medium to another in which its speed is different. This can be evaluated with the help of the wavefront diagram in Fig. 2-13. At $t = 0$, the wavefront *AB* just comes in contact with the boundary between the two media. We suppose that the speed of light, v_2, in the new medium is less than that in the first medium, which is v_1; so that secondary wavelets generated at the interface travel a shorter distance in the same time interval than do wavelets in the first medium. At $t = 0$, when the end A of the wavefront AB reaches the boundary, end B is still at a distance BC away. Since the speed of light in the first medium is v_1, end B requires time $t' = \dfrac{BC}{v_1}$ to reach the boundary at C. In this period of time the end A of the wavefront proceeds to D, where AD is smaller than BC. Evidently

$$BC = v_1 t' \text{ and } AD = v_2 t'$$

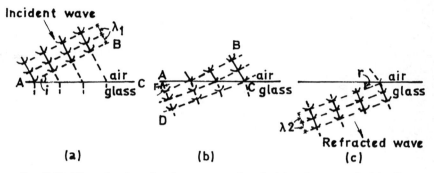

FIG. 2-13: The refraction of a plane wave as described by Huygens' principle. Note the change in wavelength on refraction. The dotted lines show the position of the successive crests.

The angle i between an approaching wavefront and the boundary between two media is called the *angle of incidence* of the wavefront and the angle r between a receding wavefront and the boundary between two media through which it has passed is called the *angle of refraction*. From Fig. 2-13(a) we see that

$$\text{Sin } i = \frac{BC}{AC} = \frac{v_1 t'}{AC}$$

and from Fig. 2-13(b)

$$\text{Sin } r = \frac{AD}{AC} = \frac{v_2 t'}{AC}$$

so that

$$\frac{\text{Sin } i}{\text{Sin } r} = \frac{v_1}{v_2} \tag{2-11}$$

Thus, *the ratio of the sines of the angles of incidence and refraction is equal to the ratio of the speeds of light in the two media*. This is known as *Snell's law** after

* Actually Snell had found that Sin i/Sin r was a constant, independent of the angle of incidence (this constant is approximately equal to 1·33 for water).

its discoverer, the seventeenth century Dutch astronomer Willebrord Snell.

We may rewrite Eq. 2-11 as

$$\frac{c}{v_1} \, \text{Sin} \, i = \frac{c}{v_2} \, \text{Sin} \, r \tag{2-12}$$

where c is the speed of light in free space. If we define,

$$\mu_1 = \frac{c}{v_1} \tag{2-13}$$

and

$$\mu_2 = \frac{c}{v_2} \tag{2-14}$$

then, μ_1 and μ_2 are known as the indices of refraction of medium 1 and of medium 2, respectively, with respect to vacuum. In terms of μ_1 and μ_2 the law of refraction can be written as

$$\mu_1 \, \text{Sin} \, i = \mu_2 \, \text{Sin} \, r \tag{2-15}$$

Table I gives a list of values of μ for a number of substances. The greater the index of refraction, the greater the extent to which a light beam is deflected upon entering or leaving that medium. It should be pointed

TABLE I

INDICES OF REFRACTION OF VARIOUS MATERIALS RELATIVE TO VACUUM
(for light of wave length $\lambda = 5 \cdot 890 \times 10^{-5}$ cm)*

Material	μ	Material	μ
Vacuum	1·0000	Quartz (fused)	1·46
Air	1·0003	Rock salt	1·54
Water	1·33	Glass (ordinary crown)	1·52
Ethyl alcohol	1·36	Glass (dense flint)	1·66
Quartz (crystalline)	1·54		

* After A. B. Aarons, *Development of Concepts in Physics*, p. 604, Addison-Wesley (1965)

out that the index of refraction decreases with increase in wavelength. However, this dependence on wavelength is very small. As an example, the variation in the refractive index with wavelength for crown glass is given in Table II.

Returning to Fig. 2-13 we note that

$$\frac{\lambda_1}{\lambda_2} = \frac{\text{Sin} \, i}{\text{Sin} \, r} = \frac{v_1}{v_2}$$

or,

$$\frac{\lambda_1}{\lambda_2} = \frac{c/v_2}{c/v_1} = \frac{\mu_2}{\mu_1} \tag{2-16}$$

TABLE II*

REFRACTIVE INDICES OF TELESCOPE CROWN GLASS AND
VITREOUS QUARTZ FOR VARIOUS WAVELENGTHS

Wavelength	Refractive Telsecope Crown	Indices Vitreous Quartz
1. $6 \cdot 562816 \times 10^{-5}$ cm	$1 \cdot 52441$	$1 \cdot 45640$
2. $5 \cdot 889953 \times 10^{-5}$ cm	$1 \cdot 52704$	$1 \cdot 45845$
3. $4 \cdot 861327 \times 10^{-5}$ cm	$1 \cdot 53303$	$1 \cdot 46318$

NOTE: The wavelengths corresponding to serial numbers 1, 2 and 3 correspond roughly to the red, yellow and blue colours. The table shows the accuracy with which the wavelengths and refractive indices can be measured.

*After F. A. Jenkins and H. E. White, *Fundamentals of Optics* (Third Edition), p. 465, McGraw-Hill (1957)

Now, if we assume that the medium above the glass is vacuum rather than air then μ_1 becomes 1 and

$$\lambda_2 = \lambda/\mu_2 \tag{2-17}$$

where λ is the wavelength of the wave in free space. This shows that the wavelength of light in a material medium is less than the wavelength of the same light in vacuum. Fig. 2-13 also shows clearly the differences in wavelength in the two media.[†]

It should be noted that reflection and refraction can also take place at a rough or a broken surface. The irregularities on the surface have to be measured in relation to the wavelength of the incident wave. A smooth surface is one in which the dimensions of the surface irregularities are much smaller than the wavelength; here the reflection and refraction take place according to the laws discussed above, and the resulting waves are said to be *coherent*. In fact it can be shown that if the surface is smooth to within $\lambda/8$, we will have regular reflection.[‡]

On the other hand, an unpolished surface destroys all phase relations between the elements on the wavefront. The secondary wavelets start from the elevated portions of the surface first, since these portions are struck first by the incident wave, and the reflected wave front, instead of being plane is corrugated in an irregular manner (Fig. 2-14). It is impossible to arrange any zone system on such a surface, for there are all possible phase differences irregularly distributed over the reflected wave front, consequently each point on the surface acts as an independent luminous source, sending light out in all directions.

[†] It is interesting to point out that the above considerations show that when a wave gets refracted into a different material, the wavelength and the speed change in such a way that the frequency remains the same.

[‡] See, for example, R. W. Wood, *Physical Optics*, Chapter I, Macmillan (1934).

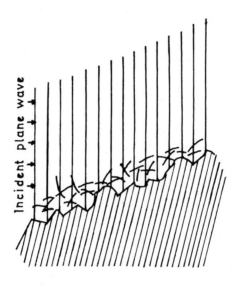

FIG. 2-14: Diffuse reflection of plane waves from
a rough boundary.

2-10 Reflection of Light from a Point Source near a Mirror

Let us consider a luminous point object, *O*, placed near a plane mirror
(Fig. 2-15). In the figure, *ABC* is a diverging spherical wavefront from
the source *O*, which has just touched the plane mirror *MM'* at the point

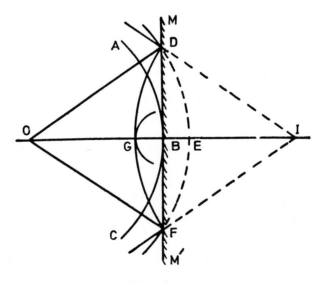

FIG. 2-15: Reflection of light from a point source near a
plane mirror.

B. If the mirror was not there, the wavefront would have, after a short time, reached a position *DEF* where $AD = BE = CF = l$, the distance travelled in the medium of propagation in this time. The secondary wavelet starting from the point *B* will, therefore, have reached a point *G*, where $BG = BE$, while those radiating from *D* and *F* will be only just starting. The new wave front is therefore *DGF*, and if other secondary wavelets are constructed, starting from points along *DF*, it will be found that it is an arc of a circle with centre *I*, where $BI = OB$. Thus the reflected wave will appear to be diverging from the point *I* and hence *I* is referred to as the image of *O* in the mirror.

2-11 Refraction of a Spherical Wave by a Spherical Surface*

A slightly more complicated case will be the refraction (or reflection) of a spherical wave by a spherical surface. Let us consider refraction by a converging surface forming a real image. In Fig. 2-16 *APC* is a spherical

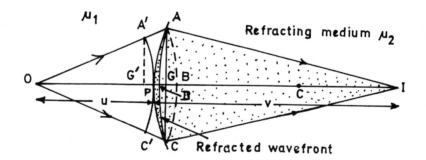

FIG. 2-16: Refraction of a spherical wave by a spherical surface.

refracting surface, convex to the less dense medium and therefore converging. A wavefront diverging from a real object *O* will meet the surface first on the axis at *P*, the position of the wavefront being *A'PC'*. If the refracting surface were not there, the wavefront would go on to a position *ABC*, but in the refracting medium the distance travelled by the light in a given time will be less than that in the first medium. If the indices of refraction of the first and the second media be μ_1 and μ_2 then the refracted wavefront, *ADC* (which is the envelope of all the secondary wavelets), will be such that

$$\frac{PB}{c/\mu_1} = \frac{PD}{c/\mu_2} \tag{2-18}$$

or
$$(PG + GB)\,\mu_1 = \mu_2\,(PG - GD) \tag{2-19}$$

* Various cases of reflection and refraction at spherical surfaces are well discussed in F. J. H. Dibdin, *Essentials of Light*, Chapter 12, Cleaver-Hume (1961).

where PG is the perpendicular on the line joining the point O and the centre of curvature of the refracting surface C. Now we assume that the refracted wave is almost spherical, converging to the point I and $PI = v$. We also assume that the distances PG and GB are very small compared to u and v; thus

$$AG^2 = (2OB - GB) \quad (GB) \approx 2u \ (GB)$$

Also
$$AG^2 = (2CP - GP) \quad (GP) \approx 2R(GP)$$

and
$$AG^2 = (2ID - GD) \quad (GD) \approx 2v(GD)$$

$$\therefore \qquad 2u(GB) = 2R(GP) = 2v \ (GD) \tag{2-20}$$

where R is the radius of curvature of the spherical surface. Using Eqs. 2-19 and 2-20 we obtain

$$\frac{\mu_1}{u} + \frac{\mu_2}{v} = \frac{\mu_2 - \mu_1}{R} \tag{2-21}$$

From the above equation we can determine the position of the image for a given object position.

2-12 Ray Treatment of Light: Geometrical Optics

The method of Huygens' secondary wavelets can be extended to cover refraction in a lens, or in a system of lenses, but this becomes very cumbersome as there are many wavefronts to be considered. However, since the wavelength of visible light is very small ($\approx 10^{-5}$ cm), it may be expected that a good first approximation to the propagation laws may be obtained by a complete neglect of the finiteness of the wavelength. It is found that for many optical problems such a procedure is entirely adequate; in fact, phenomena which can be attributed to departure from this approximate theory (e.g. so called diffraction phenomenon) can only be demonstrated by means of carefully conducted experiments.

The branch of optics which is characterized by the neglect of the wavelength, i.e. that corresponding to the limiting case $\lambda \to 0$, is known as *geometrical optics*, since in this approximation the optical laws may be formulated in the language of geometry. The energy may then be regarded as being transported along certain curves (light rays). The physical model of a pencil of rays may be obtained by allowing the light from a source of negligible extension to pass through a very small opening in an opaque screen. The light which reaches the space behind the screen will fill a region, the boundary of which (the edge of the pencil) will at first sight appear to be sharp. We are of course neglecting here the diffraction phenomenon which will make the light intensity rapidly but continuously vary from the darkness in the shadow to the lightness in the illuminated region (see Section 2-6). Therefore, as long as this magnitude is neglected in comparison with the dimensions of the opening, we may speak of a *sharply bounded pencil of rays.* On reducing the size of the opening to the dimensions of the wavelength, diffraction effects must be taken into account. If, however, one considers the limiting case of negligible wavelengths, no restriction on the size of the opening is imposed, and we may say that an opening of vanishingly small dimensions defines an infinitely thin pencil —the *light ray.* Further, within the approximation of geometrical optics, the laws of refraction and reflection established for plane waves incident upon a plane boundary remain valid under more general conditions.

2-13 Fermat's Principle

One may ask: What is the path that a light ray will travel from one point to another; i.e. in the geometrical optics approximation, what is the curve along which the energy is transported? The answer to this question is the Fermat's principle which was discovered in 1650. This principle is often expressed as follows: *A light ray travelling from one point to another will follow a path such that, compared with nearby paths, the time required is either a minimum or a maximum or will remain unchanged (i.e. it will be stationary).**

Let us derive the laws of reflection and refraction from this principle.

Let us consider two points *A* and *B* and a plane mirror *MM'* (Fig. 2-17). How do we go from *A* to *B* in the shortest time? One obvious answer is to go straight from *A* to *B*! But if we impose the rule that the light has to strike the mirror and come back in the shortest time then the answer is

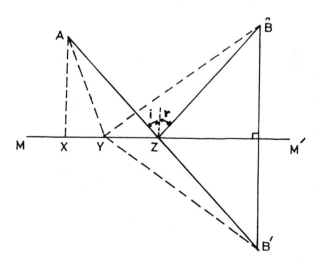

Fig. 2-17: Illustration of Fermat's principle for reflection.

* A mathematical description of light rays (from Maxwell equations) and a precise mathematical formulation of Fermat's principle (along with its proof) are given in Chapter 3 of M. Born and E. Wolf's *Principles of Optics*, Pergamon Press (1969). The mathematics is quite involved.

not very easy. One has to consider a large number of paths and find out which one takes the minimum time.

We construct on the other side of MM' an artificial point B', which is at the same distance below the plane MM' as the point B is above the plane. Now *any* point Y on the plane of the mirror will be such that $YB = YB'$. Therefore, the sum of the two distances, $AY + YB$, which is proportional to the time it will take for light to cover them if the light travels with constant velocity is also the sum of the two lengths $AY + YB'$. Therefore the problem becomes, when is the sum of these two lengths the least? The answer is: When the line goes through the point Z as a straight line from A to B'. It now becomes very easy to show that the path will be such that the angle of incidence is equal to the angle of reflection and that the incident ray, reflected ray and the normal lie in the same plane.

Next, we will derive the laws of refraction. This we shall do by using differential calculus.*

Let A and B be two points in different media and a ray AB connecting them (Fig. 2-18). The time t is given by

$$t = \frac{L_1}{v_1} + \frac{L_2}{v_2}$$
$$= \frac{\mu_1 L_1 + \mu_2 L_2}{c} \tag{2-22}$$

where c is the speed of light in vacuum. The quantity $\mu_1 L_1 + \mu_2 L_2$ is

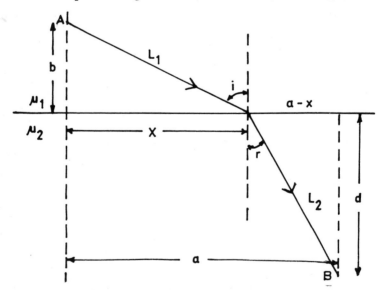

FIG. 2-18: Illustration of Fermat's principle for refraction.

* A non-mathematical proof is given in *The Feynman Lectures on Physics* by R. P. Feynman, R. B. Leighton and M. Sands, Chapter 26, Addison-Wesley (1965).

called the optical path length of the ray. Now from Fig. 2-18 we have

$$t = \frac{1}{c}\left[\mu_1\sqrt{b^2 + x^2} + \mu_2\sqrt{d^2 + (a - x)^2}\right]$$

Now Fermat's principle requires that t should be a minimum (or a maximum or must remain unchanged) which implies that $\dfrac{dt}{dx}$ must be zero. Thus

$$\frac{dt}{dx} = \frac{1}{c}\left[\frac{1}{2}\mu_1\frac{2x}{\sqrt{b^2 + x^2}} - \mu_2\frac{1}{2}\frac{2(a - x)}{\sqrt{d^2 + (a - x)^2}}\right] = 0$$

or

$$\mu_1\frac{x}{\sqrt{b^2 + x^2}} = \mu_2\frac{(a - x)}{\sqrt{d^2 + (a - x)^2}}$$

or

$$\mu_1 \, \mathrm{Sin} \, i = \mu_2 \, \mathrm{Sin} \, r \tag{2-23}$$

which is the law of refraction.

2-14 Applications of Fermat's Principle

Fermat's principle is very useful in dealing with transmission in non-homogeneous media. The phenomenon of mirage, for example, can be treated by means of it, as we will now show.

The air is very hot just above the ground and is cooler higher up. Hotter air is more expanded than cooler air, and therefore the refractive index increases in the upward direction. Consider now two light rays from A in slightly different directions (Fig. 2-19). As the lower one is always in a region of slightly lower refractive index, the velocity there is larger and therefore the upper ray must travel a shorter path to take the same time. Fig. 2-19 illustrates why the path is curved.

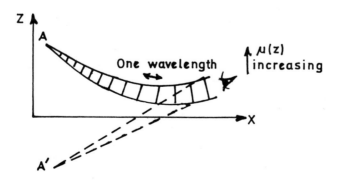

FIG. 2-19: Fermat's principle applied to a mirage.

Another interesting application is to find the direction of the ray in an anisotropic medium. For a point source in an isotropic medium the speed of the wave is the same in all directions and therefore the wavefronts are spherical and the rays are normal to the wavefronts. However, if the speed of light is different in different directions (as is indeed the case in calcite) the wavefronts are not spherical in nature. In a particular case (for calcite) the wavefronts are ellipsoid of revolution. When a narrow beam of light is incident on such a crystal the refracted wavefronts are obtained by finding a common tangent to the ellipsoid and the direction of the ray (which is now *not* normal to the wavefront) corresponds to the shortest optical path for the transfer of light energy from the point on the surface to the wavefront [see Fig. 3-27(b)]. This we will discuss in greater detail in Chapter 3.

In each of the examples that we have discussed the optical path proves to be a *minimum*. Problem 31 describes a case in which it may be a maximum, minimum, or stationary.

We return to Fig. 2-16 and show how geometrical optics and the ray diagram can be used in constructing images. In Fig. 2-20 we have shown

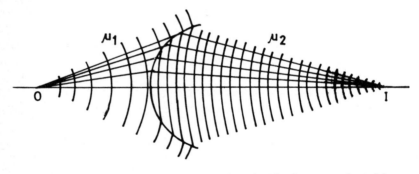

Fig. 2-20: Spherical waves originate from the point O; these are refracted by a spherical surface and the refracted waves converge approximately at I.

the successive positions of the wavefront as it refracts through the spherical surface. It should be noted that although the refracted wavefront is not exactly spherical, the direction of the rays is always normal to the wavefront. This is because the secondary wavelets are spherical (although their envelope is not) and the line joining the center of the secondary disturbance and the point of contact of the secondary wavelet and its envelope is at right angles to the envelope. Further, a very small portion of a spherical surface can be thought of as a plane surface and a ray will change its direction in such a way that

$$\mu_1 \operatorname{Sin} i = \mu_2 \operatorname{Sin} r \qquad (2\text{-}24)$$

The ray diagram corresponding to Fig. 2-20 is shown in Fig. 2-21

It is immediately obvious that it is much easier to construct images by

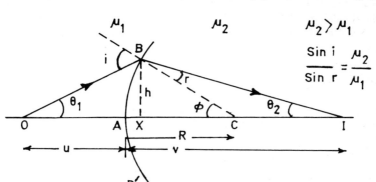

FIG. 2-21: The ray diagram corresponding to Fig. 2-20. I is the real image of O.
All angles are very small in practice but are exaggerated here for clarity.

using the ray diagram technique rather than by using Huygens' principle.

Thus we conclude that when a train of light waves passes through an optical instrument, the curvature of the wave fronts is altered at each boundary surface. However, a wavefront that is originally spherical or plane does not, in general, have a geometrically simple form after refraction at a spherical surface. This makes it practically impossible to analyze the passage of light through an optical instrument in terms of wave surfaces, and it is here that the simplifications of the ray method are most useful. A ray, in its passage through an optical instrument, is made up of a number of segments of straight lines, deviated at reflecting or refracting surfaces by angles which can be computed from the law of reflection or from Snell's law. Hence the problem of tracing the path of a ray reduces to a problem in geometry and this branch of optics is called *geometrical optics. It must, however, be pointed out again that geometrical optics is only an approximation, valid only when effects of diffraction are neglected.*

2-15 Applications of Geometrical Optics

Assuming the conditions for geometrical optics to be satisfied, we now try to construct images of objects formed by refraction at spherical surfaces.*
At this point it is necessary to adopt a convention on signs for distances. We shall use the following:

1. Draw all figures with light incident on the refracting or reflecting surface from the left.
2. Consider object distances (u) positive when the object lies at the left of the vertex of the refracting or reflecting surface.
3. Consider image distances (v) positive when the image lies at the

---

* The corresponding considerations for reflection are slightly simpler and are outlined in the problems given at the end of the chapter.

right of the vertex.

4. Consider radii of curvature (R) positive when the center of curvature lies at the right of the vertex.

5. Consider a transverse dimension positive when measured upward from the axis.

(We do not have to worry about the sign convention for angles as long as the final formula does not contain any angle.)

2-16 Refraction at a Spherical Surface

We refer back to Fig. 2-21 which shows a pencil of rays diverging from a point object O. The curve BB' denotes a section of a spherical surface of radius R separating two transparent substances of refractive indices μ_1 and μ_2. The point C is the centre of curvature of the spherical surface. A line such as OAC through the centre of curvature is called an axis. The point A, where the axis intersects the surface is called the vertex.

Now, according to our sign convention, u, v and R are all positive quantities. Further, from Fig. 2-21, it is easy to see that the angles of incidence and refraction are given by:

$$i = \theta_1 + \phi \tag{2-25}$$

$$r = \phi - \theta_2 \tag{2-26}$$

Next, the distance h can be expressed in a variety of ways, as follows:

$$h = u \tan \theta_1 = v \tan \theta_2 = R \tan \phi \tag{2-27}$$

(The point B is assumed to be near the axis so that the distance AX is very small compared to v or R.) Since the point B lies close to the axis, the angles θ_1, θ_2 and ϕ are small and if they are measured in radians we would have*

$$\tan \theta_1 \approx \theta_1, \ \tan \theta_2 \approx \theta_2 \text{ and } \tan \phi \approx \phi \tag{2-28}$$

Thus Eq. 2-27 would reduce to

$$u\theta_1 = v\theta_2 = R\phi \tag{2-29}$$

Finally, Snell's law in the small-angle approximation is

$$\mu_1 i = \mu_2 r \tag{2-30}$$

* It is worthwhile to point out the region of validity of Eq. 2-28. For example, for $\theta = 5°$,

θ (measured in radians) $= \frac{\pi}{180} \times 5 = 0 \cdot 0873$

and Sin $\theta = 0 \cdot 0872$, tan $\theta = 0 \cdot 0875$.

Thus $\theta \approx$ Sin $\theta \approx$ tan θ, within about $0 \cdot 4\%$.

Similarly, for $\theta = 10°$,

θ (measured in radians) $= \frac{\pi}{180} \times 10 = 0 \cdot 1745$

and Sin $\theta = 0 \cdot 1736$, tan $\theta = 0 \cdot 1763$.

Thus $\theta \approx$ Sin $\theta \approx$ tan θ, within about $1 \cdot 6\%$.

It is easy to eliminate all the angles from Eqs. 2-25, 2-26, 2-29, and 2-30 and obtain the following relation connecting u, v and R:

$$\frac{\mu_1}{u} + \frac{\mu_2}{v} = \frac{\mu_2 - \mu_1}{R} \tag{2-31}$$

which is the same as Eq. 2-21.

It should be pointed out that it is possible for the image distance v to turn out negative. In such a case, the refracted wave does not converge to the right of the refracting surface but *diverges* in such a way as to appear to have originated at a point to the left of the surface at a distance $-v$ (Fig. 2-22). Correspondingly, when the incident wave is a converging

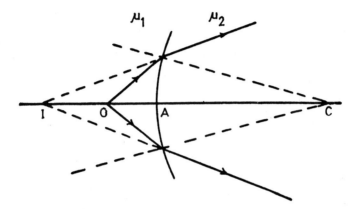

FIG. 2-22: The refracted wave appears to diverge from the point I, leading to a virtual image. The distance $AI = -v$.

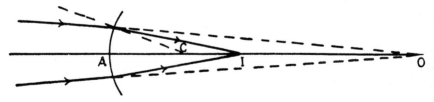

FIG. 2-23: The incident wave is a converging wave that would converge at a certain distance right of the refracting surface if the refracting surface were not there. Here u is —ive and v is +ive and the distances AO and AI are $-u$ and v respectively.

spherical wave that would converge at a certain distance to the *right* of the refracting surface if the refracting surface were not there, the object distance u is negative (Fig. 2-23). Last, the curvature of the surface may be opposite to that shown in Figs. 2-22 and 2-23. When the centre of curvature is to the left, R is a negative quantity (Fig. 2-24).

On the other hand, if a plane wave is incident on a spherical surface, it gets focused at the focal point. This defines the focus and the focal length (Fig. 2-25). In practical situations, however, spherical surfaces are usually used in pairs, in the form of *lenses*, and so we shall now discuss the situation where there are two spherical surfaces adjacent to each other.

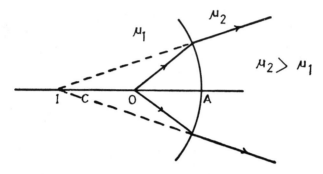

FIG. 2-24: Refraction of a spherical wave at a spherical refracting surface whose centre of curvature C, lies on the same side as the object. Therefore, the distances AO, AC and AI are u, $-R$, and $-v$ respectively.

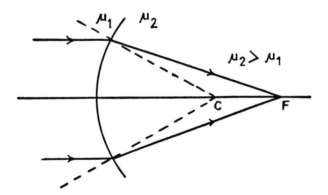

FIG. 2-25: A parallel beam of light gets focussed at the focal point F.

2-17 Thin Lenses

We now consider an optical system consisting of a material of index of refraction μ_2, with spherical surfaces of radii R_1 and R_2 respectively, embedded in a medium of index of refraction μ_1. Such a system is called a *lens*. When a lens is in vacuum, $\mu_1 = 1$; for air at ordinary temperature and pressure, $\mu = 1\cdot00028$, and the approximation $\mu_1 = 1$ is often used. It is, however, useful to include in the analysis the possibility that the lens may be immersed in some other fluid, for which μ_1 is different from unity.

The analysis proceeds as follows: If the second surface were not present, the first would form an image of any object placed to the left of it. This image can be considered as an *object* for the second surface, even though it may not actually be formed when the second surface is present. This image still exists in the sense that the direction of the rays emerging from the first surface are characteristic of a certain image position. Let u_1 and v_1 represent the object and image positions for the first surface, and u_2 and v_2 those for the second surface, with the usual sign conventions. Applying Eq. 2-31 to the two surfaces, we find

$$\frac{\mu_1}{u_1} + \frac{\mu_2}{v_1} = \frac{\mu_2 - \mu_1}{R_1} \tag{2-32}$$

and

$$\frac{\mu_2}{u_2} + \frac{\mu_1}{v_2} = \frac{\mu_1 - \mu_2}{R_2} \tag{2-33}$$

In Eq. 2-33 the roles of μ_1 and μ_2 are reversed, since for the second surface the medium characterised by μ_1 is on the right rather than on the left.

Next we note that the *image* formed by the first surface is the *object* for the second. Furthermore, if the first image distance v_1 is positive, the corresponding object distance for the second surface is negative, and conversely; thus if the thickness of the lens is negligible (as compared to the image and object distances), we have $v_1 = -u_2$. Using this relationship together with Eqs. 2-32 and 2-33 we can eliminate v_1 and u_2 to obtain a single equation relating the original object distance u_1 and the final image distance v_2. The result is

$$\frac{1}{u_1} + \frac{1}{v_2} = \left(\frac{\mu_2}{\mu_1} - 1\right)\left(\frac{1}{R_1} - \frac{1}{R_2}\right) \tag{2-34}$$

Having derived this equation, we now describe the lens as a single unit; we drop the subscripts on u and v and simply state that the first object position u and the final image position v are related by

$$\frac{1}{u} + \frac{1}{v} = \left(\frac{\mu_2}{\mu_1} - 1\right)\left(\frac{1}{R_1} - \frac{1}{R_2}\right) \tag{2-35}$$

The same sign conventions apply to Eq. 2-35 as for the single-surface equation.

When the source or object is very far away, the incident waves are practically plane waves. The value of v at which such plane waves are focussed is called the *focal length* of the lens and is denoted by f. That is, an incident plane wave, corresponding to rays parallel to the axis, is brought to focus at a distance f from the lens, where

$$\frac{1}{f} = \left(\frac{\mu_2}{\mu_1} - 1\right)\left(\frac{1}{R_1} - \frac{1}{R_2}\right) \tag{2-36}$$

Eq. 2-36 shows that, within the approximations we have made, the lens is symmetric; i.e. if it is reversed, side to side, the focal length and all the other characteristics remain unchanged.

Now, in terms of the focal length, the lens equation can simply be written as

$$\frac{1}{u} + \frac{1}{v} = \frac{1}{f} \tag{2-37}$$

Consideration of Eq. 2-37 reveals that in the usual case when $\mu_2 > \mu_1$, f is a positive quantity whenever the lens is thicker in the middle than at the edges, while f is negative if the lens is thicker at the edges. In the latter case, a beam of parallel rays is not brought to a real focus but rather *diverges* from the lens as though it originated from a virtual focus, a distance f to the left of the lens. Similarly, it is possible for v to be negative, corresponding to a virtual image. The object distance u may be negative, corresponding to a virtual object describing rays which converge as they approach the first surface of the lens.

Understanding of the formation of an image by a lens is facilitated by the construction of principal rays which are drawn as follows:*

1. A ray parallel to the optic axis† is refracted through the focus.
2. A ray passing through the focus is refracted parallel to the optic axis.
3. A ray which passes straight through the centre of the lens is not deviated at all.

Fig. 2-26 shows the principal ray diagram for two cases. In Fig. 2-26(*a*) the object is farther from the lens than the focal length and a real image results. In Fig. 2-26(*b*), the object is closer than the focal length, resulting in a virtual image. Similarly, in Fig. 2-27, the image formation is shown for a diverging lens having negative focal length.

2-18   Compound Lenses

If we have a system of several lenses, how can we possibly analyze it? It is easy. We start with some object and calculate where its image is for the first lens using Eq. 2-37 or by drawing a principal-ray diagram. Then we treat this image as the object for the next lens, and use the second lens, with whatever its focal length is, to again find an image. We simply chase the ray through the succession of lenses. That is all there is to it. It involves nothing new in principle. We will discuss a few examples later.

---

* The first two laws are valid only when f is positive. They have to be modified when f is negative.

† The optic axis is the line joining the two centres of curvature.

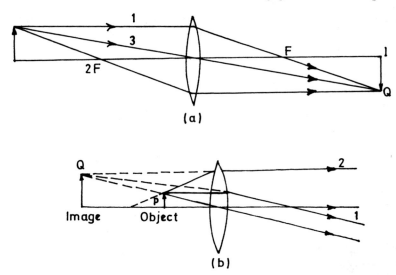

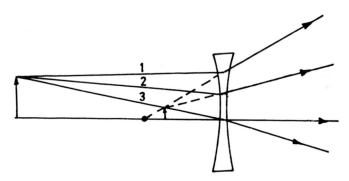

Fɪɢ. 2-26: Principal-ray diagrams showing image formation by a convex lens: (*a*) when the object distance is greater than the focal length, a real inverted image is formed; (*b*) when the object distance is less than the focal length, a virtual erect image is formed; its position is found by projecting the principal rays backward.

Fɪɢ. 2-27: Principal-ray diagram showing image formation by a diverging lens.

2-19 Aberrations

At this point we must mention that we have limited ourselves, strictly speaking, to *paraxial* rays, i.e. the rays near the axis. A real lens having a finite size will, in general, exhibit *aberrations*. For example, a ray that is *on* the axis, of course, goes through the focus; a ray that is very close to the axis will still come to the focus very well. But, as we go farther out, the ray begins to deviate from the focus, perhaps by falling short, and a ray striking near the top edge comes down and misses the focus by quite a wide margin (Fig. 2-28). So, instead of getting a point image, we get a smear. This effect is called *spherical aberration*, because it is a property of

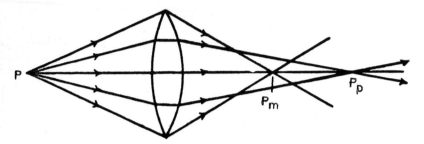

FIG. 2-28: Spherical aberration of a convex lens. The paraxial rays meet the axis at P_p whereas the marginal rays meet the axis at P_m.

spherical surfaces. This could be remedied, for any specific object distance, by using several lenses arranged so that the aberrations of the individual lenses tend to cancel each other. Lenses have another fault: lights of different colours have different speeds, or different indices of refraction, and therefore, the focal length of a given lens is different for different colours. So for a white object the image will show colours: when we focus for red, blue is out of focus and vice versa. This effect is called *chromatic aberration* (see also Example on p. 68).

There are still many other faults. For example if the object is off the axis, then the focus is really not perfect any more, when it gets far enough off the axis. Thus, if one has to design an optical instrument, including analysis of aberrations, then he has to simply trace the rays through the various surfaces using the laws of refraction from one lens to the other and to find out where they come out and see if they form a satisfactory image. People have said that this is too tedious, but today with computing machines, it is the right way to do it.* One can set up the problem and make the calculations for one ray after another very easily and then design a satisfactory optical system, a process which frequently involves a considerable amount of trial and error calculations. So the subject is really quite simple, and involves no new principles.

At this point it is worth while mentioning that all the above considerations are in fact based on Huygens' principle. For example, if we consider a thin convex lens (Fig. 2-29), a point object sends out spherical waves and B_1BB_2 denotes a portion of the spherical wave-front when it just touches the pole of the lens. Each point on this wavefront is a source of secondary disturbance and one can draw wavefronts at different places as the wave passes through the lens. Fig. 2-29 represents a particular case in which the wavefronts in the lens medium are plane. This has been assumed in order to simplify the figure. Nevertheless, it is easy

* See, for example, F. D. Smith 'How Images are Formed', *Scientific American*, September 1968.

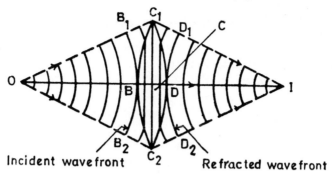

FIG. 2-29: Refraction of spherical waves through a convex lens.

to see how, in general, the successive wavefronts can be constructed. [While constructing the wavefronts one should only take care of the fact that the speed of light in the material of the lens (say glass) is different from that in air.] The refracted wavefront can be assumed to be approximately spherical, converging towards the image *I*. If the refracted wavefront is indeed spherical, then, obviously all the rays (i.e. $OB_1C_1D_1I$ or $OBDI$ or any other ray) will take the same time in converging towards the point *I*. This is usually referred to as *all the rays traverse equal optical paths*.

Now, in general, the refracted wavefront will not be truly spherical; thus the normals at different points will not pass through the same point *I*. The wavefront will therefore not converge towards a single point and will give rise to aberrations.

Another important fact is that even if the wavefront is spherical, it is only a portion of a sphere. This will give rise to diffraction effects. We shall see in Chapter 6 that because of diffraction the image will no longer be a point but a 'smear' over a region whose dimensions will be 'of the order of' $f\lambda/a$ where f is the focal length, λ, the wavelength and a the aperture of the lens.

It should also be noted that Huygens' principle will be valid for any wave phenomenon. For example, we can apply Huygens' principle to study the propagation of sound waves, and indeed, if we can form a biconvex lens of oxygen gas placed in an atmosphere of hydrogen gas, then one can focus sound waves using a ray diagram, provided the aperture of the lens is large compared to the wavelength of sound waves. However, since the wavelength of sound waves is of the order of 50 cm, the apparatus may be difficult to make (because, for reasonable focusing, the aperture of the lens should be about 500 cm), but it certainly is possible and has been done.*

---

* The lens can be formed by filling oxygen gas in a (thin) rubber 'bladder'.

$$\left(\frac{V_{H_2}}{V_{O_2}} \approx 4 \right)$$

2-20 Optical Instruments

We shall now briefly discuss a few simple applications of lenses and mirrors in the design of optical instruments. A most common use of a lens is in a camera as shown in Fig. 2-30(a). The function of the lens is

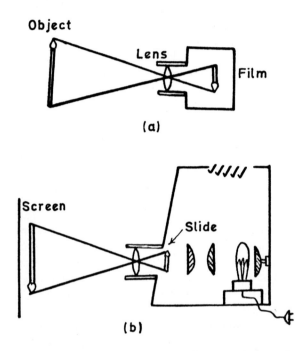

FIG. 2-30: (a) In a camera the lens forms a real inverted image (reduced in size) on the plane of the film. (b) In a projector an enlarged image of a slide is formed on a screen. The lenses between the bulb and slide serve to concentrate light from the bulb on the slide; they are known as condenser lenses.

simply to form a two-dimensional real image on the plane of the photographic film. The figure indicates that this can be accomplished with a single lens (or even a pinhole, see Fig. 7-1). In actual practice, camera lenses usually contain several elements in order to reduce various aberrations.

A similar problem is that of a slide or a movie projector. Here the function of the lens is just the opposite of that of a camera: the lens creates an enlarged real image on a screen of a slide as shown in Fig. 2-30(b). These two applications of lenses have several features in common. In both cases the ratio of object to image size is the same as the ratio of object to image distance from the lens; the image is real and is inverted with respect to the object. The same general principles are characteristic of the operation of the human eye, in which the lens produces a real inverted image on the retina.

2-21   The Magnifying Glass

Another application of a single lens is that of a simple magnifying glass. The closer an object is to the eye, the more the detail that can be seen, but most persons cannot focus their eyes comfortably on an object *closer* than about 25 cm. This is known as the *least distance of distinct vision*. A magnifying glass forms an enlarged image at a distance sufficiently far away from the eyes for comfortable viewing. Unlike the case of the camera or slide projector, the image need not be real. Fig. 2-31 shows how a magnifying glass (which is essentially a biconvex lens) with $f = 10$ cm forms an enlarged, erect virtual image of an object located between the focus and the lens. It is customary to define the magnification, Υ, of

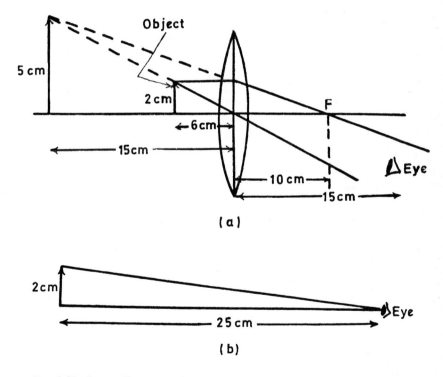

(a)

(b)

Fig. 2-31: A magnifying glass forms an enlarged, erect virtual image. The angle subtended at the eye when placed at the least distance of distinct vision (*b*) is about 2 cm/25 cm $= 1/12 \cdot 5$ radians. The angular size of the image for the case shown in (*a*) is about $5/30 = 1/6$ radian. The magnification is about

$$\left(\frac{1}{6}\right)\Big/\left(\frac{1}{12 \cdot 5}\right) \approx 2 \cdot 6.$$

an optical system as

$$\gamma = \frac{\tan u'}{\tan u} \tag{2-38}$$

where u is the angle subtended at the eye by the object when placed at the least distance of distinct vision ($= 25$ cm) and u' is the angle subtended by the image (usually at infinity) at the eye. Thus if y is the height of the object and f the focal length then

$$\tan u = \frac{y}{25} \qquad (y \text{ measured in cm})$$

and

$$\tan u' = \frac{y}{f} \text{ (when the object is at the focal point)}$$

Therefore,

$$\gamma = \frac{25}{f} \tag{2-39}$$

where f is in cm.

Thus in the example of Fig. 2-31 the magnification is about 3, often written as $3\times$. For greater magnification it is often desirable to use two or more lenses in combination. We consider here only two simplest examples of such schemes, the compound microscope and the astronomical telescope (see also the problems at the end of the chapter).

2-22 Compound Microscope

Fig. 2-32 shows the general construction of a compound microscope. The lens closest to the object, called the *objective*, forms a real enlarged,

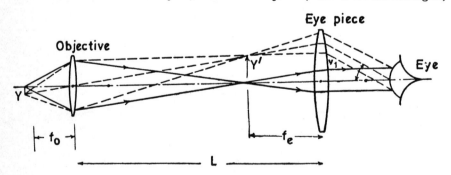

FIG. 2-32: The compound microscope.

inverted image of the object in the barrel of the microscope. The second lens, called the *eyepiece*, uses this image as its object and forms a finol virtual enlarged, erect image, which can be located at distance from the eye convenient for comfortable viewing. Compound microscopes constructed according to this general scheme but using more complex lens groupings can be made with a magnification of several hundred.

The overall magnification M of a compound microscope can easily be calculated:

$$M = \frac{\tan u}{\tan u'} \qquad (2\text{-}40)$$

where the angles u and u' have been defined earlier. As before $\tan u = y/25$ (y measured in cm) and $\tan u' = y'/f_e$, where f_e is the focal length of the eyepiece and y' is the height of its image formed by the objective. Eq. 2-40 follows from the fact that the image (formed by the objective) is at the focal plane of eyepiece (see Fig. 2-32). Thus

$$M = \frac{y'}{y} \frac{25}{f_e}$$

But

$$\frac{y'}{y} \approx \frac{L}{f_o} \qquad (2\text{-}41)$$

where f_o is the focal length of the objective and L, the length of the microscope tube.

$$\therefore \quad M \approx \frac{L}{f_o} \frac{25}{f_e}$$

It seems as if the magnification can be made as large as we wish by making the f's sufficiently small, but our approximate lens theory conceals the failure of lenses to form good images when the focal lengths are made too small. Furthermore, we shall see later that, while the magnifying power increases the image does not gain in detail, even though all lens aberrations have been corrected. This limit to the *useful* magnification is set by the diffraction of light waves, which essentially implies that the laws of geometrical optics do not hold strictly for a wavefront of limited extent. Physically, the image of a point source is *not* the intersection of *rays* from the source but the diffraction pattern of those *waves* from the source that pass through the lens system (this point will be discussed in greater detail in Chapter 6).

2-23 Astronomical Telescopes

The operation of the refracting astronomical telescope, shown in Fig. 2-33, is very similar to that of the compound microscope, except for differences in magnitudes of the various focal lengths. Like the compound microscope, it makes use of an objective lens which forms a real image in the barrel of the telescope, and an eyepiece lens which forms an enlarged virtual image of the first image. The final image, as in the case of a compound microscope, is inverted with respect to the object.* The magnifying power can easily be calculated. From Fig. 2-33.

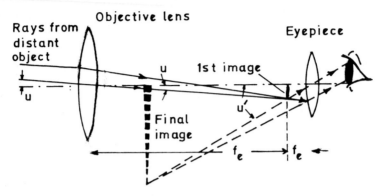

FIG. 2-33: A telescope consists of a long-focal-length objective and a short-focal length eyepiece. The figure shows the calculation of the apparent magnification. The object and the final image are far to the left.

$$\tan u = \frac{y'}{f_o}$$

$$\tan u' = \frac{y'}{f_e}$$

where y' is the height of the image formed by the objective.

Thus, the magnifying power M is given by

$$M = +\frac{f_o}{f_e} \tag{2-42}$$

(Some books write $M = -\frac{f_o}{f_e}$ indicating that the image is inverted.)

The angular magnification of a telescope is therefore equal to the ratio of the focal length of the objective to that of the eyepiece. Telescopes are used not only because of their magnifying power but also because of their 'light-gathering' power. The former property has just been mentioned. The 'light-gathering' ability depends on the area of the objective lens. In astronomy, it is important to have high light-gathering power in order to observe faint objects.

We will now show how the diameter of the objective sets a limit to the useful magnification of a telescope.

The eyepiece of a telescope forms a real reduced image of the objective lens itself in the space beyond the eyepiece as shown in Fig. 2-34. It is

---

* This is also the principle of operation of most ordinary binoculars, which use a pair of prisms for each eye, in which internal reflections invert the image back to the same orientation as the object.

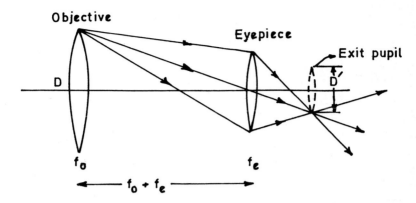

Fig. 2-34: The eyepiece of a telescope forms a real, inverted and reduced image of the objective lens itself. This image shown by the dotted curve defines the exit pupil of the eye.

easy to see that the entire light that enters the objective and refracted by the eyepiece must pass through this image of the objective, which is called the *exit pupil* of the telescope. The diameter of the transmitted beam is a minimum in the plane of the exit pupil. If all the light that is transmitted by the telescope is to enter the pupil of the observer's eye, the diameter of the exit pupil of the telescope must not be larger than the diameter of the pupil of the eye as, usually, the eye is placed at the exit pupil of the telescope.

We shall assume that the object and also the virtual image formed by the eyepiece are both at infinity. If D be the diameter of the objective and D' the diameter of its image formed by the eyepiece, then it can easily be shown that (Problem 26)

$$\frac{D}{D'} = \frac{f_o}{f_e} \tag{2-43}$$

where, as before f_o and f_e are the focal lengths of the objective and eyepiece respectively, and we have disregarded the algebraic signs. But f_o/f_e is the angular magnification, M, of the telescope. Thus, we may write

$$M = \frac{D}{D'}$$

It is worthwhile mentioning that using the above equation we can easily determine the angular magnification of the telescope. The telescope is directed toward a bright sky and a screen is moved along the axis until the minimum diameter of the transmitted beam is found. This is the image of the objective formed by the eyepiece. Measuring the diameter of this image (which is equal to D') and the diameter of the objective (which is equal to D) we can easily calculate M.

The *normal magnification* of a telescope is defined as that at which the diameter of the exit pupil is just equal to the diameter of the pupil of the eye which is usually assumed to be 2 mm.

Let us work out a simple example:

Consider a telescope whose objective has a diameter of 30 mm. Since the diameter of the pupil of the eye is assumed to be 2 mm, the normal magnification would be $30/2 = 15 \times$.

The diameter of the objective is 15 times that of the pupil of the eye and its area is therefore 225 times as great. It therefore admits 225 times as much light.

Let us first assume that the normal magnification $15 \times$, is used. The diameter of the exit pupil then equals the pupillary diameter of the eye; and all the light admitted by the objective can enter the eye. The linear dimensions of the image formed at the retina of the eye are increased by a factor of 15 and its area by a factor of 225. Hence 225 times the light is distributed over 225 times the area and the apparent brightness of the objective viewed is the same with the telescope as with the unaided eye.

If we want to have a magnification of $20 \times$, the area of the retinal image is increased by a factor of 400. The diameter of the exit pupil is only 1·5 mm. The objective therefore admits 225 times as much light as does the unaided eye and all this light can enter the eye. However, since the light is distributed over an area 400 times as great, the apparent brightness is less than that seen by the unaided eye.

On the other hand, when the magnification is $10 \times$, the area of the retinal image is increased by a factor of 100. The diameter of the exit pupil is now 3 mm and its area is 2·25 times that of the pupil. Hence about half of the light passing through the exit pupil can enter the eye. Thus, although the objective admits 225 times as much light as the unaided eye, only $1/2\cdot25$ of this light reaches the retina and therefore 100 times the light is distributed over 100 times the area. The apparent brightness remains the same.

Finally, we consider briefly the principle of the reflecting telescope. Here the objective lens is replaced by a concave mirror. When the telescope is to be used only with very distant objects, this scheme has several advantages. One is that a mirror is intrinsically free from chromatic aberration.* Another is that, by making the mirror paraboidal instead of spherical, spherical aberrations can be completely eliminated in the first image. A third is that it is usually easier and cheaper to make large mirrors than lenses of comparable size. The general layout of a reflect-

---

* Because, for all wavelengths, the angle of incidence is equal to the angle of reflection.

ing telescope is shown in Fig. 2-35.

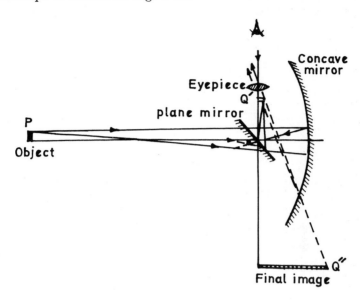

FIG. 2-35: A reflecting telescope. The image formed by the mirror is brought away from the optic axis by a small plane mirror placed at 45° to the optic axis; Q' is the (real) image formed by the concave mirror and the plane mirror. The eyepiece forms a final virtual image as shown. (After H. D. Young, *Modern Physics and Optics*, p. 132.)

Example

A hollow sphere of glass, of refractive index μ, has a small mark on its interior surface which is observed from a point outside the sphere on the side opposite the centre. The inner cavity is concentric with the external surface and the thickness of the glass is everywhere equal to the radius of the inner surface. Prove that the mark will appear nearer than it really is by a distance $(\mu - 1)R/(3\mu - 1)$, where R is the radius of the inner surface.

Solution

Considering refraction by the inner surface (see Fig. 2-36) we have

$$u = 2R, \ \mu_1 = 1, \ \mu_2 = \mu$$

Therefore, using Eq. 2-31

$$\frac{1}{2R} + \frac{\mu}{v} = -\frac{\mu - 1}{R}$$

or

$$v = -\frac{2\mu R}{2\mu - 1}$$

The image is virtual at the point I_1. For the second surface the pole is at the point B, therefore

$$u = R + \frac{2\mu R}{2\mu - 1} = \frac{(4\mu - 1)R}{(2\mu - 1)}$$

$$\mu_1 = \mu$$

$$\mu_2 = 1$$

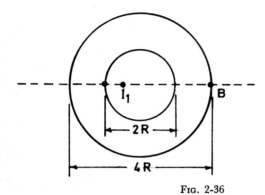

FIG. 2-36

Again if we use Eq. 2-31

$$\frac{\mu(2\mu - 1)}{(4\mu - 1)R} + \frac{1}{v} = -\frac{1 - \mu}{2R}$$

because the radius of the outer surface is $2R$. On solving we get

$$v = -\frac{(4\mu - 1)\,2R}{(3\mu - 1)}$$

The required distance is then

$$3R - \frac{(4\mu - 1)\,2R}{(3\mu - 1)} = \frac{(\mu - 1)R}{(3\mu - 1)}$$

Example

Consider an astronomical telescope (see Fig. 2-37). The focal lengths of the objective and eyepiece are in the ratio of 10:1. Both the lenses are convex. There is an object which is 1000 metres tall and is at a distance

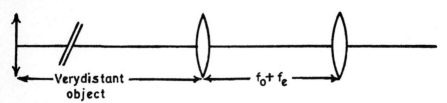

FIG. 2-37

of 0.25×10^6 metres. Show that one obtains a virtual image 100 metres tall at a distance of 2500 metres from the telescope.

Solution

If the object distance is u then the distance, v_1, at which the image is formed by objective is given by:

$$\frac{1}{v_1} + \frac{1}{u} = \frac{1}{f_o}$$

or

$$v_1 = \frac{u f_o}{u - f_o}$$

(Although in the given problem $u \ggg f_o$, yet at this point we should *not* write $v_1 \approx f_o$.) The distance of this image from the eyepiece, u_1, is given by

$$u_1 = f_o + f_e - \frac{u f_o}{u - f_o} = \frac{u f_e - f_o{}^2 - f_o f_e}{u - f_o}$$

If the final image is formed at a distance v from the eyepiece then

$$\frac{1}{v} + \frac{1}{u_1} = \frac{1}{f_e}$$

or

$$\frac{1}{v} = \frac{1}{f_e} - \frac{u - f_o}{u f_e - f_o{}^2 - f_o f_e}$$

or

$$v = -\frac{f_e\,(u f_e - f_o{}^2 - f_o f_e)}{f_o{}^2}$$

Further

$$\gamma = \frac{\text{Height of the final image}}{\text{Height of the object}} = \frac{v_1}{u} \times \frac{v}{u_1}$$

$$= -\frac{f_e}{f_o}\left[\frac{u f_e - f_o{}^2 - f_o f_e}{u f_e - f_o{}^2 - f_e}\right]$$

In the problem

$$\frac{f_o}{f_e} = 10$$

and

$$u \ggg f_o, f_e$$

$$\therefore \quad v \approx -\left(\frac{f_e}{f_o}\right)^2 u$$

and

$$\gamma \approx -\frac{f_e}{f_o}$$

For

$$u = 0.25 \times 10^6 \text{ m}$$

$$v \approx 2500 \text{ m}$$

Further

Height of the final image = (Height of the object).γ

$$= -(1000) \times \tfrac{1}{10}$$

$$= -100 \text{ m}$$

The negative sign indicates that the final image is inverted.

It should be noted that the height of the image is only 1/10th of that of the object, but the image is closer than the object by a factor of 100, so it subtends an angle ten times bigger, i.e. the image *appears* ten times bigger.

Example

For a given material of a lens, the refractive indices (and hence the focal lengths) are different for different colours. Consider two lenses (one made of crown glass and the other of flint glass) placed in contact with one another. Find the condition for the combination to have the same focal length for blue and red lights.

Solution

Let μ_b and μ_b be the refractive indices for the blue and red light for the crown glass and μ_b' and μ_r' be the corresponding values for the flint glass. If F_b and F_r represent the focal lengths of the combination for the blue and red lights then

$$\frac{1}{F_b} = \frac{1}{f_b} + \frac{1}{f_b'}$$

$$= (\mu_b - 1)\left(\frac{1}{R_1} - \frac{1}{R_2}\right) + (\mu_b' - 1)\left(\frac{1}{R_1'} - \frac{1}{R_2'}\right)$$

where the primed quantities refer to the flint glass lens and the unprimed quantities refer to the crown glass lens. Now, we may write

$$\frac{1}{F_b} = \frac{\mu_b - 1}{\mu - 1} \cdot \frac{1}{f} + \frac{\mu_b' - 1}{\mu' - 1} \cdot \frac{1}{f'}$$

where

$$\mu \equiv \frac{\mu_b + \mu_r}{2} \approx \mu_y$$

$$\mu' \equiv \frac{\mu_b' + \mu_r'}{2} \approx \mu_y'$$

$$\frac{1}{f} \equiv (\mu - 1)\left(\frac{1}{R_1} - \frac{1}{R_2}\right)$$

and

$$\frac{1}{f'} \equiv (\mu' - 1)\left(\frac{1}{R_1'} - \frac{1}{R_2'}\right)$$

Similarly

$$\frac{1}{F_r} = \frac{\mu' - 1}{\mu - 1} \cdot \frac{1}{f} + \frac{\mu_r' - 1}{\mu' - 1} \cdot \frac{1}{f'}$$

For the two focal lengths to be equal we must have

$$\frac{\omega}{f} + \frac{\omega'}{f'} = 0$$

where

$$\omega = \frac{\mu_b - \mu_r}{\mu - 1}$$

is known as the dispersive power. Since ω and ω' are both positive, the focal lengths must be of opposite sign. Thus if one of the lenses is a convex lens the other has to be concave. Such a combination is known as an *achromatic doublet* (Fig. 2-38). From the lens manufacturer's point of view, for good optical contact the curvature of the second surface of the crown glass lens must be the same as that of the first surface of the flint glass lens (Fig. 2-38). Achromatic doublets are extensively used in optical instruments.

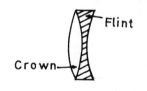

Fig. 2-38

2-24 Dispersion

Throughout this chapter we have assumed that the velocity of the light wave (or, to be more precise, of the electromagnetic wave) through matter is different from its velocity of propagation in vacuum. One may ask as to why a light wave travels in a material medium with a velocity different from that in vacuum. The answer to this question lies in the fact that when an electromagnetic wave propagates through matter it induces oscillations in the charged particles of the atoms or molecules, which then emit secondary or 'scattered' waves. These scattered waves are superposed on the original wave, giving a resultant wave. The phases of the secondary waves are, in general, different from those of the original wave, since a forced oscillator is not always in phase with the driving force. A detailed analysis* indicates that this phase difference affects the resultant wave in such a way that the wave appears to have a velocity different from the wave velocity in vacuum.

On the other hand, reflection of light from the surface of a solid or a liquid surface involves only the oscillators (electrons) located in a small pill-box-shaped volume at the surface of the material. Indeed, the radiation that makes up the reflected wave originates in a thin layer whose thickness is about half the wavelength of the incident light. The radiation adds up coherently only in a direction such that angle of incidence is equal to angle of reflection.†

SUGGESTED READING

FEYNMAN, R. P., R. B. LEIGHTON and M. SANDS, *The Feynman Lectures on Physics*, Vol. I, Addison-Wesley (1965). Chapters 26, 27, 31 and 51 are on 'The Principle of Least Time', 'Geometrical Optics', 'The Origin of Refractive Index' and 'Waves' respectively and are directly relevant to the concepts developed in this chapter.

HUYGENS, C., *Treatise on Light*, Dover Publications (1962). The treatise (in French) was communicated to the Royal Academy of Science in 1678; the work was published in 1690. Although about 300 years old, the work has withstood the test of time; and even now the exquisite skill with which Huygens applied his conception of the propagation of waves of light to unravel the intricacies of the phenomena of double refraction of crystals‡ and of the refraction of the atmosphere will excite the admiration of the student of optics.

NEWTON, ISAAC, *Optiks*, Dover Publications (1952). This is based on the fourth edition published in 1730. The first edition appeared in 1704. According to Einstein, '. . . this new edition of his [Newton's] work on Optics is nevertheless to be welcomed with warmest thanks, because it alone can afford us the enjoyment of a look at the personal activity of this unique man. . . . In one person he [Newton] combined the experimenter, the theorist, the mechanic and, not least, the artist in exposition. He stands before us strong, certain, and alone: his joy in creation and his minute precision are evident in every word and in every figure.'

PHYSICAL SCIENCE STUDY COMMITTEE, *Physics*, D. C. Heath and Co. (1965). Chapter 14 gives a good account of the particle model of light. Chapter 16 discusses refraction, dispersion and diffraction.

SMITH, F. D., 'How Images are Formed', *Scientific American*, September 1968. With modern computers and a wide variety of possible optical materials, it is possible to design and construct a lens as close to the theoretical limits as necessary for the purpose. Various methods of producing and evaluating images are discussed in this article.

SCHRODINGER, E., 'The Fundamental Ideas of Wave Mechanics', *Nobel Lectures, Physics*, 1922-1941, Elsevier Publishing Co., (1965). This is Schrodinger's Nobel Prize lecture given in December 1933. Of special interest in this article is an elegant discussion on Fermat's principle, the limitations of geometrical optics and the phenomenon of diffraction.

WEISSKOPF, V. F., 'How Light Interacts with Matter', *Scientific American*, September 1968. The article describes how absorption, refraction, reflection, or scattering of light can be explained

* See, for example, R. P. Feynman, R. B. Leighton and M. Sands, *The Feyman Lectures on Physics*, Vol. 1, Chapter 31 on 'The Origin of Refractive Index', Addison-Wesley (1965).

† See also V. W. Weisskopf, 'How Light Interacts with Matter', *Scientific American*, September 1968.

‡ Double refraction is discussed in Chapter 3.

in terms of what is known about the structure of matter and the ways in which the electrons and atoms respond to light.

PROBLEMS*

1. Show that if quantities which depend on λ^2 in Eq. 2-1 are not neglected then the slight increase in the area of the zones as we recede from the centre of the system is compensated by the increased distance.

2. Light of wavelength 6×10^{-5} cm is passed through a narrow circular aperture of radius 0.9 mm. At what distance along the aixs will be the first maximum intensity observed?
   (135 cm)

3. A strong parallel beam of monochromatic light is incident normally on a thin plate having a small circular hole of diameter 1 mm. If the screen be moved through a distance of 12.5 cm from the first position where the centre is black to the second similar position, what is the wavelength of light used? (5×10^{-5} cm)

4. In the particle model of light, assuming that when refraction takes place the component of the velocity along the boundary remains the same, show that Sin i/Sin r is a constant.
   In this model, do you expect the speed of light in water to be less or more than its speed in air?

5. In the particle model of light how would you explain diffuse and specular reflection os light.

6. In the particle model of light, how would you explain the illumination varying inversely with the square of the distance from the point source.

7. When a parallel beam of light is incident on the surface of a rarer medium, total internal reflection takes place when
$$i > \text{Sin}^{-1} \frac{1}{\mu}$$
where μ is the refractive index of the denser medium with respect to the rarer medium. Prove the above law using Huygens' theory.

8. A plane wavefront is incident on one face of a prism. Construct the transmitted wavefront using Huygens' principle. Also calculate the deviation produced by the prism in terms of the angles of incidence and emergence and the angle of prism. ($\delta = i + e - A$)

9. Consider the reflection of a spherical wave at a spherical surface. Using Huygens' principle obtain the mirror equation
$$\frac{1}{u} + \frac{1}{v} = \frac{2}{R}$$
where R is the radius of curvature. (You may consider, for example, reflection by a converging surface, i.e. by a concave mirror, forming a virtual image. The above relation is however true for all cases, with the appropriate sign convention.)

10. Derive Eq. 2-31 by considering refraction of a spherical wave at a spherical surface using Huygens' principle.

11. (a) Consider a biconvex lens of oxygen gas in an atmosphere of hydrogen gas. The radius of curvature of each of the rubber membranes is 6 metres. Calculate the focal length for sound waves. ($f = 100$ cm)
    (b) Assuming an aperture of 1 metre, do you expect it to give reasonable focusing for sound waves of frequency 1000 cps.

12. Find the condition for the formation of a real image when paraxial rays of light are refracted at a concave spherical glass-air interface of radius R ($\mu_g = 3/2$) (Fig. 2-39). ($u > 3R$)

13. An object is 1 metre in front of the curved surface of a plano-convex lens whose flat surface is silvered. A real image is formed 120 cm in front of the lens. What is focal length of the lens?

Fig. 2-39

(1200/11 cm)

* Answers when provided are in parentheses at the ends of the problems.

14. An object is placed 40 cm from the surface of a glass sphere of radius 10 cm along a dia-meter. Where will the final image be formed after refraction at both surfaces? Refractive index of glass = 1·5.   (11·4 cm from the second surface)

15. An equi-convex lens of refractive index 1·5 and focal length 10 cm is held with its axis vertical and its lower surfce immersed in water, the upper surface being in air. At what distance from the lens will a vertical beam of parallel rays falling on the lens be focused? How will the position of focus be altered if the lens is wholly immersed in water?
                                                                        (20 cm, 40 cm)

16. A thin equi-convex lens is of focal length 15 cm ($\mu = 1·5$). The lens is mounted so that one of its faces is in air and the other in a transparent liquid of refractive index 1·4. Find the position of the image of a small object which lies in the liquid at a distance of 20 cm from the lens.   (33·3 cm)

17. A plano-convex lens acts like a concave mirror of 28 cm focal length when its plane surface is silvered and like a concave mirror of 10 cm focal length when its curved surface is silvered. Calculate the refractive index of the material of the lens.   (1·56)

18. Fig. 2-40 shows a rectangular tank of water whose refractive index is 4/3. There is a small fish in this water. It's eye is at the point B. In this water there is a spherical air bubble

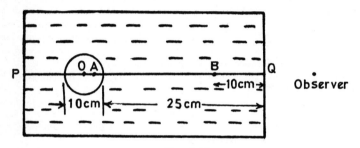

Fig. 2-40

containing a fly whose eye is at the point A. The diameter of this air bubble is 10 cm, and its centre is at O. The points O, A and B lie on the same straight line which hits the wall of the container normally. The distances OA, OB, BQ are 2 cm, 20 cm and 10 cm respect-ively. Find
(a) where the fish appears to be, to an observer outside the tank   (closer by 2·5 cm);
(b) where the fly appears to be, in relation to the fish   (1/3 cm farther);
(c) where the fly appears to be in relation to an outside observer.
                                                (25 cm from the wall of the tank)

19. Two convex lenses L_1 and L_2 are placed 39 cm apart on a common principal axis (Fig. 2-41). A small object is placed 18 cm in front of the lens L_1. Find the position, nature, and magnification of the image formed by the lens combination. The focal lengths of L_1 and L_2 are 12 cm and 4 cm respectively.

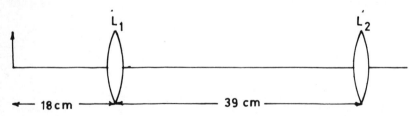

Fig. 2-41

(The final image is virtual, inverted, magnified eight times and is formed at a distance of 12 cm on the left of L_2)

20. In Problem 19, if the lens L_2 is replaced by a concave lens of the same focal length and the distance between the two lenses is reduced by 6 cm, show that the image will again be magnified 8 times but will be real, inverted and at a distance of 63 cm from the object.

21.* Consider a simple magnifier of focal length 6 cm as shown in Fig. 2-42. An object 1 mm in height is placed 5 cm from the lens. (*a*) Find and describe the image. (*b*) Calculate the magnifying power. The eye is at a distance of 10 cm from the lens.
    [(*a*) Virtual image at a distance of 30 cm from the lens; 6 mm in height. (*b*) 15/4]

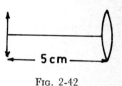

FIG. 2-42

22. Consider a compound microscope (Fig. 2-32). The focal lengths of the objective and the eyepiece are 1 cm and 10 cm respectively. Both the lenses are convex. The two lenses are separated by 18 cm. An object of height 0·01 mm is placed at a distance of 1·1 cm from the objective. (*a*) Draw the ray diagram and find the final image. (*b*) Calculate the magnifying power.
    [(*a*) At a distance of 5·3 cm from the objective. The image is inverted and of height 1/3 mm. (*b*) 33.]

23. Consider a Gallilean telescope as shown in Fig. 2-43. (This type of combination is used in an opera glass.) The height of the object may be assumed to be 2 m (about the height of a man). Show that the image is erect and is about 1 m high at a distance of 12·5 m. What is the angular magnification? (+ 2)

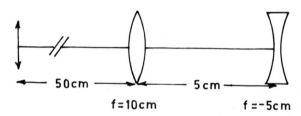

FIG. 2-43

24. A person at the bottom of a swimming pool switches on a flashlight so that the beam makes an angle of 60° with the vertical (Fig. 2-44). (*a*) Where does the beam go? (*b*) Next, let there be many layers of different kinds of oil spread on the water (Fig. 2-45), sketch the path.
    [(*b*) Total internal reflection at layer No. 4]

FIG. 2-44   FIG. 2-45

25. The (*b*) part of Problem 24 can be used to explain the phenomenon of mirage. The air is hottest near the surface of the earth and hence the refractive index is smaller. On the other hand, far from the surface, the refractive index is 1·0003. An observer sees the road surface only if he looks down at an angle of 89° or less. What is the approximate refractive

    * Assume in all problems that the least distance of distinct vision is 25 cm.

index of air at the surface? (1·0001)

[Actually, the index of refraction of air changes gradually rather than in discrete steps and then one *must* consider wave construction (why?). See Fig. 2-29.]

26. Show that for a telescope if D be the diameter of the objective and D' the diameter of its image formed by the eyepiece, then (disregarding algebraic signs)

$$\frac{D}{D'} = \frac{f_o}{f_e}$$

27. (*a*) In a simple astronomical telescope the focal length of the objective lens is 25 cm and its diameter is 2 cm. Assuming the diameter of the pupil of the eye to be 2 mm calculate
(*i*) the normal magnification of the telescope;
(*ii*) the focal length of the eyepiece which would give the normal magnification.
(*b*) What would be the diameter of the exit pupil of the eyepiece if eyepieces were used which gave magnifications of 50% more and 50% less than the normal magnification. Discuss.

[(*a*) (*i*) 10× (*ii*) 2·5 cm]
[(*b*) 0·133 cm; 0·4 cm]

28. (*a*) While discussing the magnifying power of a microscope it was mentioned that the approximate lens theory conceals the failure of lenses to form good images when the focal lengths are made too small. Duscuss this point.
(*b*) Show that for a convex lens, even under optimum conditions

$$\frac{f}{a} \gtrless 1$$

29. Show that the entire light that enters the objective of a telescope and is refracted by the eyepiece must pass through the exit pupil of the telescope.

30. In a Huygens' eyepiece the component lenses are plano-convex of focal length $3f$ and f and placed $2f$ apart. Show that the combination is convergent or divergent according as the light first falls on the first or second lens.

31. Fig. 2-46 shows two points P and Q connected by a light ray POQ. Show that, by comparison with nearby rays, such as PXQ, the ray POQ represents a minimum, a stationary, or a maximum optical path, depending on whether the distance u is respectively less than, equal to, or greater than the quantity

$$R\frac{\mu + 1}{\mu - 1}$$

where R is the radius of curvature of the spherical surface.

Hint: Express the optical path length of the path PXQ in terms of θ in Fig. 2-46; assume θ to be small and take the first and second derivatives of the path with respect to θ.

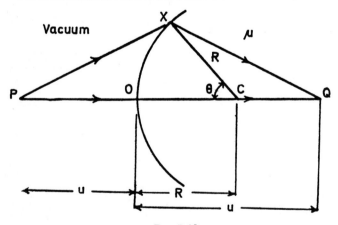

FIG. 2-46

Electromagnetic Character of Light and Polarization

Perhaps the most dramatic moment in the development of physics during the 19th century occurred to J. C. Maxwell one day in the 1860's, when he combined the laws of electricity and magnetism with the laws of the behaviour of light...Maxwell could say, when he was finished with his discovery, 'Let there be electricity and magnetism, and there is light'.

—*The Feynman Lectures on Physics*, Volume 1, Chapter 28

3-1 Introduction

Early in the nineteenth century the notion that light consists of waves, a view already expressed by Christiaan Huygens in the seventeenth century, came into ascendance. A variety of interference phenomena were observed, the most famous of which was the two-slit experiment* of Thomas Young, performed in 1802. By a brilliant interpretation of the results of his experiments in terms of a wave theory of light, Young was able to give strong support to the theory that light is fundamentally a *wave phenomena*. Further, he was able to determine with considerable precision the wavelengths of visible light, ranging from 7×10^{-5} cm at the red end of the visible spectrum to 4×10^{-5} cm at the violet end. However from these experiments the nature of the waves could not be established, the solution to that problem came from quite a new direction —the theoretical study of electricity and magnetism.

James Clerk Maxwell predicted in 1862 the existence of *electromagnetic waves*, consisting of fluctuating electric and magnetic fields which propagate through space (see Chapter 1). Maxwell showed how the speed of electromagnetic waves could be calculated from laboratory

*This experiment will be described in detail in Chapter 4.

experiments in which the same quantity of electric charge is measured by two different methods. One of the methods makes use of the Coulomb force that is proportional to the magnitude of the charge. The second method makes use of the magnetic force exerted on a charge moving in a magnetic field: the force that is proportional to the product of the magnitude of the charge and its speed (see also Appendix A).*

The necessary measurement had been performed in Germany five years earlier by Weber and Kohlrausch. Using their values Maxwell calculated that the speed of electromagnetic waves should be 310,740,000 metres per second. He was immediately struck by the fact that this number was very close to the value of the speed of light (314,858,000 metres per second) as measured by Fizeau in 1849.† The close similarity could have been a chance occurence, but Maxwell with faith in the rationality of nature believed that there must be a deep underlying reason for these two values being same, within the limits of experimental error. In the words of Maxwell himself, the speed of electromagnetic waves

...calculated from the electromagnetic measurements of Kohlrausch and Weber, agrees so exactly with the velocity of light calculated from the optical experiments of M. Fizeau, that we can scarcely avoid the inference that *light consists in the transverse undulations of the same medium which is the cause of electric and magnetic phenomena.*

In 1865, after he had shown that the equations‡ of his theory could be derived from his general principles of electromagnetism, Maxwell wrote to his cousin, Charles Cay:

...I have also a paper afloat, with an electromagnetic theory of light, which till I am convinced to the contrary, I hold to be great guns.

The synthesis of electromagnetism and optics was a great event in physics. In fact, physics had known no greater time since the 1680's when Newton was doing his monumental work on mechanics.

---

*To be more precise, Maxwell showed that the speed of the electromagnetic waves in vacuum c, is given by $c = \dfrac{I_{emu}}{I_{esu}}$ where I_{esu} is the electrostatic unit of current and I_{emu}, is the electromagnetic unit of current. The electrostatic unit of current (esu) is defined from the esu of charge: the esu of charge is that charge which when placed one cm away from an exactly similar charge in vacuum experiences a force of one dyne. Now, if an esu of charge flows through a given cross-section of a wire in one second it constitutes one esu of current in the wire. On the other hand, the electromagnetic unit of current is the current which when present in a long straight wire results in a force of 1 dyne/cm when the wire is placed one cm from a parallel conductor carrying the same current.

†Since then, a number of sophisticated experiments have been carried out to measure the speed of light (c) and it is now one of the most precisely determined fundamental physical constants. Its value in vacuum is $(2 \cdot 997924 \pm \cdot 000010) \times 10^{10}$ cm/sec.

‡ Maxwell's equations are discussed in Appendix A.

3-2 Some Basic Consequences of Electromagnetic Theory

We now state some of the basic consequences of Maxwell's electromagnetic theory:

(1) The electromagnetic waves are generated by any accelerated charge.

(2) When propagating in vacuum the waves travel with a speed of nearly 3×10^8 metres/sec.

(3) Electric waves are always accompanied by magnetic waves and vice-versa.

(4) At any one point in vacuum the electric and magnetic fields, **E** and **B** are mutually perpendicular to each other.

(5) The waves are transverse, i.e. the electric and magnetic fields are perpendicular to the direction of propagation of the waves.

(6) As the waves spread outward into empty space the amplitudes of the electric and magnetic fields decrease steadily.

(7) When two disturbances are superimposed they combine according to the superposition principle; i.e. at any given point and given instant the resultant electric field is the vector sum of the two individual fields. The same is true for magnetic fields.

(8) Electromagnetic waves of many different frequencies can exist. All such waves would be propagated through space with the speed of light. Light itself would correspond to waves of only a small range of frequencies, from 4×10^{14} cps to 7×10^{14} cps, frequencies that happen to be detectable by the human eye.

3-3 A Representation of an Electromagnetic Wave

Fig. 3-1 illustrates the simplest type of electromagnetic wave, in which

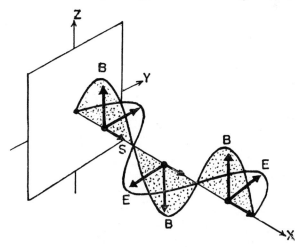

FIG. 3-1: The electric and magnetic vectors in a sinusoidal electromagnetic wave travelling along the $+x$-axis.

we have plane wavefronts; at any fixed point through which the wave-train passes, the electric and magnetic fields oscillate along straight lines (in a sinusoidal manner) at right angles to one another and to the direction of propagation.* The wave is assumed to propagate along the $+ x$ axis and the three components of the electric field E, and the three components of the magnetic field B, may be written as

$$
\left.
\begin{aligned}
E_x &= 0 & B_x &= 0 \\
E_y &= E_0 \operatorname{Sin} (\omega t - kx) & B_y &= 0 \\
E_z &= 0 & B_z &= B_0 \operatorname{Sin} (\omega t - kx)
\end{aligned}
\right\} \quad (3\text{-}1)
$$

where $k = 2\pi/\lambda$, $\omega = 2\pi\nu$ and E_0 and B_0 denote the amplitudes of the electric and magnetic fields respectively. In the MKS system of units (see Appendix A)

$$
B_0 = \frac{1}{c}\, E_0 \qquad (3\text{-}2)
$$

It is important to note from Fig. 3-1 (and also from Eq. 3-1) that the electric and magnetic components of the wave are in phase with each other, i.e. when E_y has its maximum value, B_z is also a maximum.

The wave in Fig. 3-1 is said to be *linearly polarized*, implying that at any fixed point, the tip of the electric or magnetic vector oscillates along a line. The wave is also called *plane polarized*, from the fact that each of the sine waves lie in a plane. Other types of polarization will be discussed later in the chapter.

3-4   Hertz's Experiments

We have already discussed about Maxwell's predictions regarding the production and properties of electromagnetic waves. For example, Maxwell had predicted that they are generated by any accelerated charge, that they are transverse waves, and that they travel with the velocity of light in free space.

To test these predictions, it was necessary to design some apparatus that could produce and detect electromagnetic waves of other frequencies than those of light. This was first done by the German physicist Heinrich Hertz[†] in 1888. Hertz's apparatus is shown in Fig. 3-2. Two plane brass plates are connected to a spark gap P and sparks are caused to jump

---

*To be more precise, the direction of propagation is perpendicular to the plane of the electric and magnetic vectors in the sense of advance of a right handed screw rotated from the electric to the magnetic vector through the smaller angle between their positive directions. In the language of vector analysis, the direction of propagation is along the vector $E \times B$.

[†]Heinrich Hertz (1857-1894). These experiments were carried out while he was professor of physics at the Technical High School at Karlsruhe, Germany during 1885-1889. He was then given a professorship at the University of Bonn, which he held until his death. During the course of his work on electromagnetic waves, Hertz made another discovery, the photoelectric effect, which has had a profound influence on modern physics. We will study this effect in Chapter 8.

across the gap by charging the plates to high voltage with an induction coil. Now, the discharge of the plates by the spark is an oscillatory one. Each time the potential difference between the knobs of the gap reaches the point where the air in the gap becomes conducting, a spark passes.

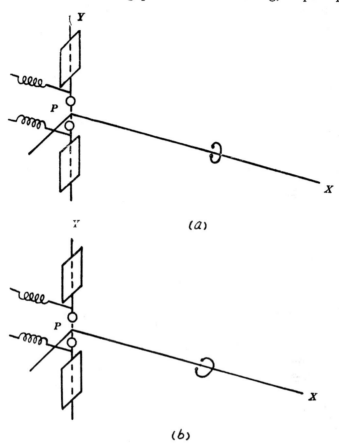

FIG. 3-2: The apparatus used by Hertz for the generation and detection of electromagnetic waves. Hertz showed that with the source and detector parallel to one another, as in (a), a signal was received. But with the detector turned at right angles with respect to the source, as in (b), no signal was received.

This represents a sudden surge of electrons across the gap, and the signs of the charges on the two plates are reversed. But since the air is still conducting, this will produce a return surge, another reversal of sign, and the process repeats until the energy is dissipated as heat. The frequency of these oscillations depends on the inductance and capacitance of the circuit. Thus we have an electric charge undergoing very rapid oscillations, and electromagnetic waves should be radiated.* Now, along the line PX, the direction of the electric field (associated with the wave) is parallel to the Y-axis and that of the magnetic field along the Z-axis (see also Appendix C). To detect the waves Hertz used a short wire, bent in a circular shape but with a small gap. This device is called a resonator. If the resonator is placed in the X-Y plane, i.e. with its plane perpendicular to the magnetic field of the wave [Fig. 3-2(a)], the varying magnetic field induces an *emf* in the resonator, resulting in sparks at its gap. On the other hand, if the plane of the resonator is parallel to the magnetic field [Fig. 3-2(b)] no *emf* is induced and no sparks are observed at the gap.

3-5 Standing Electromagnetic Waves

Hertz also produced standing electromagnetic waves. However, in order to understand the formation of such waves we must consider what happens when an electromagnetic wave strikes a metallic conductor.

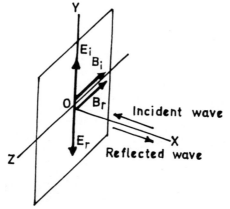

FIG. 3-3: Reflection of electromagnetic waves from a conducting surface.

Let us suppose that a plane electromagnetic wave propagating in the $-x$ direction strikes normally the plane surface of a perfect conductor (Fig. 3-3). We take the Y- and Z-axes as being parallel to the electric and magnetic fields respectively; i.e. we have a plane-polarized incident wave (see Fig. 3-1) with the electric field oscillating in the X-Y plane. The electric field is then parallel to the surface of the conductor. But a perfect conductor does not permit the existence of an electric field *tangential* to its surface.† Therefore, the electric field must always be zero

---

*See also the discussion on dipole oscillator in Appendix C.

†This follows from the fact that if an attempt is made to establish an electric field tangential to the surface, the mobile charges inside the conductor immediately readjust themselves so as to cancel it (see, for example A. Sommerfield's *Optics* Chapter 1, Academic Press)

at $x = 0$. But since the incident wave does not always have $E = 0$ in this plane, the currents which flow in the conductor must generate another wave such that the superpostion of it with the incident wave results in a total electric field which is always zero in the plane $x = 0$.

This situation is completely analogous to that of reflection of waves at a fixed point on a vibrating string discussed in Chapter 1. In that case the displacement at the fixed point was always zero, just as the conductor requires that the electric field be zero in a certain plane. The fixed point exerts forces on the string such as to generate a reflected wave, and here the currents induced in the conductor generate a reflected electromagnetic wave.

In order to provide complete cancellation at $x = 0$, the electric field of the reflected wave at $x = 0$ must have the same amplitude as that of the incident wave but must be out of phase by π radians. Thus if the electric field E_i of the incident wave (which is in the Y-direction) is given by

$$E_i = E_0 \cos \frac{2\pi}{\lambda} (x + ct) \tag{3-3}$$

then the reflected wave is given by

$$E_r = -E_0 \cos \frac{2\pi}{\lambda} (x - ct) \tag{3-4}$$

and the total electric field would be

$$E = E_i + E_r = -2E_0 \sin\left(\frac{2\pi}{\lambda} x\right) \sin \omega t \tag{3-5}$$

where

$$\omega = 2\pi\nu = \frac{2\pi c}{\lambda} \tag{3-6}$$

(Eq. 3-4 is obtained by using the method described in Sec. 1-13, where the expressions for standing waves in a string have been developed.) Thus, in the present case we obtain standing electromagnetic waves.

The corresponding expressions for the magnetic fields of the incident and reflected waves, and the total magnetic field, are

$$B_i = -B_0 \cos \frac{2\pi}{\lambda} (x + ct) \tag{3-7}$$

$$B_r = -B_0 \cos \frac{2\pi}{\lambda} (x - ct) \tag{3-8}$$

$$B = B_i + B_r = -2B_0 \cos\left(\frac{2\pi}{\lambda} x\right) \cos \omega t \tag{3-9}$$

($B_0 = E_0/c$, see Eq. 3-2)

The negative sign in Eq. 3-7 arises because, for a wave propagating in the $-X$ direction if E is in the $+Y$ direction then B must be in the $-Z$ direction (see Fig. 3-1). Similarly for a wave propagating in the $+X$ direction, if the electric field is in the $-Y$ direction then the magnetic field will again be in the $-Z$ direction. This gives rise to the negative sign in Eq. 3-8. We note that the resultant magnetic field does not vanish at the surface of the conductor, and there is no reason why it should. In fact the resultant magnetic field has maximum amplitude at the surface.

It can easily be shown (Problem 1) that for standing waves represented by Eqs. 3-5 and 3-9, the electric and magnetic fields are $\pi/2$ out of phase and that the positions of nodes for E correspond to antinodes for B, and conversely.

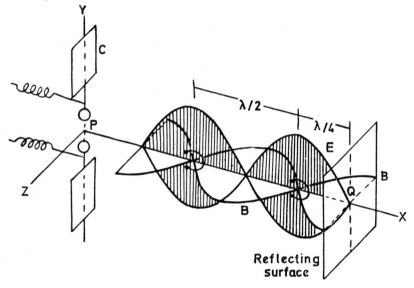

FIG. 3-4: Hertz's experiment on standing electromagnetic waves.

To produce standing electromagnetic waves, Hertz placed a reflecting surface (made of a good conductor) at Q (Fig. 3-4). The standing waves that are formed are also shown in Fig. 3-4. When the resonator is at a node of the magnetic field, no matter what its orientation is, it will show no induced emf (or sparks). At an antinode of the magnetic field, the induced emf (and hence sparking) is greatest when the resonator is oriented at right angles to the magnetic field. By moving the resonator along the line PQ, Hertz found the position of the nodes and antinodes and the direction of the magnetic field. The results obtained by Hertz agreed well with the calculated values of Eq. 3-9. By measuring the distance between two successive antinodes, Hertz could calculate the wavelength of the electromagnetic waves. The frequency of the oscillations could be calculated from the inductance and capacitance in the circuit. Thus Hertz could calculate* the velocity c of the electromagnetic waves by using the equation $c = \lambda \nu$. It was by this means that Hertz measured the speed of these electromagnetic waves and found that this speed, as Maxwell had predicted, is the same as the speed of light.

3-6 Reflection and Refraction of Electromagnetic Waves

In subsequent experiments, Hertz showed that the electromagnetic radiation has all the properties of light waves. Utilizing metal sheets through which the electromagnetic waves could not pass, Hertz was able to define a reasonably narrow beam (Fig. 3-5). With this narrow beam he was able to demonstrate that the waves he had detected do travel in straight lines, and that they can be reflected from a third metallic sheet as light reflects from a mirror; i.e. the angle of incidence i, equals the angle of reflection r.

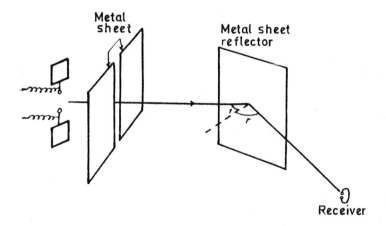

Fig. 3-5: Hertz showed that he could define a straight beam of what we now call radio waves by inserting two metal sheets, with a gap between them, in the path of the wave beam. This beam reflected from a third metal sheet as light reflects from a mirror.

It was not very difficult to prove that the electromagnetic waves reflect according to the observed laws of reflected light but (according to Hertz)†:

* In some of the experiments the frequency of these oscillations was about 10^9 cycle per second. The corresponding wavelength was about 30 cm.

† Hertz's description of his experiment on refraction of electromagnetic waves is taken from S. J. Inglis, *Physics: An ebb and flow of ideas*, p. 243, John Wiley (1970). See also M. R. Shamos (Edited) *Great Experiments of Physics*, Holt-Dryden (1959).

in order to find out whether any refraction of the ray takes place in passing from air into another *insulating* medium, I had a large prism made of so called hard-pitch, a material like asphalt* (Fig. 3-6). The

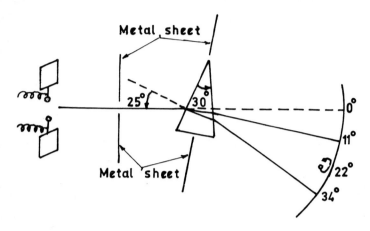

FIG. 3-6: Hertz also showed that radio waves are refracted and dispersed into a spectrum in the manner of Newton's optical spectrum.

base was an isosceles triangle 1·2 metres in the side, and with a refracting angle of nearly 30°. The refracting edge was placed vertical, and the height of the whole prism was 1·5 metres. But since the prism weighed about 12 cwt (1200 pounds), and would have been too heavy to move as a whole, it was built up of three pieces, each 0·5 metre high, placed one above the other. The material was cast in wooden boxes which were left around it, as they did not appear to interfere with its use. The prism was mounted on a support of such height that the middle of its refracting edge was at the same height as the primary and secondary spark gaps. When I was statisfied that refraction did take place, and had obtained some idea of its amount, I arranged the experiment in the following manner: The producing mirror transmitter was set up at a distance of 2·6 metres from the prism and facing one of the refracting surfaces, so that the axis of the beam was directed as nearly as possible towards the centre of mass of the prism, and met the refracting surface at an angle of incidence of 25° (on the side of the normal towards the base). Near the refracting edge and also at the opposite side of the prism were placed two conducting screens which prevented the ray from passing by any other path than that through the prism. On the side of the emerging ray there was marked upon the floor a circle of 2.5 metres radius, having

* Asphalt is a black or dark brown hard substance anciently used as a cement, and now for paving, cisterns, water pipes, etc.

as its centre the centre of mass of the lower end of the prism. Along this the receiving mirror (receiver) was now moved about, its aperture being always directed towards the centre of the circle. No sparks were obtained when the mirror was placed in the direction of the incident ray produced; in this direction the prism threw a complete shadow. But sparks appeared when the mirror was moved towards the base of the prism, beginning when the angular deviation from the first position was about 11°. The sparking increased in intensity until the deviation amounted to about 22°, and then again decreased. The last sparks were observed with a deviation of about 34°. When the mirror was placed in a position of maximum effect, and then moved away from the prism along the radius of the circle, the sparks could be traced up to a distance of 5-6 metres. When an assistant stood either in front of the prism or behind it the sparking invariably ceased, which shows that the action reaches the secondary conductor through the prism and not in any other way.

Hertz's experiments provided dramatic confirmation of Maxwell's electomagnetic theory. They showed that electromagnetic waves actually exist, that they travel with the speed of light, and that they have the familiar characteristics of light. There was now rapid acceptance of Maxwell's theory by mathematical physicists, who applied it with great success to the detailed analysis of a wide range of phenomena.

3-7 Energy and Intensity of an Electromagnetic Wave

The intensity of mechanical waves was shown in Chapter 1 to be proportional to the square of the amplitude. A similar result follows from Maxwell's electromagnetic theory. It can be shown* that in vacuum the electromagnetic field has an energy density given by:

$$\text{Energy per unit volume } E = \frac{U}{V} = \tfrac{1}{2}\varepsilon_0 E^2 + \frac{1}{2\mu_0}B^2 \qquad (3\text{-}10)$$

where E and B are the *instantaneous* values of the fields, and ε_0 and μ_0 are called the *permittivity* and *permeability* of vacuum respectively.† Their numerical values in the MKS system are

$$\left. \begin{aligned} \varepsilon_0 &= \frac{1}{36\pi \times 10^9} \text{ coul}^2/\text{nt-m}^2 = 8\cdot85 \times 10^{-12} \text{ coul}^2/\text{nt-m}^2 \\ \mu_0 &= 4\pi \times 10^{-7} \text{ coul}^2 \text{ sec}^2/\text{nt} \end{aligned} \right\} \qquad (3\text{-}11)$$

$$\frac{1}{\varepsilon_0\mu_0} = c^2 = 9 \times 10^{16} \text{ (metre/sec)}^2$$

---

* See, for example, L. Page and N. I. Adams, *Principles of Electricity*, 2nd Edition, p. 564, D. Van Nostrand Company (1949).

† ε_0 and μ_0 are defined in Appendix A.

For the simple plane waves we will study

$$E = \frac{\varepsilon_0}{2} E_0^2 \operatorname{Sin}^2 (kx - \omega t) + \frac{1}{2\mu_0} B_0^2 \operatorname{Sin}^2 (kx - \omega t) \qquad (3\text{-}12)$$

But

$$B_0 \doteq \frac{E_0}{c} = E_0 \sqrt{\varepsilon_0 \mu_0}$$

or

$$\frac{1}{2\mu_0} B_0^2 = \frac{\varepsilon_0}{2} E_0^2$$

Thus, half the energy is associated with the electric vector and half with the magnetic vector giving

$$E = \varepsilon_0 E_0^2 \operatorname{Sin}^2 (kx - \omega t) \qquad (3\text{-}13)$$

Since the optical frequencies are very large (ω is of the order 10^{15} sec^{-1}), one cannot observe the instantaneous values of any of the rapidly varying quantities, but only their time average taken over a time interval (say $-T' \leqslant t \leqslant T'$) which is *large* compared to the fundamental period $T = \frac{2\pi}{\omega}$.

Thus

$$\langle E \rangle_{av} = \frac{1}{2T'} \int_{-T'}^{T'} E \, dt$$

$$= \frac{\varepsilon_0 E_0^2}{2T'} \int_{-T'}^{T'} \operatorname{Sin}^2 (kx - \omega t) \, dt$$

$$= \frac{\varepsilon_0 E_0^2}{2T'} \cdot \frac{1}{2} \int_{-T'}^{T'} [1 - \operatorname{Cos} 2 (kx - \omega t)] \, dt$$

$$= \frac{\varepsilon_0 E_0^2}{4T'} \left[t - \frac{1}{2\omega} \operatorname{Sin} 2 (kx - \omega t) \right]_{-T'}^{+T'}$$

$$= \frac{\varepsilon_0 E_0^2}{4} \left[2 - \frac{1}{4\pi} \left(\frac{T}{T'} \right) \{ \operatorname{Sin} 2 (kx - \omega T') - \operatorname{Sin} 2 (kx + \omega T') \} \right]$$

$$\therefore \quad \langle E \rangle_{av} \approx \frac{\varepsilon_0 E_0^2}{2} \qquad (3\text{-}14)$$

because $T' \gg T$ and the expression within the braces has to be between -2 and $+2$.

The intensity of the wave will merely be the product of the above expression by the velocity c, since this represents the volume of the wave that will stream through unit area per second. We therefore have

$$I = \tfrac{1}{2}\ \varepsilon_0 c E_0{}^2 \tag{3-15}$$

Substituting the values of ε_0 and c we obtain

$$I = \tfrac{1}{2} \times (8\text{·}85 \times 10^{-12}\ \text{coul}^2/\text{nt-m}) \times (3 \times 10^8\ \text{m/sec})\ E_0{}^2$$

or

$$I \approx (1\text{·}33 \times 10^{-3}\ \text{watt/volt}^2)\ E_0{}^2 \tag{3-16}*$$

Example
The light from a 220 watt lamp uniformly spreads out in all directions. Find the intensity of the electromagnetic waves (and the amplitude E_0) at a distance of 10 metres from the lamp.

Solution
The intensity at a distance of 10 metres is given by

$$I = \frac{220}{4\pi(10)^2} = 1\text{·}75 \times 10^{-3}\ \text{watt/m}^2$$

$$= 1\text{·}75 \times 10^{-7}\ \text{watt/cm}^2$$

Thus, if we consider one square centimetre of area placed normally to the direction of the propagation of the beam, then $1\text{·}75 \times 10^{-7}$ Joules of energy will cross this area in one second.

Now, from the relationship between the intensity and the wave amplitude, E_0, given by Eq. 3-16, we obtain

$$E_0{}^2 = \frac{1\text{·}75 \times 10^{-3}\ \text{watt/m}^2}{1\text{·}33 \times 10^{-3}\ \text{watt/volt}^2}$$

or

$$E_0 \approx 1\text{·}15\ \text{volt/m}$$

(This amplitude will fall of roughly as $1/r$ with distance, whereas the intensity will fall of as $1/r^2$. We are, of course, assuming an approximately point source of light.)

Example
Lasers (described in Appendix D) are light sources which are almost perfectly parallel of high intensity. Suppose a 2000 watt laser beam was concentrated by a lens into cross-sectional area of about 10^{-6} sq. cm. Find the corresponding intensity and the amplitude of the electric field.

---

* Although Eq. 3-16 is exactly true only for plane waves in vacuum, it represents an excellent approximation in many situations.

Solution

The intensity is given by

$$I = \frac{2000 \text{ watt}}{10^{-6} \text{ cm}^2}$$

$$= 2 \times 10^9 \text{ watt/cm}^2 = 2 \times 10^{13} \text{ watt/m}^2$$

Further, from Eq. 3-16

$$E_o{}^2 = \frac{2 \cdot 0 \times 10^9 \text{ watt/m}^2}{1 \cdot 33 \times 10^{-3} \text{ watt/volt}^2}$$

or

$$E_o \approx 1 \cdot 2 \times 10^8 \text{ volt/m}$$

Because a laser beam has a high power and can be brought to a very sharp focus, it can be used as an exceedingly effective drill to burn through a target.

3-8 Momentum of an Electromagnetic Wave

Maxwell had also predicted that electromagnetic waves transport *momentum*. This momentum is in the direction of the propagation of the wave and is equal to the energy divided by c. Thus if p denotes the momentum per unit volume associated with the electromagnetic wave then

$$p = \frac{E}{c} \tag{3-17}$$

where E denotes the energy density.

Since electromagnetic waves carry momentum, they give rise to a certain pressure when they are reflected or absorbed at the surface of a body. We illustrate this by considering some examples.

Suppose that a plane electromagnetic wave falls perpendicularly on a perfectly absorbing surface [Fig. 3-7(a)]. If the incident momentum per unit volume is p then the amount of momentum in the radiation falling per unit time on the surface of area A is obtained by multiplying p by the volume cA; i.e. pcA. Since the radiation is assumed to be completely absorbed by the surface, this is also the momentum absorbed by the surface A; that is the force on A. Dividing by the area A, we get the pressure due to the radiation.

$$P_{\text{rad}} = cp = E \qquad \begin{bmatrix} \text{Perfect Absorber:} \\ \text{Normal Incidence} \end{bmatrix} \tag{3-18}$$

Thus for normal incidence the radiation pressure* on a perfect absorber is equal to the energy density in the wave. On the other hand, if the

---

* See, G. E. Henry, 'Radiation Pressure', *Scientific American*, June 1957.

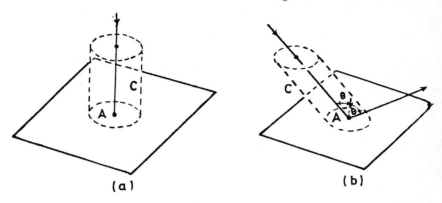

FIG. 3-7: (a) Radiation pressure at normal incidence. (b) Radiation pressure at oblique incidence.

surface is a perfect reflector, the radiation after reflection has a momentum equal in magnitude but opposite in direction to the incident radiation. The change in momentum per unit volume is thus $2p$, and the radiation pressure is accordingly

$$P_{\text{rad}} = 2pc = 2E \qquad \begin{bmatrix} \text{Perfect Reflector:} \\ \text{Normal Incidence} \end{bmatrix} \qquad (3\text{-}19)$$

These results can be generalized to the case of oblique incidence [Fig. 3-7(*b*)] in which the change in momentum per unit volume at the perfectly reflecting surface is $2p \cos\theta$, and the corresponding radiation pressure is

$$P_{\text{rad}} = 2cp \cos^2\theta = 2E \cos^2\theta$$

$$\begin{bmatrix} \text{Perfect Reflector:} \\ \text{Oblique Incidence} \end{bmatrix} \qquad (3\text{-}20)$$

where θ is the angle of incidence.

Example
A parallel beam of light with an intensity of 50 watt/cm² falls normally for one hour on a perfectly reflecting plane mirror of area 2 cm². Calculate the momentum that is delivered to the mirror in this time and the force that acts on the mirror.

Solution
The energy that is reflected from the mirror is given by

$$U = (50 \text{ watt/cm}^2) \ (2 \text{ cm}^2) \ (3600 \text{ sec})$$
$$= 3 \cdot 6 \times 10^5 \text{ Joule}$$

The momentum delivered after one hour's illumination is given by

$$p = 2\,\frac{U}{c} = \frac{2 \times 3\cdot6 \times 10^5 \text{ Joules}}{3 \times 10^8 \text{ m/sec}}$$

$$= 2\cdot4 \times 10^{-3} \text{ kg m/sec}$$

Now, from Newton's second law, the average force on the mirror is equal to the average rate at which momentum is delivered to the mirror, i.e.

$$F = \frac{p}{t} = \frac{2\cdot4 \times 10^{-3} \text{ kg m/sec}}{3600 \text{ sec}}$$

$$= 0\cdot667 \times 10^{-6} \text{ Newton}$$

Thus the radiation pressure P_{rad} would be given by

$$P_{\text{rad}} = \frac{0\cdot667 \times 10^{-6}}{2 \text{ cm}^2} \text{ Newton}$$

$$= 0\cdot334 \times 10^{-6} \text{ Newton/cm}^2$$

$$= 3\cdot34 \times 10^{-3} \text{ Newton/m}^2$$

(This is to be compared with the atmospheric pressure which is about 10^5 Newton/m^2.)

The radiation pressure being very small, experimentally it is extremely difficult to measure. The technical difficulties involved in testing this prediction were not solved until 1899, when Lebedev in Russia and two years later, Nichols and Hull in the United States finally confirmed the existence of radiation pressure. Nichols and Hull used a torsion balance technique to measure radiation pressure (Fig. 3-8). They allowed light to fall on mirror M and the radiation pressure caused the balance arm to turn through a measured angle θ, twisting the torsion fibt   . Assuming a suitable calibration for their torsion fiber, the actual pressure could be calculated. The intensity of the light beam was measured by allowing it to fall on a blackened metal disc of known absorptivity and by measuring the temperature rise of this disc. In a particular run, the measured value of the radiation pressure was $7\cdot01 \times 10^{-6}$ Newton/metre^2 which was in excellent agreement with the value $7\cdot05 \times 10^{-6}$ Newton/metre^2 predicted by Maxwell's theory (i.e. by using Eq. 3-19).

Thus, at the beginning of the twentieth century, Maxwell's electromagnetic theory stood on the same level as Newton's laws of mechanics, as an established part of the foundations of physics.

3-9   The Electromagnetic Spectrum

Hertz used radiation with a wavelength of about a million times the wavelength of visible light. One may ask: is this an isolated set of wavelengths or a small segment of a spectrum of much greater extent? Experiments show that there is a wide and continuous variation in the

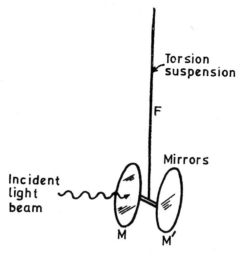

Fig. 3-8: Torsion balance used to measure radiation pressure. The amount of twist in the horizontal beam is proportional to the pressure that radiation exerts on the mirror.

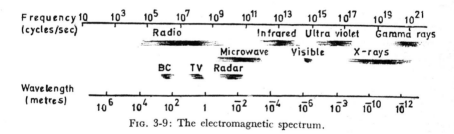

Fig. 3-9: The electromagnetic spectrum.

wavelength (and frequency) of electromagnetic waves; the entire possible range is called the electromagnetic spectrum (Fig. 3-9). The wavelengths of visible light lie between about 4×10^{-5} cm for the extreme violet and 7×10^{-5} cm for the deep red. Just as the ear becomes insensitive to sound above a certain frequency, so do the eyes fail to respond to light vibrations of frequencies greater than that of the extreme violet or less than that of the extreme red. Electromagnetic radiation of wavelength shorter than that of the visible, down to about 5×10^{-7} cm is known as 'ultraviolet light'. Beyond this is the X-ray region which extends to a wavelength of about 6×10^{-10} cm. Still shorter wavelengths correspond to the gamma rays which are emitted by radioactive substances. On the long wavelength* side of the visible region lies the

---

* Electromagnetic waves with a wavelength as long as 1,86,000 miles have been detected. The origin of these waves is unknown, but they promise to be useful in geophysics. (See, for example, J. H. Heirtzler, 'The Longest Electromagnetic Waves', *Scientific American*, March 1962.)

infrared, which may be said to merge into the radio waves at a wavelength of about 4×10^{-2} cm. As we all know radiowaves can have as long a wavelength as 10^5 cm.

Although all the radiations that are listed in Fig. 3-9, are similar in nature, differing only in wavelength, the term 'light' is conventionally referred to the region extending between the violet and the red end of the spectrum. It is seen that 'light' covers an almost insignificant fraction of the electromagnetic spectrum. The divisions between the different types of radiation are purely formal and are roughly fixed by the fact that in the laboratory the different types of waves are generated and detected in different ways. For example, light may be perceived through its effect on the retina of the eye, but to detect radio waves we need electric equipment to select and amplify the signal. On the other hand to detect gamma rays we require 'nuclear counters'.

All the waves in the electromagnetic spectrum, although produced and detected in various ways, behave as predicted by Maxwell's theory. All electromagnetic waves travel through vacuum at the same speed. They all carry energy; when they are absorbed, the absorber is heated. Electromagnetic radiation, whatever its frequency, can be emitted only by a process in which energy is supplied to the source of radiation. There is now overwhelming evidence that electromagnetic radiation originates from accelerated charges. This acceleration may be produced in many ways by heating materials to increase the vibrational energy of charged particles, by varying the charge on an electric conductor (antenna), or by causing a stream of charged particles to change its direction.* In these and other processes the work done by the force applied to the electric charges becomes the energy of radiation.

3-10 Describing an Electromagnetic Disturbance

One could choose either the electric or the magnetic wave to describe an electromagnetic disturbance. In the interference or the diffraction phenomena, which we will discuss in Chapters 4-6, the electric waves will superpose with each other in exactly the same way as the magnetic waves. We could, therefore, assume any one of them to represent the 'displacements' that we have been using in Chapter 1. However, it has been found that it is the electric vector that affects the photographic plate and presumably the retina of the eye. In this sense, therefore the electric wave is the part that really constitutes 'light', and the magnetic wave though no less real is less important.

In all what follows, we shall describe an electromagnetic disturbance by *just the electric waves*. We shall further make the following two assumptions:

* For example X-rays are commonly produced by the sudden deflection or stopping of electrons when they strike a metal target.

(i) The wavetrain is monochromatic, i.e. it involves only a single wavelength.

(ii) One is concerned with only a small region of space and that this region is so far away from the light source that the amplitude throughout the region is practically constant.

Accordingly, one can depict the wavetrain at a given instant by sinusoidal curve similar to that shown in Fig. 3-1. One can thus, specify the disturbance as

$$E = a \; \text{Sin} \; (\omega t - kz) \qquad (3\text{-}21)$$

where, a is the amplitude of the wave, ω, the angular frequency and $k = 2\pi/\lambda$. The expression, represented by E, is usually referred to as the displacement at any given time t and any given position z along the axis of propagation.

In order to indicate the displacement (or more specifically, the electric field strength), one may write it explicitly as in Eq. 3-1 or, one may write the symbol i or j in front of the algebraic expressions and mention that i and j are unit vectors along the positive X-and positive Y-axes respectively.

3-11 Superposition of two Disturbances

Let us suppose that we want to superpose two beams of light that have the same wavelength and same phase constant and both the waves are propagating along the Z-axis. We shall also assume that the waves have identical directions of displacement (say along the Y-axis) and that the two amplitudes, a_1 and a_2 are not equal. According to the electromagnetic theory, when the two beams are combined, the resulting disturbance E, corresponds to the simple addition of the two initial disturbances. Thus

$$\begin{aligned} E &= a_1 \; \text{Sin} \; (\omega t - kz) + a_2 \; \text{Sin} \; (\omega t - kz) \\ &= a \; \text{Sin} \; (\omega t - kz) \end{aligned} \qquad (3\text{-}22)$$

where

$$a = a_1 + a_2$$

Thus, the resulting disturbance is the same as either of the initial disturbances except that the amplitude is greater, being the sum of the two initial amplitudes.

The slightly more complicated case in which the two disturbances have the same phase but having displacements that are parallel to X- and Y-axes can also be easily discussed. We may have

$$E_1 = i \; a_1 \; \text{Sin} \; (\omega t - kz) \qquad (3\text{-}23a)$$

and

$$E_2 = j \; a_2 \; \text{Sin} \; (\omega t - kz) \qquad (3\text{-}23b)$$

Thus, the x- and y-components of the resultant E ($=E_1 + E_2$) are given by

$$E_x = a_1 \operatorname{Sin}(\omega t - kz) \qquad (3\text{-}24)$$
$$E_y = a_2 \operatorname{Sin}(\omega t - kz)$$

Now what kind of electric field is made up of an x-component and a y-component which oscillate at the same frequency and in the same phase? If one adds to an x-vibration a certain amount of y-vibration at the same phase, the result is a vibration in a new direction in the x-y plane. Fig. 3-10 illustrates the superposition of different amplitudes for

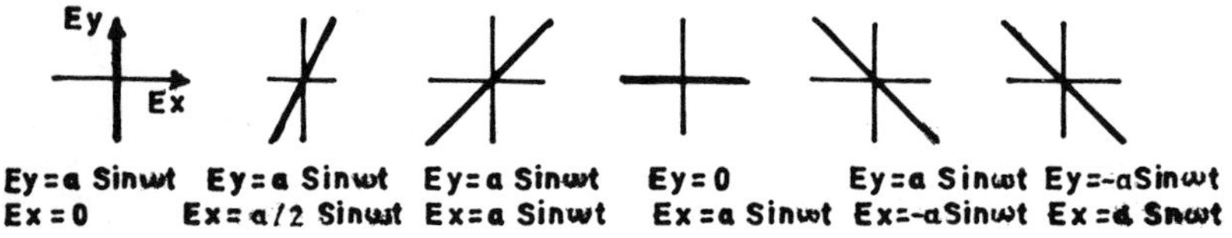

Ey=a Sinωt Ey=a Sinωt Ey=a Sinωt Ey=0   Ey=a Sinωt Ey=-aSinωt
Ex = 0   Ex=a/2 Sinωt Ex=a Sinωt Ex=a Sinωt Ex=-aSinωt Ex=a Snωt

FIG. 3-10: Superposition of two electric fields at right angles to each other and oscillating in phase.

the x- and the y-vibrations. That the resultant vibration is also linearly polarized also follows from the fact that

$$\frac{E_y}{E_x} = \frac{a_2}{a_1} \qquad \text{(independent of } t\text{)} \qquad (3\text{-}25)$$

which is the equation of a straight line in the E_x-E_y plane.

In all the cases that have been shown in Fig. 3-10 we have assumed that the x-vibration and the y-vibration are *in phase*, but it need not have to be that way. Indeed, when the x-vibration and the y-vibration are not in phase*, the electric field vector moves around in an ellipse, and we can understand this by considering the following simple example: If we suspend a small mass from a support by a long string, so that it can swing freely in a horizontal plane, it will execute sinusoidal oscillations. If we imagine horizontal x-and y-coordinates with their origin at the rest position of the ball, the ball can swing in either the x- or y-direction with the same frequency. By making the proper initial displacement and initial velocity, we can set the ball in oscillation along either the x- or

* How do we obtain the experimental conditions so that we superpose two plane polarized beams (of the same frequency) with their electric vectors at right angles to each other? It turns out that the experimental conditions are not too difficult to obtain. It will be shown later that if light is passed through a calcite crystal there are, in general, two emergent rays. Both these rays are plane polarized and their planes of vibration are at right angles to one another. These two plane polarized beams can be superposed after being made to traverse different optical paths. It is possible to rotate their planes of vibration before superposing them but the way in which this is done will be discussed later. In this way it is possible to study the effects of superposing plane polarized disturbances under various conditions.

the y-axis, or along any straight line in the x-y plane. These motions of the ball are analogous to the oscillations of the electric field vector illustrated in Fig. 3-10. In each case, since the x-vibrations and the y-vibrations reach their maxima and minima at the same time, the x- and y-oscillations are in phase. But we know that the most general motion of the ball corresponds to the case when the ball moves on the circumference of an ellipse (in the horizontal plane). This motion in an ellipse corresponds to oscillations in which the x- and y- directions are *not* in the same phase. The superposition of x- and y-vibrations which are *not* in phase is illustrated in Fig. 3-11 for various values of the phase

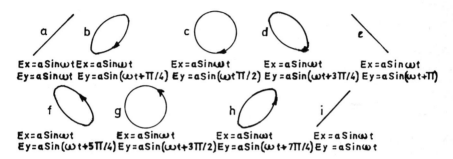

Fig. 3-11: Superposition of two electric fields (of equal amplitude) at right angles to each other and oscillating with various phase differences.

difference between the x- and the y- component of vibration. The general result is that the electric vector moves around an ellipse. The motion in a straight line is a particular case corresponding to a phase difference of zero, or an integral multiple of π. On the other hand the motion in a circle corresponds to the superposition of x-vibrations and y-vibrations of equal amplitudes with a phase difference of $\pi/2$ or any odd integral multiple of $\pi/2$. We can arrive at the same conclusion from simple mathematical considerations. Let the x- and y-components of the electric vector be given by

$$E_x = a_1 \ \text{Sin} \ \omega t \tag{3-26}$$
$$E_y = a_2 \ \text{Sin} \ (\omega t + \delta) \tag{3-27}$$

where a_1 and a_2 are the amplitudes, ω is the angular frequency and δ, the phase difference between the two vibrations.

If E_x and E_y are imagined as components of the motion of a particle, the path of the resultant motion of the particle is obtained by eliminating t from Eqs. 3-26 and 3-27.

Now

$$\begin{aligned} E_y &= a_2 \ \text{Sin} \ (\omega t + \delta) \\ &= a_2 \ [\text{Sin} \ \omega t \ \text{Cos} \delta + \text{Sin} \delta \ \text{Cos} \omega t] \\ &= a_2 \ [\text{Sin} \ \omega t \ \text{Cos} \delta + \text{Sin} \delta \ \sqrt{1 - \text{Sin}^2 \ \omega t}] \end{aligned}$$

Thus, in order to obtain the resultant vibration we substitute the value of Sin ωt (in terms of E_x) in the above equation. The result is

$$E_y = a_2 \left[\frac{E_x}{a_1} \, \mathrm{Cos}\,\delta + \mathrm{Sin}\,\delta \, \sqrt{1 - \left(\frac{E_x}{a_1}\right)^2} \, \right]$$

After carrying out simple agebraic manipulations, we obtain

$$\frac{E_x{}^2}{a_1{}^2} + \frac{E_y{}^2}{a_2{}^2} - \frac{2E_xE_y}{a_1a_2} \, \mathrm{Cos}\,\delta = \mathrm{Sin}^2\delta \tag{3-28}$$

[If we replace E_x and E_y by x and y respectively, we find that the curve represented by Eq. 3-28 (in the x-y plane) is, in general, an ellipse.] Thus in the E_x-E_y plane the tip of the E moves on the circumference of an ellipse. In general, the angle θ between the coordinate axes and the axes of the ellipse is given by:

$$\tan 2\theta = \frac{2a_1a_2 \, \mathrm{Cos}\,\delta}{a_1{}^2 - a_2{}^2} \tag{3-29}$$

Some special cases are given below:

(i) If $\delta = 2n\pi$ where $n = 1, 2, 3,\ldots$.
Eq. (3-28) becomes

$$\frac{E_x{}^2}{a_1{}^2} - \frac{2E_x \, E_y}{a_1a_2} + \frac{E_y{}^2}{a_2{}^2} = 0$$

or

$$E_y = \frac{a_1}{a_2} \, E_x \tag{3-30}$$

which is the equation of a straight line. On the other hand, for $\delta = (2n + 1)\pi$, it can be easily seen that

$$E_y = -\frac{a_1}{a_2} \, E_x \tag{3-31}$$

In either case, we have linearly polarized light.
(ii) If $\delta = (n + \tfrac{1}{2})\pi$ where $n = 0, 1, 2,\ldots$.
Eq. (3-28) becomes

$$\frac{E_x{}^2}{a_1{}^2} + \frac{E_y{}^2}{a_2{}^2} = 1 \tag{3-32}$$

which is the equation of an ellipse with the axes of the ellipse parallel to the E_x- and E_y- axes. This ellipse degenerates into a circle when a_1 becomes equal to a_2.
  Now, for $\delta = \pi/2, 5\pi/2, 9\pi/2,\ldots$.
$$E_x = a_1 \, \mathrm{Sin}\,\omega t$$
and
$$E_y = a_2 \, \mathrm{Cos}\,\omega t$$

and the ellipse is described *clockwise* with respect to an observer towards whom the wave travels and the light is said to be left-handed elliptically polarized. This can easily be verified by plotting the above equations. For example, we may proceed as follows:

(*a*)   When $t = 0$, $E_x = 0$ and $E_y = a_2$

(*b*)   When $t = \dfrac{\pi}{4\omega}$, $E_x = \dfrac{a_1}{\sqrt{2}}$ and $E_y = \dfrac{a_2}{\sqrt{2}}$

(*c*)   When $t = \dfrac{\pi}{2\omega}$, $E_x = a_1$ and $E_y = 0$

The plot can easily be drawn and the tip of the electric vector rotates in the clockwise direction.

On the other hand, if $\delta = 3\pi/2$, $7\pi/2, \ldots$ the ellipse is described in the anticlockwise direction and the light is said to be right-handed elliptically polarized light (Fig. 3-12).*

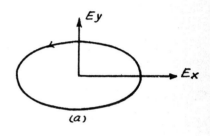

(a)

We conclude this section by noting that light is *linearly polarized* (sometimes called *plane polarized*) when the electric field oscillates on a straight line. When the end of the electric field vector travels in an ellipse the light is *elliptically polarized*. When this ellipse degenerates into a circle, we have *circularly polarized light*.† If the end of the electric vector, when we look at it as the light comes straight toward us, goes around in clockwise direction, we call it left-hand circular polarization. Fig. 3-11(*c*) shows left-hand circular polarization and Fig. 3-11(*g*) illustrates right-hand circular polarization.

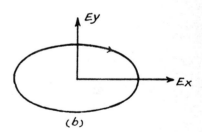

(b)

FIG. 3-12: (a) If $\delta = \dfrac{3\pi}{2}, \dfrac{7\pi}{2}, .$ the tip of the electric vector rotates in the anticlockwise direction.

(b) If $\delta = \dfrac{\pi}{2}, \dfrac{5\pi}{2}, \ldots$ the tip of the electric vector rotates in the clockwise direction.

---

* Our convention for labelling left-hand and right-hand circular (or elliptical) polairzation is consistent with the one used by Feynman and also with that which is used today for all the other particles in physics which exhibit polarization (e.g. electrons). However, in some books on optics the opposite conventions are used, so one must be careful!

† It is worthwhile pointing out that in the rope experiment of Chapter 1 (Fig. 1-1) if we rotate one end of the string in a circle we will generate a circularly polarized wave.

3-12　Unpolarized Light

So far we have considered linearly, circularly, and elliptically polarized light, which covers everything except for the case of *unpolarized light.* Indeed light from ordinary sources are unpolarized. This is due to the fact that one atom emits radiation for a time of the order of 10^{-9} sec and if one atom emits a certain polarization, and then another atom emits light with a different polarization, the polarization will change every 10^{-9} sec. If the polarization changes more rapidly than we can detect it, then we call the light unpolarized, because all the effects of the polarization average out.　Thus, light is unpolarized only if we are unable to find out whether the light is polarized or not.

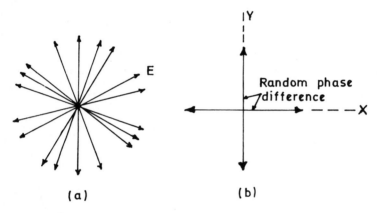

(a)　　　　　　　　　　　　　(b)

FIG. 3-13: (a) An unpolarized electromagnetic wave can be thought of as a random superposition of many plane polarized waves.

(b) Another equivalent description of the unpolarized wave consists in viewing the unpolarized wave as the superposition of two linearly polarized waves (with vibrations at right angles to each other) and with a random phase difference.

Fig. 3-13(*a*) describes the unpolarized light. The planes of vibration are randomly oriented about the direction of propagation. Another completely equivalent description of an unpolarized light wave is shown in Fig. 3-13(*b*); here the unpolarized wave is viewed as two plane-polarized waves with a *random phase difference.* The orientation of the *x*-and *y*-axes about the direction of propagation is completely arbitrary. It is because of this random orientation of the planes of vibration that on casual study, the true transverse nature of the waves is concealed. Thus, to study the transverse nature of the wave we must try to produce polarized light.

3-13 Production of Polarized Light

In the laboratory, the usual method of producing polarized light essentially consists of dividing unpolarized light (actually or at least in principle) into two orthogonally polarized components and then eliminating one of the components. The component that remains is essentially a polarized beam.

The device which divides the unpolarized light into two components and discards one is known as a polarizer.* If the discarding process is not perfect, so that the unwanted component is not completely eliminated, the polarizer is then known as a partial polarizer and the light that comes out is known as 'partially polarized light'. Clearly, a polarizer does not create transverse vibrations, but merely divides the existing vibrations into two components and selects one. Surely, there can be many types of manipulations which may be used in a polarizer in order to select (or discard) one of the components. For example, either of the processes involving absorption, reflection, scattering, etc., may be used to produce a polarized beam. In what follows we will consider the most easily understood class of polarizer — that which employs assymmetry of absorption, or *dichroism*. Perhaps the simplest example is the 'wire grid polarizer'.

3-14 The Wire Grid Polarizer

This device consists of an array of thin wires arranged parallel to one another as shown in Fig. 3-14. Let us suppose that an electromagnetic wave is propagating in the $+ z$ direction (the z-axis is perpendicular to the plane defined by the wires). Let the electric field in the incident electromagnetic radiation has both x- and y-components. We will consider the effect of the wires on these two components separately.

First, let us consider the effect of the y-component along the wires. The electric field of the incident radiation drives the electrons along the wire.† The electric field does work on the electrons; they transfer some of their energy to the copper lattice through collisions. The energy associated with the field is therefore converted to heat because of the small but *significant* electrical resistance of the wires. Thus the wire grid eliminates the y-component of the electric field.

Next, let us consider what happens to the x-component of the electric field. The electrons are not free to move along the x-axis, because they cannot leave the wire. Instead of reaching a steady terminal velocity (as they do for motion along the y-axis), the electrons soon build up a

---

* If the device separates the two components and preserves both, it is called a 'polarizing beam splitter.'

† The wires are usually made of copper or gold or any good metallic conductor. The wires being good conductors provide high conductivity for electric fields parallel to the wires.

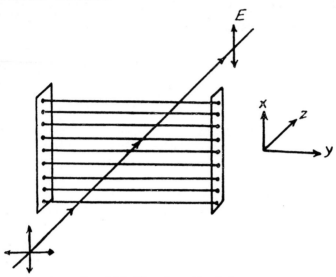

FIG. 3-14: The wire grid polarizer.

surface charge along the $+x$ and $-x$ edges of the wire. When the field due to the surface charge is sufficient to cancel the incident field (inside the wire), the electrons stop moving. Thus the electrons are always in a sort of static equilibrium (or nearly so) with no velocity or acceleration. They do not absorb energy. Thus the x-component of the radiation is unaffected.

It should be noted that although surface charges do build up along the length of the wires (i.e. along the y-axis), the resulting field from these end charges can be made as small as we wish in the region of interest (near the centre of the grid) by making the wires sufficiently long in the y-direction.

Thus, the energy associated with one of the components of the un-polarized beam is almost completely lost, whereas the other component passes through with almost no decrease in intensity at all. The transmitted component is the one whose electric vector oscillates in a direction perpendicular to the wires. (If the wires were arranged randomly in all directions, both components would be absorbed and the device would transmit no light at all.)

Although easy to understand, the wire grid polarizer is difficult to construct. Indeed no one succeeded in making a really statisfactory wire-grid polarizer until 1960, when Bird and Parrish succeeded in produc-ing wires of less than a wavelength of light in diameter and separating them by gaps of less than a wavelength wide.* Such a construction is

* See *Polarized Light* by W. A. Shurcliff and S. S. Ballard, p. 28, Van Nostrand. The details of the procedure for making this wire grating are also discussed in the book. The original work of G. R. Bird and M. Parrish was published in the *Journal of Optical Society of America*, Vol. 50, 886 (1960).

indeed rather difficult since about 30,000 wires are needed to be put in about one inch.

The wire-grid performs very badly if the wires are thick and the spaces between them are large. A thick wire permits some flow of current transveresly across the individual wire, and thus even the wanted component of the light is absorbed to some extent. Wide spaces between wires permit some of the unwanted component to leak through. Wire grids for polarizing 3-cm microwaves are easy to make, since it is only necessary that the wire diameter and spacing be small compared to three centimetres. For visible light, however, the diameter and spacing must be small compared to the wavelength of, say, green light, which makes the manufacturer's task a difficult one.

It is easy to understand that if we place two wire grid polarizers, one with horizontal wires and the other with vertical wires, one after the other (as in Fig. 3-14) then we will have almost no transmitted light.* This is due to the fact that the first polarizer eliminates the y-component and the second one the x-component of the electric field.

The above type of experiment leads one to believe that light waves were transverse and not longitudinal. In order to understand this let us carry out a simple experiment with waves propagating on a rope.

Let us generate transverse waves in a rope by oscillating one of its ends (Fig. 13-15). A slit P can be used to confine the vibration to one plane after which they can be transmitted or obstructed by a second slit A. The slit P 'polarizes' the wave, for it allows only those vibrations which are parallel to the slit. The second slit A, 'analyses' the polarized wave and if it is placed parallel to the slit P it will allow the wave to pass through and if it is placed perpendicular to the slit P, no wave is propagated through. On the other hand, if the rope were replaced by a coiled spring, compressional (longitudinal) waves in it would pass through the slit regardless of their orientation. Thus

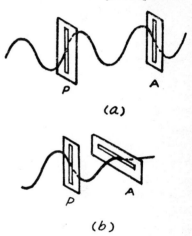

FIG. 3-15: Mechanical analogue of polarization. The slit P 'polarizes' the wave, for it allows only those vibrations to pass through which are parallel to the slit. The slit A 'analyzes' the wave. In (a), the slit A is parallel to the slit P and hence it allows the wave to pass through. In (b), the slit A is perpendicular to the slit P and it does not allow any wave to pass through.

* This point will be considered in greater detail when we discuss Malus' law.

polarizability is a characteristic of *transverse waves*.

The above analogy, when applied to electromagnetic waves, should be used with some caution. For, in the wire grid polarizer it is the electric field vibrating parallel to the *length* of the wires which get absorbed and the field at right angles to the length of the wires get transmitted. Thus the rope analogy may be helpful to beginners, but it is wrong by just 90°.

3-15 H-Sheet: World's Most Popular Polarizer

The commonest type of polarizer is an *H*-sheet, invented by E. H. Land in 1938. It may be regarded as a chemical version of the wire grid. Instead of long, thin wires it employs long chain polymeric molecules that contain many iodine atoms. These long straight molecules are aligned almost perfectly parallel to one another, and because of the conductivity provided by the iodine atoms, they strongly absorb the electric vibration component parallel to the molecules. The component perpendicular to the molecules passes through, with very little absorption.

The aligning of the 'conducting' molecule is not very difficult and the experimental details of producing the polarizer are given by Shurcliff and Ballard.*

The long, thin iodine chains can act like wires. They absorb electric vibrations *parallel* to their alignment axis and transmit freely the vibrations perpendicular to that axis. Thus the transmission axis is perpendicular to the stretched direction. Ideally, about 50% of the power of the incident beam is transmitted. Typically in one popular brand of the polarizer (Polaroid HN-38) about 38% of light is transmitted.

3-16 Percentage Polarization and Malus' Law

When light is incident on a polarizer as in Fig. 3-16 the transmitted light is plane polarized. The lines on the polarizer indicate the direction of the electric vector in the transmitted light, i.e. the sheet will transmit only those wavetrain components whose electric vectors vibrate parallel to this direction and will absorb those that vibrate at right angles to this direction. The transmitted light is allowed to fall on a photocell,[†] and the current in the ammeter is proportional to the intensity of light incident on it.

If the incident light is unpolarized, then as the polarizer is rotated about the incident ray as an axis, the reading of the ammeter does not change. The polarizer transmits only those wavetrain components whose electric fields vibrate parallel to the direction of transmission of the polarizer, and by symmetry the components are equal for all azimuths.

On the other hand, if the incident light is partially polarized, the

---

* W. A. Shurcliff and S. S. Ballard, *Polarized Light*, Van Nostrand.

† Photocell is a device which measures the light intensity, something like the one seen on a camera.

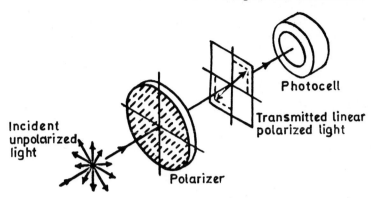

FIG. 3-16: A polarizer produces plane polarized light from unpolarized light. The dotted lines, which are actually not visible on the sheet, denote the characteristic polarizing direction of the sheet. If the polarizer is rotated about the axis of the incident ray, the intensity of the transmitted light remains the same.

ammeter will show variation in intensity as the polarizer is rotated about the incident ray as an axis. If I_{max} and I_{min} represent the maximum and minimum values of the ammeter readings,* then the percentage polarization of the incident light is defined as

$$\text{Percentage polarization} = \frac{I_{max} - I_{min}}{I_{max} + I_{min}} \times 100 \qquad (3\text{-}33)$$

Next, let us insert a second polaroid between the first polarizer and the photocell as shown in Fig. 3-17. Let the second polaroid be placed

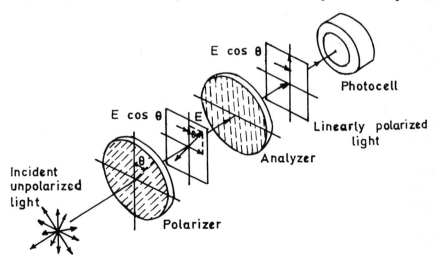

FIG. 3-17: The analyser transmits only that component of the linearly polarized light which is parallel to its transmission direction.

* The ammeter readings will be proportional to the light intensity.

in such a way so that the polaroid transmits only those wavetrain components whose electric vectors vibrate vertically. Further, let the transmission direction of the first polaroid make an angle θ with the vertical. The linearly polarized light transmitted by the polarizer may be resolved into two components, one parallel and the other perpendicular to the transmission direction of the analyzer (Fig. 3-17).

Evidently only the vertical component, of amplitude $E \cos\theta$, will be transmitted by the second polaroid.* Thus the photocell will record maximum current when $\theta = 0$ and the current will be zero when $\theta = 90°$. In the latter case (i.e. when $\theta = 90°$) the polarizer and the analyzer are said to be *crossed*. At intermediate angles, since the current is proportional to the square of the electric intensity, we will have

$$I = I_{max} \cos^2\theta \qquad (3\text{-}34)$$

where I_{max} is the maximum amount of light transmitted and I, the amount of light transmitted at an angle θ. Eq. 3-34 which was discovered by Malus in 1809, is called Malus' law. It should be noted that if *either* the analyzer or the polarizer is rotated, the amplitude of the transmitted beam varies in the same way with the angle between them.

Eq. 3-34 describes precisely the lack of symmetry about the direction of propagation that must be exhibited by plane polarized waves. Longitudinal waves could not possibly show such effects.

3-17   Polarization by Reflection

Let us consider nonpolarized light to be incident on a dielectric like glass, as shown in Fig. 3-18. The electric vector for each wavetrain in the beam

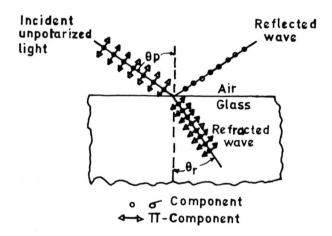

FIG. 3-18: For an unpolarized light incident on a dielectric like glass, if the angle of incidence is equal to $\tan^{-1}\mu$ then the reflected wave is completely polarized and the refracted wave is partially polarized.

* The second polarizer, when so used, is usually called an analyzer.

can be resolved into two components, one perpendicular to the plane of incidence, (i.e. perpendicular to the plane of the paper) and one lying in the plane. The first component is called the σ-component; whereas the second component (represented by arrows) is called the π-component. For an unpolarized light, the two components are of almost equal amplitude.

Experimentally, for dielectric materials like glass, there is a particular angle of incidence, called the polarizing angle ($\approx 57°$ for ordinary glass), at which the reflection coefficient for the π-component is zero. This means that the beam reflected from the glass, although of low intensity, is plane polarized with its plane of vibration at right angles to the plane of incidence. That the reflected light is plane polarized can easily be verified by analysing it with a polaroid.

Brewster discovered that at the polarizing angle the reflected and refracted rays are just 90° apart. This remarkable discovery enables one to correlate polarization with the refractive index; indeed, one can eailsy show that (Problem 11)

$$\tan \theta_p = \mu \qquad (3\text{-}35)$$

This is known as Brewester's law. The angle θ_p varies somewhat with wavelength, but for ordinary glass the dispersion is such that the polarizing angle θ_p does not change much over the whole visible spectrum.

At the polarizing angle, the π-component is entirely refracted. Therefore, the transmitted beam, which is of high intensity, is only partially polarized. If we use a stack of glass plates (Fig. 3-19) rather than a single

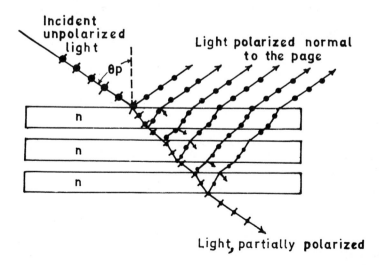

FIG. 3-19: Polarization of light by a pile of glass plates. The angle of incidence is equal to the Brewster's angle.

plate, reflections from successive surfaces occur and the intensity of the emerging reflected beam can be increased; and the transmitted beam becomes more completely polarized (the π-component).

3-18 Interpretation of Brewster's Law*

We have already discussed that an oscillating charge acts as the source of electromagnetic waves which are radiated in the form of spherical wavelets. Further, the intensity is maximum in a direction perpendicular to the direction of the oscillating charge, and zero along the line of oscillation (see Fig. 1-17).

Now consider a linearly polarized plane wave incident on the surface of a dielectric with its electric vector in the plane of incidence (Fig. 3-20).

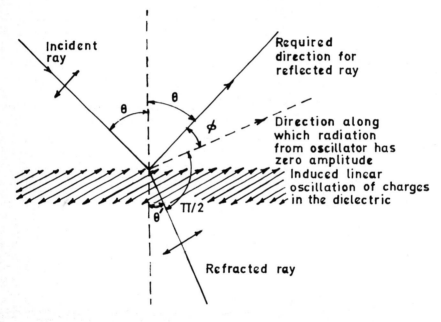

FIG. 3-20: Motion of charge near the surface of a dielectric due to a linearly polarized wave with its *E*-vector in the plane of incidence.

Let the angle of incidence be θ. The charge within the dielectric is subject to sinusoidally varying forces due to the electric vector of the refracted wave which propagates in a direction making an angle θ′ with the normal. The charge will therefore execute harmonic oscillations along a direction perpendicular to the refracted ray. Each charge is the source of a spherical wavelet. This might lead us to suppose that the reflected wave contains energy travelling in all possible directions.

* May be skipped in the first reading.

However, since the motion of all oscillating charges is induced by the field of the refracted wave, these oscillations have definite phase relationships with each other. Thus we must take account of the possibility that there are certain directions for which the contributions from the individual oscillators will completely cancel each other. It can be shown* that there is only one possible direction in which the contributing wavelets interfere constructively, which is the direction of the reflected wave. In all other directions the individual wavelets add up so as to produce a zero resultant.

Now, we already know that the intensity of electromagnetic waves, radiating from an oscillating charge is maximum in a direction perpendicular to the direction of the oscillating charge and zero along the line of oscillation (see Fig. 1-17). In Fig. 3-20 the dotted line is parallel to the direction of the oscillating charges and therefore in this direction the radiation from oscillators have zero amplitude. If the angle of incidence is chosen so that $\phi = 0$, the amplitude of the radiation from the oscillators is zero in the direction of the reflected ray and therefore, there is no reflected wave. From Fig. 3-20, we see that for $\phi = 0$ the reflected and refracted rays be at right angles to each other, which as we know is the condition for the polarizing angle. We may conclude that the reason for zero reflection of linearly polarized light whose electric vector is in the plane of incidence is that the charges in the dielectric are oscillating parallel to the reflected ray and produce no radiation in this direction.

3-19   Polarization of Scattered Light

Next, let us consider a light wave passing through a gas. The oscillating electrons in this case being separated by relatively large distances and not being bound together in a rigid structure, act independently of one another. Thus the rigid cancellation of wave disturbances do not occur.

If an unpolarized light beam falls on a gas and if we are situated at right angles to the beam, we will see polarized light. This follows immediately from the fact that although the electrons can oscillate in any direction at right angles to the beam, only the component oscillating along the X-axis gives rise to the wave propagating in the Y-direction (Fig. 3-21).

A familiar example is the scattering of sunlight by the molecules of the earth's atmosphere. If the atmosphere were not present, the sky would appear black except when we look directly at the sun.[†] We can easily check with a polarizer that the light from a cloudless sky is partially polarized. Another important example is the double scattering of X-rays.

---

* See for example, D. H. Towne's *Wave Phenomena*, Section 12-11, Addison-Wesley (1967).

† This indeed happens on the moon, which does not have any atmosphere. The shadows on the moon are also perfectly dark.

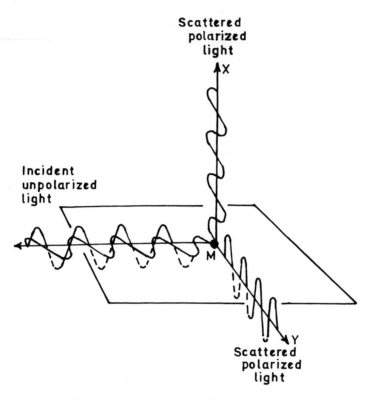

FIG. 3-21: The radiation scattered at right angles to the incident beam
is linearly polarized.

When X-rays were discovered in 1898, there was much speculation
whether they were waves or particles. In 1906 Barkla established X-
rays as transverse waves by means of a *polarization* experiment.

When an unpolarized X-ray beam strikes the scattering block S_1 (Fig.
3-22) they set the electrons into oscillatory motion. The X-rays scattered
towards S_2 are plane polarized. This plane polarized beam after getting
scattered from S_2, are detected by a detector D which is rotated in a plane
at right angles to the line joining the blocks. The electrons will oscillate
parallel to each other and the positions of maximum and zero intensity
will be as shown in Fig. 3-22. A plot of the detector reading as a function
of the angle ϕ supports the hypothesis that X-rays are transverse waves.

3-20 Birefringence

Uptil now we have assumed that the speed of light, and thus the index of
refraction, is independent of the direction of propagation of the electro-

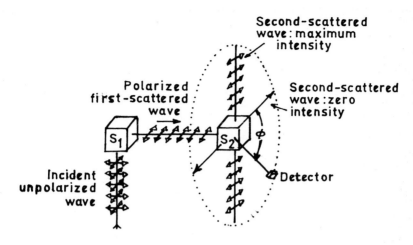

Fig. 3-22: Double scattering of X-rays.

magnetic wave and of the state of the polarization of the wave. Liquids, amorphous solids like glass and certain crystalline solids usually show this behaviour and are said to be optically isotropic. For such a medium the progress of a wavetrain can be determined by Huygens' construction. The secondary wavelets in such a medium are spherical surfaces.

Many other crystalline solids are optically anisotropic, i.e. their optical properties are different in different directions. Let us suppose we have some material which consisted of long non-spherical molecules, longer than they are wide, and suppose that these molecules were arranged in the substance with their long axes parallel. Then, what happens when a plane polarized electromagnetic wave passes through this substance? Suppose that because of the structure of the molecule, the electrons in the substance respond more easily to oscillations in the direction parallel to the axes of the molecules than they would respond if the electric field tries to push them at right angles to the molecular axis. Thus we should expect a different response for polarization in one direction than for polarization at right angles to that direction. Let us call the direction of the axes of the molecules as the *optic axis*. When the polarization is in the direction of the optic axis, the refractive index is different than it would be if the direction of polarization were at right angles to it. Such a substance is called *birefringent*. Thus in a birefringent substance there must be a certain amount of lining up, for one reason or another of unsymmetrical molecules. It has two refractive indices, depending on the direction of the polarization inside the substance.

Next, let us suppose that a polarized light beam is incident on a plate of birefringent substance. If the polarization is parallel to the optic axis, the light will go through with one velocity; if the polarization is perpendicular to the axis, the light is transmitted with a different velocity. What would happen if light is linearly polarized at an angle θ to the optic axis? We have already shown that this state of polarization can be represented as a superposition of the *x*- and the *y*-polarizations of suitable amplitudes oscillating in phase as shown in Fig. 3-10. Since the *x*- and *y*-polarizations travel with different velocities, their phases change at a different rate as the light passes through the substance. So, although at the start the *x*- and *y*-vibrations are in phase, inside the material the phase difference between *x*- and *y*-vibrations is proportional to the depth in the substance.

The phenomenon of birefringence can easily be illustrated with a piece of cellophane (or even with a piece of adhesive tape). Cellophane is made of long, fibrous molecules, and is not isotropic, since the fibres lie in a certain direction. To demonstrate the birefringence of cellophane, we use two sheets of polaroid, as shown in Fig. 3-23. The first gives us a linearly

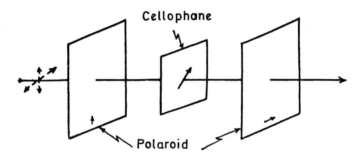

FIG. 3-23: An experimental demonstration of the birefringence of cellophane. The incident beam is unpolarized. The pass axes of the polaroid sheets and optic axis of the cellophane are indicated by arrows.

polarized beam which passes through the cellophane and then through the second polaroid sheet, which serves to detect any effect the cellophane may have had on the polarized light passing through it. If we first remove the cellophane and set the axes of the two polariod sheets perpendicular to each other, no light will be transmitted through the second polaroid. If we now introduce the cellophane between the two polaroids and rotate the cellophane about the beam axis, we observe that in general the cellophane makes it possible for some light to pass through the second polaroid. However for two orientations of the cellophane sheet at right angles to each other, no light will pass through the second polaroid. These orientations must be the directions parallel and perpendicular to the optic axis of the cellophane sheet.

We suppose that the light passes through the cellophane with two different speeds in these two different orientations, but it is transmitted without changing the direction of polarization. When the cellophane is turned halfway between these two orientations, as shown in Fig. 3-25, the light transmitted through the second polaroid is bright.

3-21 Applications of Birefringence

A. Full wave plate, Half wave plate and Quarter wave plate

Suppose that the thickness of the material is just right to introduce a phase difference of $2m\pi$, where m is an integer. In this case we get back our original polarization. Such a slab is called a *full wave plate*. The principle of the full wave plate can easily be understood if we recall that the wave with **E** vector along the optic axis travels with a different speed than the speed of the other component wave. Since both waves have the same frequency, the faster one has a longer wavelength. If the slab has a thickness t, such that

$$t = 82\lambda_{fast} = 84\lambda_{slow} \qquad (3\text{-}36)$$

then the slow wave has gone two wavelengths farther than the fast wave (equivalent to a phase difference of 4π) but they emerge in phase and so appear *unchanged*. Since the wavelength in a medium is related to the corresponding wavelength in vacuum by the relation

$$\lambda_{medium} = \lambda_{vacuum}/\mu \qquad (3\text{-}37)$$

therefore, the proper thickness for a full wave plate is given by

$$(\mu_{slow} - \mu_{fast})\, t = m\lambda \qquad (3\text{-}38)$$

where m is an integer and λ, the wavelength in vaccum.

The above discussion leads to an obvious definition of a *half wave plate*:

$$(\mu_{slow} - \mu_{fast})\, t = (2m + 1)\frac{\lambda}{2} \qquad (3\text{-}39)$$

This implies that if the two components go into the slab in phase, they emerge from the slab 180° out of phase (Fig. 3-24). Thus, if the electric vector of the incident plane polarized beam lies along an axis, there is no effect. On the other hand if **E** makes an angle θ with, say the fast axis then we have to take components along the slow and the fast axes. The emergent wave will thus have components of the same amplitude as the incident one, but they are now *out of phase* by π. A relative phase change of π is equivalent to reversing one of the vectors (Fig. 3-25). By recombining, we find that the polarization direction has been rotated through twice the angle θ [Fig. 3-25 (a), (b) and (c)]. It should be noted that this is independent of which component we chose to reverse, since the polarization direction only specifies the line along which **E** oscillates [Fig. 3-25(d)]. Mathematically speaking, before the wave enters the medium we may write

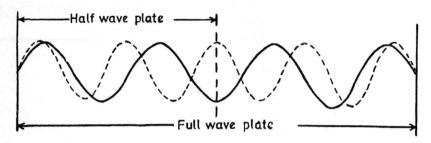

FIG. 3-24: Half wave plate and full wave plate.

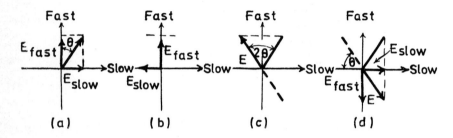

FIG. 3-25: Rotation of the plane of polarization by a half wave plate. (a) The *E*
before entering the half wave plate; (b) the slow and the fast components of the
E after emerging from the half wave plate; (c) the resultant of the vectors shown
in (b); and (d) if the fast component were reversed the resultant would remain
the same.

$$E_x = E_s = E_0 \, \text{Sin}\theta \, \text{Sin}\,\omega t$$

and

$$E_y = E_f = E_0 \, \text{Cos}\theta \, \text{Sin}\,\omega t$$
$$(3\text{-}40)$$

where E_0 is the amplitude of the incident electric field and the subscripts
s and *f* refer to the slow and fast components respectively. [The resultant
electric vector will have a direction as shown in Fig. 3-25(*a*).] Now, the
emergent beam will be given by

$$E_x = E_s = E_0 \, \text{Sin}\theta \, \text{Sin}\,(\omega t + \pi)$$
$$E_y = E_f = E_0 \, \text{Cos}\theta \, \text{Sin}\,\omega t$$

or

$$E_x = - E_0 \, \text{Sin}\theta \, \text{Sin}\,\omega t$$
$$E_y = E_0 \, \text{Cos}\theta \, \text{Sin}\,\omega t$$
$$(3\text{-}41)$$

The emergent wave will again be plane polarized but the resultant *E*-
vector will oscillate along the direction shown in Fig. 3-25(*c*). We may
thus say that the half wave plate rotates the polarization direction, but
otherwise leaves the wave unaffected:

It should be clear that a *quarter wave plate* will be one that has a thickness
given by

$$(\mu_{slow} - \mu_{fast}) \ t = (2m + 1) \frac{\lambda}{4} \qquad (3\text{-}42)$$

which would change the relative phases of the fast and slow components of E by $\pi/2$. (In the Eq. 3-42, λ again refers to the wavelength in vacuum.) In general, the result of such a phase change will be elliptical polarization. Linear polarization is a special case of elliptic polarization and occurs when E lies along an axis. The other special case is circular polarization. This is obtained from linearly polarized light by passing it through a quarter wave plate whose axes are at 45° to E of the wave (Fig. 3-26).

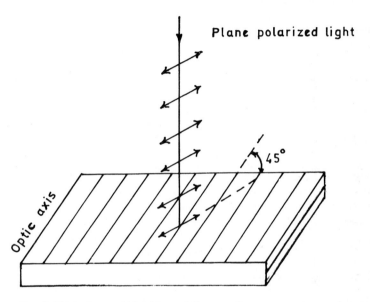

FIG. 3-26: A plane polarized beam falls normally on a quarter wave plate. The optic axis is parallel to the face of the crystal. The electric vector of the incident polarised light is assumed to make 45° with the optic axis. The emergent beam is circularly polarized.

Thus the incident wave will be given by

$$\left.\begin{array}{l} E_x = E_s = E_0 \ \text{Sin } 45 \ \text{Sin } \omega t = \dfrac{E_0}{\sqrt{2}} \ \text{Sin } \omega t \\[3mm] E_y = E_f = E_0 \ \text{Cos } 45 \ \text{Sin } \omega t = \dfrac{E_0}{\sqrt{2}} \ \text{Sin } \omega t \end{array}\right\} \qquad (3\text{-}43)$$

After the wave comes out of the quarter wave plate, the fast *component* will be ahead of phase by 90° (see also Fig. 3-24). Thus for the emergent beam

$$E_y = \frac{E_0}{\sqrt{2}} \, \text{Sin} \, (\omega t + \pi/2) = + \frac{E_0}{\sqrt{2}} \, \text{Cos} \, \omega t$$

$$E_x = \frac{E_0}{\sqrt{2}} \, \text{Sin} \, \omega t$$

$$(3\text{-}44)$$

Eq. 3-44 represents a circularly polarized beam whose electric vector is rotating in the *anticlockwise* direction. If two quarter waveplates are put together, the combination will behave as a half wave plate.

B. Photoelasticity

Another interesting application of aligned molecules is the phenomenon of *photoelasticity*. Certain plastics are composed of very long and complicated molecules, all twisted together. When the plastic is solidified carefully the molecules are all twisted in a mass, so that there are as many aligned in one direction as in another, and so the plastic is not particularly birefringent. If, we now apply tension to a piece of this plastic material, it is, quoting Feynman,* 'as if we are pulling a whole tangle of strings, and there will be more strings preferentially aligned parallel to the tension than in any other direction.' So when a stress is applied to certain plastics, they become birefringent, and one can see the effects of the birefringence by passing polarized light through the plastic. If we examine the transmitted light through a polaroid sheet, patterns of light and dark fringes will be observed. The patterns move as stress is applied to the sample, and by counting the fringes and seeing where most of them are, one can determine what the stress is. This phenomenon of photoelasticity is used to obtain stresses in odd shaped pieces that are difficult to calculate.

C.   Kerr Effect

Another interesting example of obtaining birefringence is by applying an electric field to certain liquids. Consider a liquid composed of long asymmetric molecules which carry opposite charges near the ends of the molecule, so that the molecule is an electric dipole. In general, the molecules will be randomly oriented, with as many molecules pointed in one direction as in the other. If we apply an electric field the molecules will tend to line up and the liquid will become birefringent. This phenomenon is known as the *Kerr effect*. The existence of the Kerr effect makes it possible to construct an electrically controlled "light valve". A cell with transparent walls contains the liquid between a pair of parallel plates. The cell is inserted between crossed polaroids. Light is transmitted when an electric field is set up between the plates and is cut off when the field is removed.

---

* *The Feynman Lectures on Physics* by R. P. Feynman, R. B. Leighton and M. Sands, Chapter 33, Addison-Wesley (1965).

3-22 Anomalous Refraction

Finally we will discuss a phenomenon* which was actually one of the first to be discovered: anomalous refraction. If we place a polished crystal of calcite ($CaCO_3$) on a newspaper, the image of each letter will appear double. This came to the attention of Huygens and played an important role in the discovery of polarization. As is often the case, the phenomena which are discovered first are the most difficult to explain.

Anomalous refraction is a particular case of birefringence and comes about when the optic axis (the long axis of the asymmetric molecules) is *not* parallel to the surface of the crystal. In Fig. 3-27, the

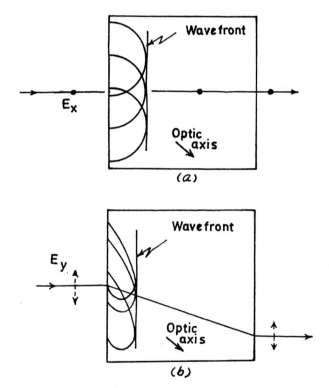

Fig. 3-27: (a) The electric vector of the incident wave is oscillating at right angles to the plane of the paper. The figure shows the path of the ordinary ray through a doubly refracting crystal.

(b) The electric vector of the incident wave is oscillating in the plane of the paper (which contains the optic axis). The figure shows the path of an extraordinary ray.

* Another interesting effect of polarization is the phenomenon of optical activity. This is discussed in Appendix E.

incident beam falling on the calcite crystal is linearly polarized in a direction perpendicular to optic axis. When this beam strikes the surface of the crystal each point on the surface acts as a source of a wave which travels into the crystal with velocity $v_\perp$ which is the velocity of light in the crystal when the direction of polarization is perpendicular to the optic axis. The wavefront is just the envelope of all these *spherical* wavelets, and this wavefront moves straight through the crystal as shown in Fig. 3-27(*a*). This is just the behaviour we would expect and this ray is called the *ordinary ray*. In Fig. 3-27(*b*) the incident beam falling on the crystal is polarized in a direction such that the optic axis lies in the plane of polarization. Now the secondary wavelets originating at any point on the surface of the crystal do not spread out as spherical waves. Light travelling along the optic axis travels with velocity $v_\perp$ because the polarization is perpendicular to the optic axis, whereas the light travelling perpendicular to the optic axis travels with velocity $v_\parallel$ because the polarization is parallel to the optic axis. In a birefringent material $v_\parallel \neq v_\perp$ and in the figure $v_\parallel < v_\perp$. A more detailed analysis shows that the waves spread out on the surface of an ellipsoid,* with the optic axis as major axis of the ellipsoid. The envelope of all these elliptical waves is the wavefront which proceeds through the crystal in the direction shown. Further, when the beam emerges out of the crystal, it will be deflected just as it was on the front surface, so that the light emerges parallel to the incident beam, but displaced from it as shown in Fig. 3-27(*b*). Clearly, this beam does not obey Snell's law and is therefore called the *extraordinary ray.*

If however, an unpolarized beam is incident on an anomalously refracting crystal, it will split into an ordinary ray, which travels through in the normal manner (obeying Snell's laws) and an extraordinary ray which is displaced as it passes through the crystal. Thus, when the crystal is rotated about an axis parallel to the incident beam (which falls normally on the crystal, see Fig. 3-28), the spot produced by the ordinary rays remains fixed, while the extraordinary rays revolve around it in a circle. The two emergent rays are linearly polarized at right angles to each other. That this is true can easily be demonstrated with a sheet of polaroid to analyze the polarization of the emergent rays. We can also demonstrate that our interpretation of this phenomenon is correct by sending linearly polarized light into the crystal. By polarization of the incident beam, we make this light go straight without splitting, or we can make it go through without splitting but with a displacement.

Example

If a crystal of quartz is cut parallel to the optic axis, one may take advantage of the maximum difference in speed of the ordinary and extraordinary rays as they enter normally and progress through the crystal.

---

* An ellipsoid is a surface obtained by rotating an ellipse about its major or minor axis.

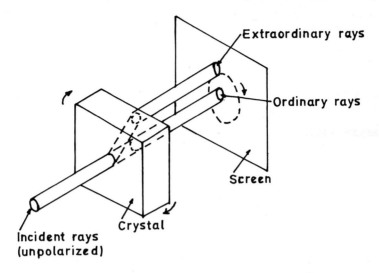

FIG. 3-28: A narrow beam of unpolarized light is split into two plane-polarized beams by a doubly refracting crystal like calcite. The ordinary rays obey the laws of refraction, whereas the extraordinary rays do not.

The refractive indices of crystalline quartz for light of wavelength 6000 Å are $\mu_0 = 1\cdot544$ and $\mu_e = 1\cdot553$, for the ordinary and extraordinary rays respectively.* What thickness of the crystal is required to shift the relative phases of these two rays by 90°, for light of the above wavelength?

Solution

Let the crystal thickness be t; then the number of wavelengths of the e-wave contained in the crystal will be

$$\frac{t}{\lambda_s} = \frac{t\mu_e}{\lambda}$$

where λ is the wavelength of the incident light in air, (the light is assumed to be monochromatic), and λ_e, the wavelength of the e-wave in the crystal. Similarly, the number of wavelengths of the e-wave in the crystal will be

$$\frac{t\mu_0}{\lambda}$$

The difference between the two expressions is in the number of waves that one wavetrain lags behind the other on emergence, since the waves are in phase when they are incident on the crystal. Each wavelength

* Values of μ_0 and μ_e for some doubly refracting crystals are tabulated in Table I given at the end of the chapter.

corresponds to a phase angle of 2π radians, so the phase difference between the emergent waves is

$$\phi_e - \phi_o = \frac{2\pi t}{\lambda} (\mu_e - \mu_o) \tag{3-45}$$

In order to introduce a phase difference of $\pi/2$ (i.e. for a quarter wave plate)

$$\phi_e - \phi_o = \frac{\pi}{2} = \frac{2\pi t}{\lambda} (\mu_e - \mu_o)$$

or

$$t = \frac{\lambda}{4 (\mu_e - \mu_o)} \tag{3-46}$$

Thus for $\lambda = 6000 \ \overset{\circ}{\text{A}}$

$$t = \frac{6 \times 10^{-5}}{4 (1 \cdot 553 - 1 \cdot 544)} \text{ cm}$$

$$= 0 \cdot 0167 \text{ mm}$$

This plate is rather thin; most quarter wave plates are made from mica splitting the sheet to the correct thickness by trial and error.

Further, if the thickness is made twice the above value, the sheet will become a half wave plate.

Example
A plane polarized light wave of amplitude E_0 falls on a quartz quarter wave plate with its plane of vibration making 45° with the optic axis (Fig. 3-29). What will be the emerging light? (The direction of propagation is out of the page.)

Solution
The x- and y-components of the incident plane polarized beam can be written as

$$\left. \begin{array}{l} E_x = E_o \text{ Cos } 45 \text{ Sin } \omega t \\ E_y = - E_o \text{ Cos } 45 \text{ Sin } \omega t \end{array} \right\} \tag{3-47}$$

(It is obvious that E_x and E_y can be chosen in a variety of ways; for example, one may choose

$$E_x = - E_o \text{ Cos } 45 \text{ Cos } \omega t$$

but then E_y would be given by

$$E_y = + E_o \text{ Cos } 45 \text{ Cos } \omega t)$$

Now in a quartz crystal since $\mu_o < \mu_e$, the ordinary wave (abbreviated as o-wave) travels faster than the extraordinary wave (abbreviated as e-wave). Therefore after it emerges from the quarter wave plate the

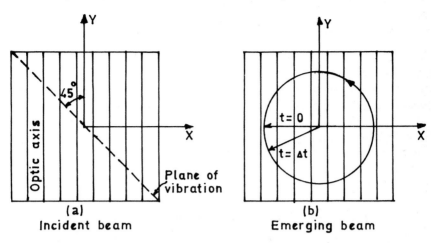

Fig. 3-29: Plane polarized light (with the E making an angle of 45° with the optic axis) falls from behind on a quartz quarter wave plate. The emerging light beam (b) is circularly polarized with the E rotating in the clockwise direction.

electric field of the o-wave (i.e. E_x, because for the o-wave the vibrations of the E is always at right angles to the optic axis) will be ahead of the e-wave by 90°. Hence for the incident beam represented by Eq. 3-47, the emergent beam would be

$$E_x = E_o \; \text{Cos } 45 \; \text{Sin } (\omega t + \pi/2)$$

or

$$E_x = \frac{E_o}{\sqrt{2}} \; \text{Cos } \omega t$$

and

$$E_y = -\frac{E_o}{\sqrt{2}} \; \text{Sin } \omega t$$

(3-48)

which represents a circularly polarized beam.

To determine the direction of rotation of the emergent circularly polarized wave, let us locate the tip of the rotating vector at $t = 0$ and at $t = \Delta t$.

At $t = 0$,

$$E_x = \frac{E_o}{\sqrt{2}}$$

$$E_y = 0$$

and at $t = \Delta t$,

$$E_x \cong \frac{E_o}{\sqrt{2}}$$

$$E_y \cong - \frac{E_o}{\sqrt{2}} \, \omega \Delta t$$

Thus it rotates in theclockwise direction and by our convention it is left circularly polarized.

It should be noted that in the experimental set-up of Fig. 3-29, if the incident plane of vibration is *not* making an angle of 45° with the optic axis then the two amplitudes in Eq. 3-47 will not be equal and in general, we will obtain elliptically polarized light. (What will happen if the angle is 90°?) Further, if the thickness of the plate does not correspond to a quarter wave plate then also we obtain elliptically polarized light (see Problem 19).

3-23   Analysis of Polarized Light

If one has a beam of light that is linearly polarized or elliptically polarized, it will appear to the eye to be no different from ordinary unpolarized light. By using simple auxiliary apparatus, however, its nature of polarization can be easily determined (see Table II). For example, if a beam is thought to be circularly polarized, then this can be verified by inserting a quarter wave plate. If the beam is circularly polarized, the two components will have a phase difference of 90° between them. The quarter wave plate will introduce a further phase difference of $+ 90°$ so that the emerging light will have a phase difference of either zero or 180°. In either case the light will now be plane polarized and can be made to suffer complete extinction by rotating a polarizer in its path.

TABLE I

PRINCIPAL INDICES OF REFRACTION 'OF SEVERAL DOUBLY REFRACTING CRYSTALS (FOR SODIUM LIGHT, $\lambda = 5890$ Å)*

Crystal	Formula	μ_o	μ_e	$\mu_o - \mu_e$
Ice	H_2O	1·309	1·313	$+ 0·004$
Quartz	SiO_2	1·544	1·553	$+ 0·009$
Wurzite	ZnS	2·356	2·378	$+ 0·022$
Calcite	$CaCO_3$	1·658	1·486	$- 0·172$
Dalomite	$CaO. MgO. 2CO_2$	1·681	1·500	$- 0·181$
Siderite	$FeO. CO_2$	1·875	1·635	$- 0·240$

* After D. Halliday and R. Resnick, *Physics*, Part II, p. 1158, John Wiley (1966).

Table II

ANALYSIS OF POLARIZED LIGHT

(1) If there is *complete* extinction by rotating a polarizer in its path then the light is plane polarized.

(2) If there is *no intensity variation* when a polarizer is rotated in the path of the beam then one has either unpolarized light or circularly polarized or a mixture of the two. Now, if with a quarter wave plate in front of the rotating polaroid,

(a) one has no intensity variation then the incident light is *natural unpolarized* light.	(b) one position of the polaroid gives zero intensity (i.e. complete extinction), one has *circularly polarized light*.	(c) no position of the polaroid gives zero intensity (but there is a variation of intensity), one has a mixture of circularly polarized light and unpolarized light.

(3) If there is some intensity variation when a polarizer is rotated, but if no position of polaroid gives zero intensity then one has either elliptically polarized light, or a mixture of plane polarized light and unpolarized light, or a mixture of ellipitcally polarized light and plane polarized light. Now, if with a quarter wave plate in front of the polarizer with optic axis *parallel* to position of maximum intensity,

(a) one gets zero intensity with the polaroid then one has *elliptically polarized* light.	(b) one gets no zero intensity	
	(i) but the same polarizer setting as before gives the maximum intensity, one has *mixture of plane polarized light and unpolarized light*.	(ii) but some other polarizer setting as before gives a maximum intensity, then one has a *mixture of elliptically polarized light and unpolarized light*.

SUGGESTED READING

American Institute of Physics, *Polarized Light: Selected Reprints*, p. 103 American Institute of Physics, New York (1963).

Feynman R. P., R. B. Leighton and M. Sands, *The Feynman Lectures on Physics*, Vol. I, Addison-Wesley (1965). Chapters 28 and 33 are directly relevant to the concepts developed in this chapter.

Halliday D. and R. Resnick, *Physics*, Part II, John Wiley (1962). Of particular interest is a one dimensional mechanical model for double refraction in Chapter 46.

Heirtzler J. H., "The Longest Electromagnetic Waves", *Scientific American*, March 1962. The author has discussed the detection of electromagnetic waves with wavelength as long as 1,86,000 miles. The origin of these waves is unknown but they promise to be useful in geophysics.

Jenkins F. A. and H. E. White, *Fundamentals of Optics*, McGraw-Hill (1957). This book should be an excellent reference for students who are interested to learn more on double refraction and optical activity.

Shurcliff W. A. and S. S. Ballard, *Polarized Light* Van Nostrand Momentum Book No. 7, Van Nostrand (1964). This book presents a wide variety of new applications of polarized light in a non-mathematical language.

WATERMAN T. H., "Polarized Light and Animal Navigation", *Scientific American*, July 1955.

WOOD E. A., *Crystals and Light*, Van Nostrand Momentum Book No. 5, Van Nostrand (1964). This book has discussed the behaviour of light in crystals, especially their appearance in cross-polarized light.

PROBLEMS

1. Show that for standing waves represented by Eqs. 3-5 and 3-9, the electric and magnetic fields are $\pi/2$ out of phase and that the positions of nodes for E correspond to antinodes for B and conversely.

2. In a typical experimental set-up of Fig. 3-4, the distance between two successive antinodes for B is 11·5 cm (see also Fig. 1-25). Calculate the frequency of the electromagnetic wave.

3. A commercially available He-Ne laser continuously emits $4·0 \times 10^{-3}$ watt of red light in a beam with a diameter of 0·14 cm. Find the beam intensity and the amplitude of the electric field in the beam.   ($2·6 \times 10^3$ watt/m²;
                                                                                   $1·4 \times 10^3$ volt/m)

4. The frequencies of the radio waves in the 'broadcast band' range from $0·55 \times 10^6$ cps to $1·6 \times 10^6$ cps. What are the longest and shortest wavelengths?   (550 m to 190 m)

5. A parallel beam of light with an intensity of 20 watt/cm² falls normally for one hour on a perfectly absorbing surface of area 2 cm². Calculate the momentum that is delivered to the surface and the radiation pressure that acts on the surface.   ($4·8 \times 10^{-4}$ kg m/sec,
                                                                                       $6·7 \times 10^{-4}$ Newton/m²)

6. When a beam of plane polarized light strikes a polaroid sheet, a fraction β of the intensity is transmitted if the polaroid axis is parallel to the polarization axis, and fraction δ is transmitted if the two axes are at right angles (For an ideal polaroid $\beta = 1$ and $\delta = 0$). Unpolarized light of intensity I_0 is normally incident on a pair of polaroid sheets with an angle θ between their axes. Calculate the transmitted intensity (Ignore reflection effects).

$$\left[I_0\{\tfrac{1}{2}(\beta^2 + \delta^2)\cos^2\theta + \beta\delta\sin^2\theta\}\right]$$

7. Discuss the intensity and polarization of the radiation emitted by an electron moving at constant speed in a circular path for points (i) on the axis of the circle and (ii) in the plane of the circle.

8. Two polarizing sheets have their polarizing directions parallel so that the intensity I_m of the transmitted light is maximum. Through what angle must either sheet be turned if the intensity is to drop by one-half?   ($\pm 45°$, $\pm 135°$)

9. We wish to use a plate of glass ($\mu = 1·50$) as a polarizer. What is the polarizing angle? What is the angle of refraction?   ($56·3°$, $33·7°$)

10. A quartz quarter wave plate is to be used with sodium light ($\lambda = 5890$ Å). What must be its thickness?
    $[\mu_e = 1·486, \ \mu_0 = 1·658]$

11. Assuming that at the polarizing angle the reflected and refracted rays are 90° apart, derive Brewster's Law.

12. A plane-polarized light wave of amplitude E_0 falls on a calcite quarter wave plate with its plane of vibration at 45° at the optic axis of the plate, which is taken as the y-axis (see Fig. 3-30). The emerging light will be circularly polarized. In what direction will the rotating electric vector appear to rotate? The direction of propagation is out of the page.

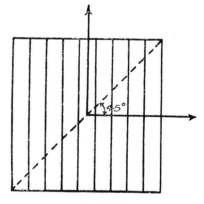

FIG. 3-30

13. Describe the state of polarization represented by these sets of equations:
    (a) $E_x = E \sin(kz - \omega t)$; $E_y = E \cos(kz - \omega t)$; (b) $E_x = E \cos(kz - \omega t)$; $E_y = E \cos\left(kz - \omega t + \frac{\pi}{4}\right)$; and (c) $E_x = E \sin(kz - \omega t)$; $E_y = -E \sin(kz - \omega t)$

14. Light is incident on a water surface ($\mu = 4/3$) at such an angle that the reflected light (ray 1 in Fig. 3-31) is completely linearly polarized.

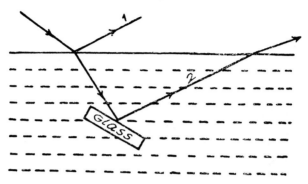

FIG. 3-31

    (a) What is the angle of incidence?

    (b) A block of glass ($\mu = 3/2$) having a flat upper surface is immersed in the water as indicated in Fig. 3-31. The light reflected from the surface of the glass (ray 2) is completely linearly polarized. Find the angle between the surface of the water and the surface of the glass.   [(a) 53°; (b) 11·4°]

15. (a) A beam of circularly polarized light is passed normally through a quarter wave plate. What is the state of polarization of the light after it emerges from the plate?

    (b) A beam of circularly polarized light is passed normally through an eighth-wave plate. What is the state of polarization of the light after it emerges from the plate?

    [(a) Linearly polarized; (b) Elliptically polarized.]

16. Calculate the thickness of a quartz half wave plate for the Fraunhofer C-line (wavelength of C line is 6563 Å). The μ_e and μ_o values for quartz are 1·55085 and 1·54181 respectively.   ($3·629 \times 10^{-3}$ cm)

17. Plane polarized light is incident on a piece of quartz cut with faces parallel to the axis. Find the least thickness for which the o-and e- rays combine to form plane polarized light, given $\mu_o = 1·5442$, $\mu_e = 1·5533$, and $\lambda = 5 \times 10^{-5}$ cm.   ($2·748 \times 10^{-3}$ cm)

18. A beam of linearly polarized light is changed into circularly polarized light by passing it through a slice of crystal 0·003 cm thick. Calculate the difference in the refractive indices of the two rays in the crystal assuming this to be minimum thickness that will produce the effect and that the wave length is 6×10^{-5} cm.   (0·005)

19. In the set up of Fig. 3-29, if the thickness of the plate is less than that of a quarter wave plate, such that it introduces a phase difference of $\pi/3$, discuss the state of polarization of the emerging light beam.

20. Give the reasons of all the analyses made in Table II.

21. We know that it is not possible to send a beam of light through two polaroid sheets with their axes crossed at right angles. But if we place a third polaroid sheet between the first two, with its 'pass' axis at 45° to the crossed axes, some light is transmitted. We know that polaroid absorbs light; it does not create anything. Then, how does the addition of the third polaroid allows more light to get through? Discuss.

22. In the set up of Fig. 3-26, suppose that the refractive indices for light of $\lambda = 4100$ Å are

$\mu_0 = 1 \cdot 557$ and $\mu_e = 1 \cdot 567$ and that the crystal is cut as a quarter wave plate for $\lambda = 6000$ Å. Describe fully the state of polarization of emergent light of this shorter wavelength which is linearly polarized (with the **E** making an angle of 45° with the optic axis, see Fig. 3-26) before entry into the crystal.

23. How can a right circularly polarized beam of light be converted into plane polarized in a specific plane into left circularly polarized?

   (Quarter wave plate with axes at 45° to the desired direction of plane polarization; two quarter wave plates with corresponding axes aligned, or one half wave plate)

24. Unpolarized light of intensity I_0 goes through three devices as shown in Fig. 3-32 and emerges with final intensity I. In Fig. 3-32 P is a polarizer with polarization axis vertical;

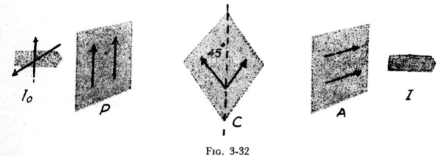

FIG. 3-32

the axes of C, if any, is always at 45° to the vertical; A is also a polarizer with polarization axis harizontal. Calculate I/I_0 when C is a (i) quarter wave plate; (ii) half wave plate; and (iii) linear polarizer.   ($\frac{1}{4}, \frac{1}{2}, \frac{1}{4}$)

25. The maximum electric field in the vicinity of a certain radio transmitter is 3×10^{-3} volt/m.

   (a) What is the maximum magnetic field?

   (b) How does this compare in magnitude with the earth's magnetic field ($B = 0.5 \times 10^{-4}$ weber/m²).

        [(a) 10^{-11} weber/m². (b) Much weaker]

26. For a 50,000 watt radio transmitting station, find the amplitudes of the electric and magnetic fields at a distance of 100 km from the antenna, if it radiates equally in all directions.   ($E_0 = 1 \cdot 73 \times 10^{-2}$ volt/m; $B_0 = 0 \cdot 577 \times 10^{-10}$ weber/m²)

27. What is the radiation pressure 100 cm away from a 500 watt light bulb? The surface on which the pressure is exerted faces the bulb and is perfectly absorbing. Assume that the bulb radiates uniformly in all directions.   ($1 \cdot 3 \times 10^{-7}$ Newton/m²)

Interference

Light + Light does not always give more light, but may in certain circumstances give darkness.

— Max Born

4-1 Introduction

One of the most striking results of the experiments with waves on a rope, described in Chapter 1 was that, two pulses travelling in opposite directions passed right through each other. The shape of the displacement of the rope was explained by adding the displacements of the individual impulses (the principle of superposition).

Let us again consider a rope (of almost infinite length) which is tied down at one end. When a periodic wave* travels along the rope, every individual pulse is reflected upside down. Now we know that a reflected pulse superposes on every oncoming pulse it meets. Suppose we first consider only two of the pulses, a and b, separated by wavelength λ, as they travel toward the reflecting end [Fig. 4-1(i)]. Some time after the first pulse is reflected it will meet the second pulse and there will be a cancellation at the midpoint P between them [Fig. 4-1(ii)]. In Fig. 4-1(iii)

* By a periodic wave we imply that the disturbance at a particular point repeats itself, say after every T seconds ($\nu = 1/T$ is the frequency, see Chapters 1 and 2). Further, at a particular instant, the displacement of two points separated by one wave length (or any integral multiple of wavelength) are *identical*.

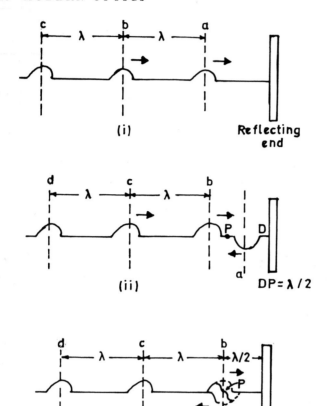

FIG. 4-1: (*i*) Three pulses travel toward the reflecting end of a string.
(*ii*) The string after one of the pulses has been reflected. The reflected pulse *a* is upside down and travelling toward the incident pulse *b*. A third pulse *c* is approaching at a distance λ behind *b*.
(*iii*) Superposition of pulses *a* and *b*.

the pulses are shown as they meet. Because they were originally a distance λ apart we can see that the point P is at a distance $\lambda/2$ from the reflecting end. The next pulse *c*, which reaches P later, will be superposed with the reflected pulse *b* so that the same cancellation occurs again. Because the wave is periodic, this will happen everytime a pulse passes P, and although the motion of the string as a whole is complicated, the point P always remains at rest. We call such a point a *node*. There are other nodes spaced $\lambda/2$ apart, as you can see by working out where the next few must be (for example, the cancellation of *a* and *c*).

It is clear that we could have obtained nodes just as easily by sending appropriate periodic waves from opposite ends of a long string. The use of the fixed end as a means of producing a wave moving in the oppposite direction is only a convenience. Indeed, in a sonometer we have waves reflected from the ends and these waves superpose on the incoming waves to form stationary waves (see Chapter 1).

The phenomenon we have just described—the superposition of two periodic waves to produce a series of nodes—is called *interference*. Rather than try to find the corresponding effect in light at once, we will first study interference between water waves in a ripple tank. Then we will look for interference in light.

4-2 Interference of Waves Originating from two Point Sources

Let us consider two vertical needles which barely touch the surface of a body of water. If the needles are identically driven (by a common source) in vertical oscillations, they periodically disturb the surface of water and send out two sets of concentric circular waves. Since both the needles are driven by a common source they are said to vibrate in phase. This type of vibration can easily be realized by attaching the two vertical

needles to an electrically driven elastic bar which operates on much the same principle as an ordinary electric bell (Fig. 4-2). When the bar starts vibrating the two needles also move up and down in the same manner and each needle produces an identical train of waves. The superposition of these two sets of waves produces the phenomenon of interference. The surface of the liquid breaks up into a number of narrow segments of alternately disturbed and calm water. An explanation of this phenomenon is given in Fig. 4-3 which shows the two points sources S_1 and S_2 from

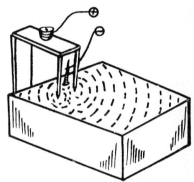

FIG. 4-2: Two electrically operated needles produce waves on the surface of a water tank.

where the wave trains emanate. (In fact S_1 and S_2 are the points at which the vertical needles touch the body of water.) Now since the needles vibrate in phase, each source produces a crest at the same instant. The waves that are produced by the sources can be represented by drawing two sets of concentric circles side by side with centres at S_1 and S_2 (Fig. 4-3). The circles represent the crests of the waves (at a particular instant) expanding from each source. Since the sources are periodic, the crests are always the same distance apart, viz. one wavelength. The distance between crests is the same in both sets of circles because the wavelengths are the same for both sources. The radii of corresponding circular crests in each set are equal because the sources of disturbance vibrate in phase.

What will happen when the waves from the two sources overlap? Let

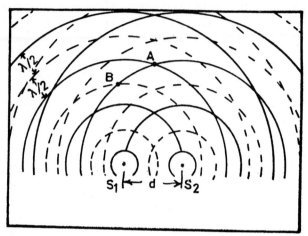

FIG. 4-3: The full circles represent (at a particular instant)
the crests of waves from two sources S_1 and S_2 which are
at a distance d apart. The dotted lines are the
corresponding troughs. The distance between a trough
and the consecutive crest is $\lambda/2$.

us try to predict the resulting wave pattern by using the principle of
superposition. Consider a point A (Fig. 4-3) which is equidistant from
both the sources, i.e. $S_1A = S_2A$. Since the distances S_1A and S_2A are
equal and since the sources vibrate in phase, if S_1 produces a crest at A
then S_2 will also produce a crest at A; and if S_1 produces a trough at A
so will S_2. We say that the waves arrive in phase at A and it is apparent
that the resulting motion will be increased. A little thinking will reveal
that if the point A is such that the ·distances S_1A and S_2A differ by $n\lambda$
($n = 0, 1, 2, 3, \ldots$), the same phenomenon will occur.

On the other hand, if we select a point B (Fig. 4-3) in such a way that
the difference between S_1B and S_2B is equal to $\lambda/2$, then the waves from
the two sources will always arrive out of phase, i.e. if S_1 produces a crest
at B then S_2 will produce a trough there (and vice versa), or, more
generally, whatever is the displacement produced by S_1 an equal and
opposite displacement will be produced by S_2. It is clear that in this case
the resultant displacement will be zero *for all times*. Once again, if the
point B is such that the distances S_1B and S_2B differ by $(n + \frac{1}{2})\lambda$ (where
$n = 0, 1, 2, 3, \ldots$), then the resultant displacement will again be zero.

It is worth while pointing out that a point like A (such that $S_1A \sim$
$S_2A = n\lambda$)* keeps on oscillating (with the same frequency as that of each
of the sources) with twice the amplitude produced by each source
independently. (For, *at a particular instant*, if the source S_1 produces zero
displacement at A, then S_2 will also produce zero displacement there.)

* Read $x \sim y$ as x difference y, i.e. $x \sim y = x - y$ (if $x > y$) and $x \sim y = y - x$ (if $y > x$).

Thus, we may conclude that for a general point P, the amplitudes of vibration will be maximum when

$$S_2P \sim S_1P = n\lambda \qquad \text{(Maxima)} \qquad (4\text{-}1)$$

or zero when

$$S_2P \sim S_1P = (n + \tfrac{1}{2})\lambda \qquad \text{(Minima)} \qquad (4\text{-}2)$$

where $n = 0, 1, 2, 3, \ldots$

4-3 Nodal Lines

The loci of points for which the displacement is always zero are known as nodal lines. These are drawn in Fig. 4-4 It can be easily seen from Fig. 4-4 that although the nodal lines are curved near the sources, they soon become quite straight (see Plate 2).

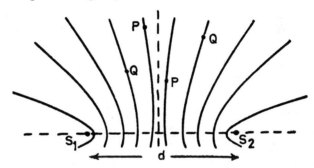

FIG. 4-4: The pattern of nodal lines formed by two point sources S_1 and S_2. The dotted line shows the line of zero path difference ($S_1P \sim S_2P = \tfrac{1}{2}\lambda$; $S_1Q \sim S_2Q = \tfrac{3}{2}\lambda$). The nodal lines are hyperbolae.

Now any point P on the first nodal line is such that

$$PS_1 \sim PS_2 = \tfrac{1}{2}\lambda \qquad (4\text{-}3)$$

Similarly any point Q on the second nodal line is such that

$$QS_1 \sim QS_2 = \tfrac{3}{2}\lambda$$

and so on. However, there will be only a finite number of nodal lines, for a little thinking will reveal that there cannot be any point such that the distance from this point to S_1 differs from the distance to S_2 by more than d, where d is the distance between the two sources. Further, it is also not difficult to show that the nodal lines are hyperbolae. For this the argument runs as follows:

Consider the line joining the sources as the x-axis and the midpoint of S_1S_2 as the origin. If (x, y) denote the coordinates of a point on the nodal line corresponding to a path difference of $(n + \tfrac{1}{2})\,\lambda$, then

$$\sqrt{\left(x+\frac{d}{2}\right)^2+y^2} - \sqrt{\left(x-\frac{d}{2}\right)^2+y^2} = (n+\tfrac{1}{2})\,\lambda = \Delta \quad \text{(say)}$$

or

$$x^2 + dx + \frac{d^2}{4} + y^2 = \Delta^2 + 2\Delta\,\sqrt{\left(x-\frac{d}{2}\right)^2+y^2} + x^2 - dx + \frac{d^2}{4} + y^2$$

or

$$(2dx - \Delta^2)^2 = 4\Delta^2\left[\left(x-\frac{d}{2}\right)^2 + y^2\right]$$

On simplification we obtain

$$\frac{x^2}{\tfrac{1}{4}\Delta^2} - \frac{y^2}{\tfrac{1}{4}(d^2-\Delta^2)} = 1 \tag{4-4}$$

which is the equation of a hyperbola. If we plot Eq. 4-4 for various values of Δ we would obtain the curves given in Fig. 4-4. (For a moment, it appears that a path difference greater than d can exist—which we know is impossible—and indeed if we substitute a value of Δ greater than d we obtain the equation of an ellipse. Certainly there is some fallacy· somewhere. Try to figure this out.)

4-4 Determination of Wavelength and Fringe Width

In the ripple tank that we have been considering, we can measure the path lengths to any point on a nodal line and using

$$PS_1 - PS_2 = (n - \tfrac{1}{2})\lambda$$

we can find the wavelength λ. We do not need to stop the waves to make such a measurement. The nodal lines stand still as we measure the distances PS_1 and PS_2. It is often convenient or even necessary to make our measurements at a point P which is far away from S_1 and S_2. Because, for any point P far away from the sources, the difference in the path lengths $PS_1 - PS_2$, depends on the angle between PS_1 and S_1S_2. Considei Fig. 4-5, which shows the two sources S_1 and S_2, and a point P very far away compared to the source separation d. The distance PA is made the same as PS_2 so that the angles PAS_2 and PS_2A are equal and $PS_1 - PS_2 = AS_1$. The farther away P is, the more nearly parallel the lines PS_1 and PS_2 become. We shall consider only points P that are far enough from S_1 and S_2 so that for all practical purposes PS_1 and PS_2 are parallel. Then the angles PAS_2 and PS_2A are almost right angles and $\triangle S_1AS_2$ is a right angled triangle. Therefore

$$\frac{AS_1}{d} = \text{Sin } \theta_n$$

where θ_n is shown in Fig. 4-5. Recalling that AS_1 is the path difference, we find that

$$PS_1 - PS_2 = AS_1 \approx d \, Sin \, \theta_n$$

The above equation expresses the path difference in terms of source separation and angle. Now when P is on the nth nodal line

$$PS_1 - PS_2 = (n - \tfrac{1}{2}) \, \lambda$$

Consequently

$$(n - \tfrac{1}{2}) \, \lambda = d \, Sin \, \theta_n$$

or

$$Sin \, \theta_n = (n - \tfrac{1}{2}) \, \lambda/d$$

as long as P is far away from S_1 and S_2. Incidentally, this result tells us that far away from the sources the direction of a nodal line does not change. It is given by the angle θ_n. Far away from the sources therefore, the nodal lines must be straight. (Actually, if these straight portions of the nodal lines are extended back toward the sources, they pass through the midpoint of the line between the sources.)

To make an accurate determination of λ, we can find the direction of the nth nodal line, i.e. the angle θ_n, and calculate λ from the equation

$$Sin \, \theta_n = (n - \tfrac{1}{2}) \, \lambda/d$$

However, it may be difficult to measure θ_n, so we shall look for a way to determine $Sin \, \theta_n$ directly, without measuring the angle θ_n itself. Since the point P (on the nth nodal line) is far away from S_1 and S_2, the lines OP and S_1P are practically parallel to each other and both are almost perpendicular to AS_2. Further, the centre line is perpendicular to S_1S_2, so $\theta_n' = \theta_n$. But from Fig. 4-5,

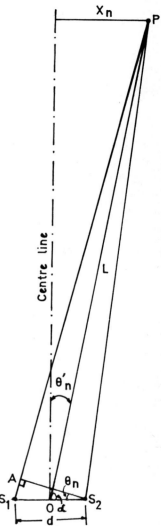

FIG. 4-5: When P is far from the sources, (i.e. $L >> d$),

$$\theta_n + \alpha \approx \frac{\pi}{2}, \quad \theta_n \approx \theta'_n.$$

Thus

$$Sin\theta_n \approx Sin\theta_n' = \frac{x_n}{L}$$

$$Sin \, \theta_n' = \frac{x_n}{L}$$

where L is the distance PO and x_n, the distance from P to the centre line. Therefore,

$$(n - \tfrac{1}{2}) \; \lambda/d = \mathrm{Sin}\,\theta_n = \mathrm{Sin}\,\theta_n' = \frac{x_n}{L}$$

or

$$\lambda = \frac{d\,(x_n/L)}{(n - \tfrac{1}{2})} \qquad\qquad (4\text{-}5)$$

Let us consider a particular example. Suppose we are working with sources 5 cm apart, i.e. $d = 5$ cm. We may choose a point P on the third nodal line, measure its distance L from the midpoint O, and measure the distance x from the centre line. Suppose we find that $L = 100$ cm and $x = 20$ cm; therefore $x/L = 0\cdot2$. We can now test the accuracy of our ratio by measuring values of L and the corresponding values of x for other points on the third nodal line. If all the points are far away, x/L value will remain close to $0\cdot2$.

Now because we are working with the third nodal line we must use $n - \tfrac{1}{2} = 5/2$ and with $d = 5$ cm we find

$$\lambda = \frac{d\,(x_n/L)}{(n - \tfrac{1}{2})}$$

$$= \frac{5 \times 0\cdot2}{5/2} = 0\cdot4 \text{ cm}$$

Using the various nodal lines, we can get several evaluations of λ. Agreement between the values obtained gives a check on our reasoning and on our measurements.

It is worth while pointing out that instead of measuring distances from the central line we could measure the distances between two consecutive nodal points lying on lines parallel to S_1S_2. This distance, β, would be equal to $x_{n+1} - x_n$; and since

$$\lambda = \frac{d\,x_n/L}{(n - \tfrac{1}{2})}$$

so

$$x_n = (n - \tfrac{1}{2})\,\lambda\,\frac{L}{d}$$

and

$$x_{n+1} = (n + \tfrac{1}{2})\lambda\,\frac{L}{d}$$

$$\therefore \; \beta \equiv x_{n+1} - x_n = \frac{\lambda L}{d} \qquad \text{(independent of } n) \qquad (4\text{-}6)$$

4-5 Phase

We have assumed the two sources of disturbance to vibrate *in phase*, i.e. both the needles dip into the water together, producing crests at the same instant. However, it is not necessary for two sources *with the same period* to be in phase. For example, one of the sources may always dip into the water somewhat later than the other, say after a time delay t. If the time period is denoted by T, the quantity p defined by

$$p = \frac{t}{T}$$

can be introduced to measure the delay. For example, if each source dips every 1/6 second and if S_2 always dips 1/18 second after S_1, then the fraction p is $\frac{1}{3}$. When two sources of the same frequency do not dip together we say that they are *out of phase*. The fraction p describes the *phase delay* of one source with respect to the other. There are no delays which are longer than the period T because we always measure the delay of the second source from the most recent dip of the first source, and its dips come a time interval T apart. Consequently the value of p is always between 0 and 1.

More conventionally, the phase delay δ is defined

$$\delta = 2\pi \, p$$

In other words, a value of $p = \frac{1}{2}$ implies a phase difference of π; and there are no phase delays more than 2π (see also Chapter 1).

Let us now use two point generators of waves operating so that S_2 has a phase delay of $2\pi \, p$ with respect to S_1. What will the interference pattern look like? We can again try to discover this graphically by

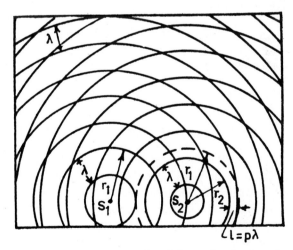

FIG. 4-6: The two sources S_1 and S_2 are vibrating in such a way that the source S_2 has a phase delay of $2\pi p$ with respect to S_1. The solid circles show the crests from the two sources at a particular instant. Clearly, the difference between the radii of corresponding crests is the distance $p\lambda$ ($= r_1 - r_2$).

drawing two sets of concentric circles representing the wave crests from each source (Fig. 4-6). As in Fig. 4-3, the crests in each set are always one wavelength, apart; however, this time the sources are not in phase, and the radii r_1 and r_2 of corresponding crests from the two sources are not equal (Fig. 4-6). The radii of the delayed crests from S_2 are shorter than those of the corresponding crests from S_1 by a distance $l = p\lambda$. Thus, if S_2 is delayed by one third of a period, the circles centered on S_2 are smaller than the corresponding ones around S_1 by $\lambda/3$.

As an example let us see what happens when one of the sources is half a period behind the other. Then the distance l equals $\frac{1}{2}\lambda$ and the phase delay is π. The conditions for maxima and minima now become

$$\left.\begin{aligned} S_2 P \sim S_1P &= (n + \tfrac{1}{2})\lambda \quad \text{(Maxima)} \\ &= n\lambda \qquad\quad \text{(Minima)} \end{aligned}\right\} \quad (4\text{-}7)$$

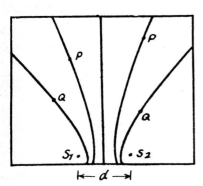

FIG. 4-7: The pattern of nodal lines formed by the point source, which are vibrating with a phase difference $\pi(p=\frac{1}{2})$. Note that the central line is a node.
$$[S_1P \sim S_2P = \lambda$$
$$S_1Q \sim S_2Q = 2\lambda]$$
(The nodal lines are hyperbolae)

(compare Eqs. 4-7 with Eqs. 4-1 and 4-2). The corresponding nodal lines are shown in Fig. 4-7.

We have given examples of the interference pattern for two particular choices of phase delay, $\delta = 0$ and $\delta = \pi$. Actually we could have chosen any phase delay from 0 to 2π, *and in each case the interference pattern would have been different.* Changing the phase delay causes the whole pattern of nodal lines to shift. As the phase delay of S_2 increases, the radii of the crests from S_2 fall behind those from S_1 by an increasing distance $l = p\lambda$. Consequently, to maintain a fixed interference pattern, the phase delay must also remain constant. For two sources which run continuously at the same frequency, the phase delay will remain constant.* But if each of the two sources is turned on and off in an irregular fashion, the phase delay will vary and with it the interference pattern. This shifting of the interference pattern will be of great importance when we discuss the interference pattern obtained by light sources.

It should however be pointed out that although our discussions have been centered on water waves, we really have not used any special property of water waves to obtain our results. *We used only the principle of superposition which is common to all wave phenomena.* The results will therefore be equally applicable to all waves.

---

* If two sources vibrate with the same frequency with their phase delay remaining constant they are said to be *coherent sources*.

4-6 Interference in Light

In Section 4-5 we studied the interference patterns produced in a ripple tank by two vibrating needles. Now we wish to do similar experiments with light. However, in the case of light the observation of interference is not so easy as for other types of wave motion. Thus, for instance, it is not sufficient to place two candles in front of a screen in order to get an interference pattern on it. If we use two conventional sources of light we fail to observe any reinforcement or weakening of light intensity in the illuminated space. In order to make the interference visible, *coherent sources* of light must be used, i.e. the sources of light must emit waves whose phase difference remains constant with time. If the phase difference changes rapidly and in an irregular way, the resulting waves are non-coherent, as of course are their sources.*

In order to explain coherence more fully, we shall examine in some detail the process of emission of light. Consider two separate sources of light. The light from each source comes from a large number of individual atoms; each atom sends out a burst of light waves only during a very short time. A typical period during which an atom emits light has been found to be about 10^{-9} seconds. Even if all the atoms of the same kind radiate under similar conditions, waves from different atoms differ in their initial phases. Thus in a light wave the phase may be regarded as constant during a period of about 10^{-9} seconds only. For the same reason, the phase difference of radiation from two different sources (say from two sodium discharge lamps) can only be constant during the same short interval, which also determines how long the interference pattern will stay on a screen illuminated by both sources. In one second, the interference pattern changes one thousand million times. The eye can only detect the intensity changes which last at least one tenth of a second. If the frequency of the changes is higher, the eye perceives only the mean value of light intensity which is the same over the whole screen. For this reason interference of light waves arriving from two different sources cannot be observed. (However, if we have a camera whose time of shutter opening can be made less than 10^{-9} seconds, then the film will record an interference pattern!†)

Thus, interference pattern may only be obtained if two light waves

---

* We may here recall from Section 4-5 that if either of the two vibrating needles vibrating in a ripple tank is turned on and off in an irregular fashion, then the phase delay will vary, and with it the interference pattern will also change. And if the phase difference changes rapidly in a random way we will not see any interference pattern at all.

† The interference pattern will be a set of dark and bright bands only if the light waves have the same state of polarization. Indeed using two independent laser beams (which have a much larger coherence time; see Chapter 5) it has been possible to obtain inteference pattern [see, for example, G. Magyar and L. Mandel, 'Interference Fringes Produced by Superposition of two Maser Light Beams', *Nature* Vol. **198**, p. 255, (1963).]

are coherent, i.e. if their phase difference remains constant. Young found a simple way to lock together the phases of two light sources so that the interference pattern will not change with time. The trick is to start with a single light source and split the light from it into two parts in phase with each other. These two parts, which act as if they came from two sources locked in phase, are then allowed to come back together to interfere with each other. With this method, we do indeed get the expected interference pattern on a distant screen.

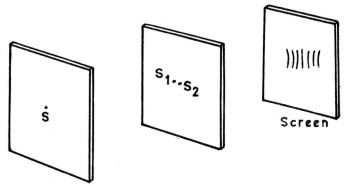

FIG. 4-8: Young's experiment for producing interference pattern. The light reaching the screen from the pinholes S_1 and S_2 comes through the pinhole S. An interference pattern is visible on the screen. The fringes are almost straight lines near the centre but become curved farther away (See Problem 2).

Young's experimental arrangement is shown in Fig. 4-8. Monochromatic light* was passed through a pinhole, and the light spreading out from this pinhole source fell upon an opaque barrier which contained two pinholes placed very close together and located equidistant from the source. Light originating from the pinhole source passed through the other two pinholes and the light at these two pinholes was then always in phase. With such an arrangement, the interference pattern of the light emerging from the pair of pinholes did not shift and could be observed. Fig. 4-9 is diagram of the wave pattern producing the interference pattern. The similarity to a ripple tank pattern is very clear. If you hold the page at the level of your eye and sight from the right edge of Fig. 4-9 toward the source (at grazing incidence) the nodal lines can be seen distinctly. These are regions of no wave disturbance and hence of no light; they have been marked 'dark' where they intersect the screen. On the screen, we would see an interference pattern consisting of alternate bright and dark bands. The bands are called interference fringes.

* Monochromatic light means light of one wavelength, i.e. of one particular colour. Different colours correspond to different wavelengths. As we go from red to violet, the wavelength changes from about 4×10^{-5} cm to roughly 8×10^{-5} cm. Light from a sodium discharge lamp is almost monochromatic with wavelength approximately $5 \cdot 89 \times 10^{-5}$ cm.

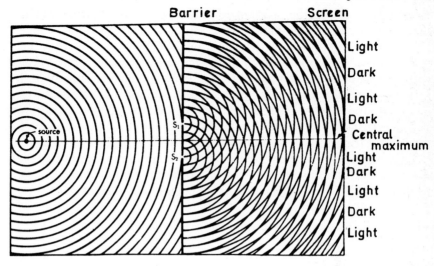

FIG. 4-9: Waves from a line source of light passing through the slits S_1 and S_2 interfere to give alternative bright and dark bands on a screen (from PSSC course on *Physics*)

Let us now try to figure out the shape of the interference fringes that will be seen on the screen. For this we must know the locus of the point P such that

$$S_1P \sim S_2P = \text{some fixed number, say } \Delta$$

(Obviously Δ has to be less than d.) In Section 4-5 we have shown that the locus of such a point is a hyperbola; and it is easy to visualize that if we rotate this hyperbola about the axis S_1S_2 all the points on the surface of this hyperbola of revolution will satisfy Eq. 4-4. Some hyperbolae of revolutions are shown in Fig. 4-10.

We are now interested in finding the curve of intersection between the screen and the hyperbola of revolution. Now in a typical experiment:

the distance $S_1S_2 \approx 10^{-2}$ cm

$$\lambda \approx 5 \times 10^{-5} \text{ cm}$$

and the distance between source and screen ≈ 50 cm.

For such distances and for a path differ-

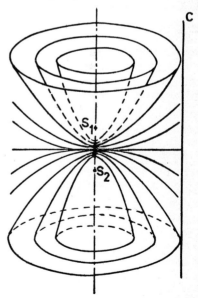

FIG. 4-10: Loci of points of constant path difference; they are, of course, hyperbolae of revolution.

ence $(S_1P \sim S_2P)$ of the order of 10λ, it can easily be shown (Problem 2) that the hyperbolae of revolution are similar to a very flat bowl and the curves of intersection with the plane of the screen will be very nearly straight lines. (Imagine yourself holding a flattish dinner plate vertically with its axis of symmetry along S_1S_2. Now if you cut this plate by a vertical plane, the plane will meet the plate at approximately a straight line.) The above series of arguments leads us to a fringe pattern shown in Fig. 4-8. It should be noted that for large values of n (i.e. far away from the central fringe) the shape of the fringes will indeed become curved.

Next, we determine mathematically the condition necessary to obtain maxima and minima on the screen. Consider a point P on the screen (Fig. 4-11). The line OO' is the central line and the distance between the

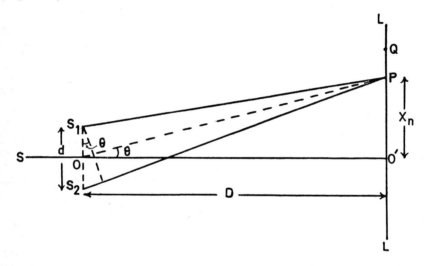

FIG. 4-11: The two 'coherent sources' S_1 and S_2 produce interference pattern on the screen LL' If $S_2P \sim S_1P = n\lambda$, we have brightness and, if $S_2P \sim S_1P = (n+\frac{1}{2})\lambda$, we have darkness.

screen and the sources is assumed to be D $(D \gg d)$. If the wavelength of the light is denoted by λ, then for the point P to correspond to the nth bright fringe we must have

$$S_2P - S_1P = n\lambda$$

Now

$$S_2P^2 - S_1P^2 = D^2 + \left(x_n + \frac{d}{2}\right)^2 - D^2 - \left(x_n - \frac{d}{2}\right)^2$$

$$= 2x_n d$$

$$\therefore \quad S_2P - S_1P = \frac{2x_n d}{S_2P + S_1P} \tag{4-8}$$

If the distances OP and $S_1 S_2$ are very small compared to D then only a very little error is introduced in replacing $S_2 P + S_1 P$ by $2D$. This can easily be understood by considering a specific example. In a typical interference experiment (with light) $d = 0.02$ cm; $D = 50$ cm, and $OP = 0.5$ cm.

For the above values

$$S_2 P = \sqrt{(50)^2 + (0.51)^2} = 50.0025 \text{ cm}$$

and

$$S_1 P = \sqrt{(50)^2 + (0.49)^2} = 50.0025 \text{ cm}$$

$$\therefore \quad S_2 P + S_1 P = 100.005 \text{ cm}$$

Thus, if we replace $S_2 P + S_1 P$ by $2D$ the error involved is 0.005%. Hence, making this approximation, we have

$$S_2 P - S_1 P = \frac{d x_n}{D} \tag{4-9}$$

Since P is the nth bright fringe, the path difference must be equal to $n\lambda$:

$$S_2 P - S_1 P = n\lambda$$

or

$$x_n = \frac{n\lambda \, D}{d} \tag{4-10}$$

If the $(n + 1)^{\text{th}}$ bright fringe is at Q, and if the distance OQ is denoted by x_{n+1} then

$$x_{n+1} = \frac{(n + 1) \, \lambda \, D}{d} \tag{4-11}$$

and the fringe width β, is given by

$$\beta = x_{n+1} - x_n$$

or

$$\beta = \frac{\lambda D}{d} \tag{4-12}$$

This is independent of the distance of the fringes from O and so we would obtain equally spaced dark and bright fringes.* For the point P to be the nth dark fringe $S_2 P - S_1 P$ must be equal to $(n + \frac{1}{2})\lambda$. The

---

* It should be pointed out that the point P (Fig. 4-11) need not be in the plane of the paper. Imagine a plane which is perpendicular to the line OO' and passing through O. On this plane, if the point P is displaced to a point which is a distance y above (or below) the plane of paper then

$$S_2 P^2 - S_1 P^2 = D^2 + \left(x_n + \frac{d}{2}\right)^2 + y^2 - D^2 + \left(x_n - \frac{d}{2}\right)^2 + y^2 = 2 x_n d$$

Thus, as long as $S_1 P + S_2 P$ is replaceable by $2D$, we will obtain straight line fringes with fringe separation equal to $\lambda D / d$. However, if y is made too large (i.e. comparable to D) the fringes will become curved (see also Problem 2).

rest of the argument will be similar and it will immediately follow that the distance between two consecutive dark fringes will also be equal to $\lambda D/d$.

A permanent record of the interference pattern produced by this system can be made by putting a camera in the place of the screen to record the result on film. The picture will show equally spaced dark and bright fringes. (see Plate 3)

4-7 Intensity Distribution

In order to calculate the intensity distribution we first note that the superposition principle for electric and magnetic fields states that at any instant, the field at any point in space arising from several sources is the vector sum of the contributions that each source would have produced if it were acting alone.* Now let us consider the resultant electric field E at the point P due to the two sources S_1 and S_2 which send out electromagenetic waves in phase (Fig. 4-11). Thus, if the source S_1 contributes a fluctuating field E_1 at the point P and S_2 contributes a field E_2, then, from the superposition principle

$$E = E_1 + E_2 \qquad (4\text{-}13)$$

The resultant electric field **E** fluctuates with the same frequency as that of the contributions E_1 and E_2. In general, the fields at the point P will be in different directions; but if the point P is very far† from S_1 and S_2 then to a good approximation we can say that the vectors E_1 and E_2 lie along the same line.

Thus, if

$$E_1 = E_{01} \; \text{Cos} \cdot (\omega t - \frac{2\pi}{\lambda} S_1 P) \left.\right\}$$

and $\qquad\qquad\qquad\qquad\qquad\qquad\qquad\qquad\qquad\qquad$ (4-14)

$$E_2 = E_{02} \; \text{Cos} \; (\omega t - \frac{2\pi}{\lambda} S_2 P) \left.\right]$$

then

$$E = E_1 + E_2$$
$$= E_{01} \; \text{Cos} \; (\omega t - \frac{2\pi}{\lambda} S_1 P)$$
$$+ E_{02} \; \text{Cos} \; (\omega t - \frac{2\pi}{\lambda} S_2 P) \qquad (4\text{-}15)$$

The resultant intensity I, would be given by (see Sec. 3-7)

---

* This principle is also a consequence of Maxwell equations (Appendix A).
† Very far compared to the distance $S_1 S_2$.

$$I(t) = \varepsilon_0 \, c \, E^2$$

$$= \varepsilon_0 \, c \left[E_{01}{}^2 \, \text{Cos}^2 \, (\omega t - \frac{2\pi}{\lambda} \, S_1 P) \right.$$

$$+ E_{02}{}^2 \, \text{Cos}^2 \, (\omega t - \frac{2\pi}{\lambda} \, S_2 P)$$

$$+ E_{01} \, E_{02} \, \{\text{Cos} \, [\frac{2\pi}{\lambda} \, (S_2 P \sim S_1 P)]$$

$$\left. + \text{Cos} \, [2\omega t - \frac{2\pi}{\lambda} \, (S_2 P + S_1 P)]\} \right]$$

For a light wave since $\omega \approx 10^{14}$ radians per second all the quantities containing ωt vary with extreme rapidity, hence any detector can only measure its average value. The average may be carried out over only one period (see Section 3-7); and one obtains

$$\left. \begin{aligned} <\text{Cos}^2 \, (\omega t - \frac{2\pi}{\lambda} \, S_1 P)> \; &= \tfrac{1}{2} \\[2mm] <\text{Cos}^2 \, (\omega t - \frac{2\pi}{\lambda} \, S_2 P)> \; &= \tfrac{1}{2} \\[4mm] <\text{Cos} \left[2\omega t - \frac{2\pi}{\lambda} \, (S_2 P + S_1 P) \right]> \; &= 0 \end{aligned} \right\} \qquad (4\text{-}16)$$

and

Therefore

$$<I \, (t)> \; = I_1 + I_2 + 2\sqrt{I_1 I_2} \, \text{Cos} \, \delta \qquad (4\text{-}17)$$

where

and

$$\left. \begin{aligned} I_1 &= \tfrac{1}{2} \, \varepsilon_0 c E_{01}{}^2 \\[2mm] I_2 &= \tfrac{1}{2} \, \varepsilon_0 c E_{01}{}^2 \end{aligned} \right\} \qquad (4\text{-}18)$$

$$\delta = \frac{2\pi}{\lambda} \, (S_2 P \sim S_1 P) \qquad (4\text{-}19)$$

Physically, I_1, is the intensity produced by the source S_1 if the hole S_2 were closed, and I_2 is the intensity produced by the source S_2. The quantity δ is the phase difference between the two wave motions at the point P. The term $2\sqrt{I_1 I_2} \, \text{Cos} \, \delta$ represents the interference term and if the point P is so far away that we may write

$$I_1 \approx I_2 = I_0 \qquad (4\text{-}20)$$

then the expression for I becomes

$$I = 2I_0 + 2I_0 \, \text{Cos} \, \delta$$
$$= 4I_0 \, \text{Cos}^2 \, \delta/2 \qquad (4\text{-}21)$$

Thus the intensity is maximum $(= 4I_0)$ when

$$\left. \begin{aligned} \delta &= 2n\pi \\[4mm] S_2 P \sim S_1 P &= n\lambda \end{aligned} \right\} \quad \begin{aligned} &\text{(Constructive} \\ &\text{Interference)} \end{aligned} \qquad (4\text{-}22)$$

or

and the intensity will be zero when

$$\delta = (2n + 1)\pi$$

or

$$S_2P \sim S_1P = (n + \tfrac{1}{2})\lambda$$

(Destructive
Interference)   (4-23)

where $n = 0, 1, 2, 3, \ldots$. The corresponding intensity distribution is shown in Fig. 4-12. Further, from Eq. 4-9 we know that

$$S_2P - S_1P \cong \frac{xd}{D}$$

In other words when $x = 0, \dfrac{\lambda D}{d}, \dfrac{2\lambda D}{d}, \ldots$ we have maximum intensity

and when $x = \dfrac{\lambda D}{2d}, \dfrac{3\lambda D}{2d}, \ldots$ we have zero intensity (see Plate 3).

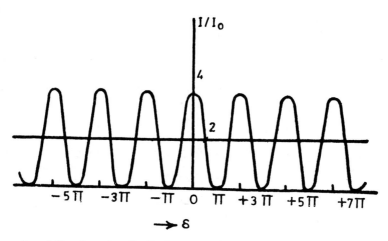

FIG. 4-12: Intensity distribution as a funciton of the phase difference $\delta \left(= \dfrac{2\pi}{\lambda} \dfrac{xd}{D} \right)$ for the interference fringes between two coherent beams.

Next, let us consider the holes S_1 and S_2 to be illuminated by different sources. As already discussed the phase difference will remain constant for a period of about 10^{-9} seconds (which will contain about a million cycles). Thus, we may carry out the averaging and obtain

$$<I(t)> = I_1 + I_2 + 2\,I_{12} <\text{Cos}\,[\delta + \phi(t)]>$$ (4-24)

where

$$I_{12} = \tfrac{1}{2}\,\varepsilon_0 c\,\boldsymbol{E}_{01} \cdot \boldsymbol{E}_{02}$$

and $\phi(t)$ is the phase difference between the two sources. (We have assumed the light to be plane polarized but not, in general, in the same direction.) Since $\phi(t)$ changes very rapidly* ($\sim 10^9$ times in a second)

* This change is extremely slow when we consider the variation of the electric field due to the term ωt.

a detector will only measure the average of $\text{Cos}\,[\delta + \phi(t)]$; this will of course be zero.

Thus, if in the Young's experiment we illuminate S_1 by one sodium lamp and S_2 by another sodium lamp, the eye or a photographic plate records merely a uniform intensity across the screen, i.e.

$$I = I_1 + I_2 \approx 2\,I_1 \qquad (4\text{-}25)$$

The pattern of alternate bright and dark bands is no longer apparent and no interference is observed. In such cases we say the sources are *incoherent*. We may therefore write that *for incoherent sources the intensities add up and no interference pattern is observed.*

Thus, except in the case of lasers (which we will discuss in Chapter 5), for two different sources we will obtain a succession of different interference patterns, shifting after intervals of approximately 10^{-9} second.* A detector which responds only to averages of time intervals like $1/10$ second will observe the average of some hundred million different patterns, and in this average no interference pattern will be observed.†

The theory that has been developed above has dealt with point sources, while, in the experiments usually two narrow slits are used. But the two slits may be regarded as a set of pairs of point sources, any one pair being in the same horizontal line. Any two pairs may be assumed to be incoherent and the intensities due to each of the pairs will add up. As long as each pair produces a set of alternate dark and bright straight line fringes, the sole effect of substituting slits for points is to get *brighter fringes*. (The effect of a wide slit will be discussed in Chapter 5.)

4-8   Determination of Wavelength

We will now discuss in some detail one of the most extensively used arrangements for producing coherent sources and then outline the method for measuring the wavelength of light. The method is due to Fresnel and consists of using a biprism, which is a single prism, one of whose angles is of the order of $179°\ 20'$ and the other two are then $20'$ each. It is used as two prisms, each of refracting angle $20'$. A horizontal cross-section of the apparatus is shown in Fig. 4-13, in which S represents a vertical slit, illuminated with monochromatic light, placed some 5 cm from the biprism B, whose two refracting edges are vertical. Each half of the biprism produces an image of the slit S and these two images S_1 and S_2 can be seen with the naked eye, if the eye is placed on the other side of the biprism from the slit S. The distance of S from the biprism is adjusted until the images are very close together (about $0\cdot02 - 0\cdot04$ cm) and the slit S is rotated in its own plane until the two images are parallel to one another.

---

* The interference pattern (even for 10^{-9} second) will be a set of dark and bright bands only if the two beams have the same state of polarization.

† See, however, the discussion on optical beats in Chapter 5.

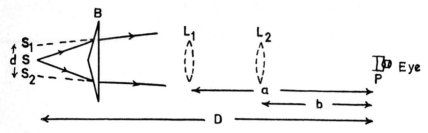

Fig. 4-13:   A cross-section of Fresnel's biprism apparatus. S denotes the source, B the biprism, P the eyepiece, L_1 and L_2 are the two positions of the lens used for measuring d.

Equally spaced interference fringes can be seen on a vertical screen placed about 30 cm from the biprism or they can be observed on a travelling eyepiece P. The fringe width increases if the eyepiece is moved further from the biprism or if the sources S_1 and S_2 are moved closer together by moving the slit S nearer to the biprism (obvious from Eq. 4-12).

Eq. 4-12 also provides a means of calculating the wavelength of light from a measurement of these fringes. We determine the fringe width by means of the travelling eyepiece, finding the distance between as large a number of dark fringes as we can see clearly and dividing this distance by the number of fringes. We work with dark rather than the bright fringes because they are the narrower, so that it is possible to set the vertical cross-wires of the eyepiece more accurately on the centre of a dark than a bright fringe (see Fig. 4-12 and Plate 3). The distance d, between the two sources and their distance D from the cross-wires of the eyepiece are found by putting a convex lens in the position L_1 between S and P (Fig. 4-13) and adjusting it so that the images of S_1 and S_2 fall on the cross-wires of the eyepiece and so can be seen in focus. The distance between these images, d_1, is measured by the eyepiece and also the distance a from the lens to the cross-wires. The lens is now moved to the second position L_2, which will bring again sharply focussed images of S_1 and S_2 in the eyepiece. Then distance d_2 between these images and the distance b from the lens to the cross-wires are measured as before. It can then easily be shown that $d = \sqrt{d_1 d_2}$ and $D = a + b$ (Problem 3). The wavelength of light can then be calculated by substituting the values of β, d and D in Eq. 4-12. A typical set of results for sodium light with $\lambda = 5 \cdot 90 \times 10^{-5}$ cm. is: $d = 0 \cdot 04$ cm, $D = 50 \cdot 0$ cm and $\beta = 0 \cdot 737$ mm.

4-9   Interference with White Light

What will happen if we replace monochromatic light with white light in the above cases of interference? What effects can we expect to observe? The screen will be found to be covered by several coloured bands since the white light is composed of many colours. Each component colour forms its own interference pattern, coloured according to the corresponding wavelength. The individual fringes will coincide only in the centre of the screen, where the combination of colours gives a white band (Fig. 4-14). To understand what will happen as we go further out from

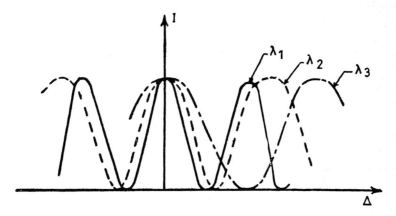

FIG. 4-14: Intensity distribution for various colours as a function of path difference $\Delta (\lambda_1 < \lambda_2 < \lambda_3)$. All the wavelengths constructively interfere at $\Delta = 0$; λ_1, λ_2 and λ_3 may be assumed to correspond to the blue, green and red regions of the spectrum respectively.

the centre to places of greater path difference, it is best to consider some numerical examples. Taking the limits of the visible spectrum as 8×10^{-5} cm (red end) and 4×10^{-5} cm (violet end), we see that the violet will be absent at the place where the path difference $(S_2P \sim S_1P)$ is 2×10^{-5} cm, but complete destructive interference does not occur for any other colour. So we shall see white light deprived of violet, which is a reddish tint. Further out, where the path difference is 4×10^{-5} cm, violet will be a maximum while red is absent, so that we shall see violet here. The central white fringe is therefore bordered by a red fringe, followed by a violet fringe on either side. At a place where the path difference is 1.5×10^{-4} cm, wavelengths given by $(2m + 1)$ $\lambda/2 = 1.5 \times 10^{-4}$ cm will be absent and the wavelengths lying within region given by $m = 2$ and $m = 3$ are visible. These wavelengths are 6×10^{-5} cm and 4.3×10^{-5} cm and correspond to the orange and blue lines respectively. The resultant colour here is white minus orange and blue; red and green will be particularly strong (why?) with some violet and the colour will be a greenish yellow. Similarly, at the place where the path difference is 4×10^{-4} cm it can easily be shown that the light of wavelengths 7.3×10^{-5}, 6.2×10^{-5}, 5.3×10^{-5}, 4.7×10^{-5}, 4.2×10^{-5} cm (in the visible region) will be absent. When five wavelengths, which are fairly evenly distributed throughout the spectrum are absent from the light the unaided eye cannot distinguish it from the white light, and the greater the path difference the more such wavelengths are absent and more nearly the colour appears to be white. So we see that with white light we get a white fringe at the point of zero path difference and a few coloured fringes on either side, fading off into uniform white light after about some ten fringes.

4-10 Displacement of Fringes

Let us now investigate the effect on the interference fringes of introducing a thin transparent plate in the path of one of the two interfering beams of monochromatic light. Let e be the thickness of the plate and μ its refractive index for the monochromatic light employed. From Fig. 4-15 it is clear that a light wave travelling from S_1 to P has to

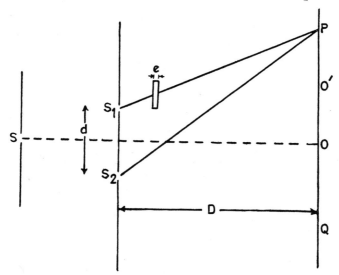

FIG. 4-15:   The displacement of the fringe pattern with the introduction of a thin plate in the path of one of the beams. After the introduction of the plate, the central fringe is formed at O'.

traverse a distance e in the plate while for the rest of the distance $(S_1P - e)$ it traverses in air. Thus, the time required to traverse this path is given by

$$t = \frac{S_1P - e}{c} + \frac{e}{v}$$

$$= \frac{1}{c}\left[S_1P - e + \frac{c}{v}e\right]$$

$$= \frac{1}{c}\left[S_1P + (\mu - 1)e\right] \tag{4-26}$$

where v is the speed of light in the plate which is equal to $\dfrac{c}{\mu}$. Eq. 4-26 tells us that due to the introduction of the plate, the effective path from S_1 to P becomes $S_1P + (\mu - 1)\, e$ in air. Similarly, the effective path from S_1 to O, the point equidistant from S_1 and S_2, becomes $S_1O + (\mu - 1)\, e$ in air, and since

$$S_1O + (\mu - 1)\, e > S_2O$$

the *central* bright fringe (corresponding to zero path difference) is *not* formed at O, which is the normal position of the central fringe in the absence of the plate. Let the new position of the central fringe be O', then

$$S_1O' + (\mu - 1)\, e = S_2O'$$

but, we know (see Eq. 4-9)

$$S_2O' - S_1O' \approx \frac{d}{D}\cdot OO'$$

$$\therefore \quad (\mu - 1)\, e = \frac{d}{D}\cdot OO'$$

Thus, if we denote the distance through which the central fringe shifts by Δ, then it is given by

$$\Delta = \frac{D\,(\mu - 1)e}{d} \tag{4-27}$$

If we use a white light source then we know that the central fringe is easily distinguishable because we get a white fringe and some coloured fringes on either side. If we can measure the shift in the central fringe then from Eq. 4-27 we can accurately measure the thickness of a thin transparent material.

It can be shown in a straightforward manner that, if we use a monochromatic source, the spacing of the interference fringes remains unaffected by the introduction of the plate and that the entire fringe system shifts laterally through a distance $\dfrac{D\,(\mu - 1)\, e}{d}$ towards the side of which the plate is placed (Problem 8). Let us now work out a simple example.

Example
A thin mica sheet ($\mu = 1\cdot58$) is used to cover one slit of a double slit arrangement. The central point on the screen is occupied by what used to be the seventh bright fringe. If $\lambda = 5\cdot5 \times 10^{-5}$ cm, what is the thickness of the mica sheet?

Solution
Let Q be the position of the seventh bright fringe, in the absence of any mica sheet (Fig. 4-15). Then

$$S_1 Q - S_2 Q = 7\lambda.$$

When the mica sheet is introduced, the fringe which previously occured at Q now occurs at O. Therefore,

$$S_2 O - S_1 O + (\mu - 1) e = 7\lambda$$

But
$$S_2 O = S_1 O$$

$$\therefore \quad (\mu - 1) e = 7\lambda$$

or
$$e = \frac{7 \times 5 \cdot 5 \times 10^{-5}}{0 \cdot 58} = 6 \cdot 6 \times 10^{-4} \text{ cm}$$

Example

The stars Betelguse and Rigel are equally bright. Light from the two stars falls on a double slit arrangement as shown in Fig. 4-16. F is a filter which passes only the wavelength λ. Obtain an expression for the intensity distribution.

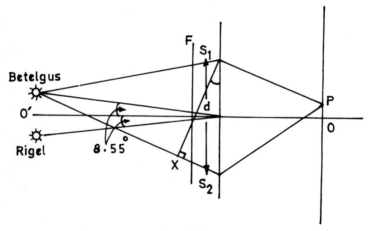

FIG. 4-16

Solution

We must first note that light from the star Rigel falls on the two slits and they form a pair of coherent sources. The same is true for light from Betelguse. But the two stars are not coherent with respect to each other. So we have a two slit interference pattern due to each star centered on the line joining the star and the slit. Since the two stars are incoherent, the resultant intensity is the sum of the two intensities produced by each star.

Now, if the sources were on the axis, then the intensity distribution would have been

$$I = I_0 \cos^2 \delta/2$$

where
$$\delta = \frac{2\pi}{\lambda} (S_2 P - S_1 P) \cong \frac{2\pi}{\lambda} \left(\frac{xd}{D} \right)$$

Thus the intensity distribution due to the star Betelguse would be given by

$$I_B = I_0 \cos^2 \frac{\delta_B}{2}$$

where

$$\delta_B = \frac{2\pi}{\lambda} [BS_2 + S_2P - BS_1 - S_1P]$$

$$= \frac{2\pi}{\lambda} [S_2P - S_1P + XS_2]$$

$$= \frac{2\pi}{\lambda} [S_2P - S_1P + d \sin 8\cdot55°]$$

or

$$I_B = I_0 \cos^2 \left[\frac{2\pi}{\lambda} \left\{ \frac{xd}{D} + 0\cdot15 \, d \right\} \right]$$

Similarly for the star Rigel

$$I_R = I_0 \cos^2 \left[\frac{2\pi}{\lambda} \left\{ \frac{xd}{D} - 0\cdot15d \right\} \right]$$

and the resultant intensity is given by (see Eq. 4-25):

$$I = I_B + I_R$$

Example
Discuss the interference pattern produced by Young's double slit arrangement in which there is a half wave plate in front of the slit S_1 with its fast axis along the slit and another half wave plate in front of the slit S_2 whose slow axis is along the slit (Fig. 4-17). The incident light is unpolarized.

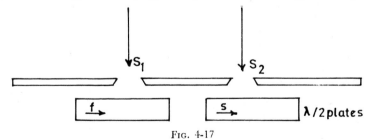

FIG. 4-17

Solution
Although the incident light is unpolarized the birefringent material changes the interference pattern. Let us first consider the component which is polarized vertically; the path through the half wave plate over slit S_2 contains, say, n wavelengths. The path through slit S_1 then contains $(n + \frac{1}{2})$ wavelengths with the result that the central fringe is now dark and the fringe system shifts half a fringe toward S_1.

The intensity pattern due to the horizontal polarization component

(*H*) also shifts, this time half a fringe toward S_2. The result (shown in Fig. 4-18) is a restored fringe pattern with a dark fringe at the center.

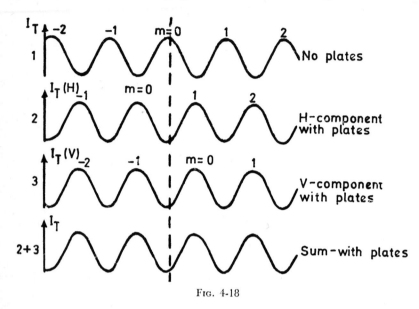

Fig. 4-18

4-11   Other Apparatus for the Production of Interference Pattern

We have discussed the Young's double slit arrangement and the Fresnel's biprism arrangement for obtaining an interference pattern. We will now discuss two more methods which may be used to produce similar interference pattern.

In the arrangement known as *Fresnel's mirrors*, light from a slit is reflected by two plane mirrors, which are slightly inclined to each other. The mirrors produce two virtual images of the slit, as shown in Fig. 4-19(*a*). They act in every respect like the images formed by the biprism, and interference fringes are observed in the region *bc*, where the reflected beams overlap. It will be noted that the angle 2θ subtended at the point of intersection *M* by the two sources is twice the angle between the mirrors.

An even simpler device, known as Lloyd's mirror [shown in Fig. 4-19(*b*)] produces interference between the light coming directly from the source without reflection. In this arrangement, the quantitative relations are similar to those in the foregoing cases, with the slit and its virtual image constituting the double source. Interference fringes are observed only in the region *bc* of the screen. An important feature of the Lloyd's mirror arrangement lies in the fact that when the screen is placed in contact with the end of the mirror [in the position *MN*, Fig. 4-19(*b*)], the edge *O* of the reflecting surface comes at the center of a dark fringe, instead of a

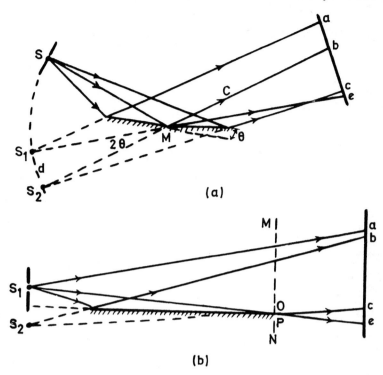

Fig. 4-19: (a) Fresnel's double mirror arrangement. Interference fringes are produced in the region *bc*.

(b) Lloyd's mirror arrangement. Interference fringes are produced in the region *bc*.

bright one as might be expected. This means that one of the two beams has undergone a sudden phase change of π. Since the direct beam could not have changed phase, this experimental observation is interpreted to mean that the reflected light has changed phase at reflection (compare this with the reflection of a pulse on a string, Fig. 4-1). The central fringe can also be observed without placing the screen in contact with the end of the mirror. This can be done by introducing a thin mica sheet in the path of the direct beam so that the central fringe is shifted somewhere in the region *bc* [Fig. 4-19(*b*)]. Now if we used a white light beam, we would find the central fringe to be dark. Since a phase difference of π corresponds to a path difference of $\lambda/2$, so for a point P (on the screen) to correspond to a bright fringe we must have (from Eq. 4-7):

$$S_2P - S_1P = (m + \tfrac{1}{2})\lambda \tag{4-28a}$$

and for P to correspond to a dark fringe we must have

$$S_2P - S_1P = m\lambda \tag{4-28b}$$

where $m = 0, 1, 2, \ldots$.

4-12 Phase Change on Reflection

Stokes used the principle of *optical reversibility* to investigate the reflection of light at an interface between two media. The principle states that if there is no absorption of light, then a light ray that is reflected or refracted will retrace its original path if its direction is reversed.*

Fig. 4-20(a) shows a wave of amplitude a reflected and refracted at a surface separating media 1 and 2, with $\mu_2 > \mu_1$. The amplitude of the reflected wave is ar, where r is the amplitude reflection coefficient. The amplitude of the refracted wave is at, where t is the amplitude transmission coefficient.

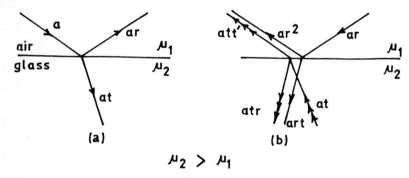

FIG. 4-20: (a) A ray is reflected and refracted at an air-glass interface.
(b) The optically reversed situation; the two rays in the lower left must cancel each other.

We consider only the possibility of phase changes of 0 or π. For example, if $r = 0.5$, we have a reduction in amplitude on reflection by one-half and no change in phase. Similarly, for $r = -0.5$ we have a phase change of π because

$$\text{Sin}\,(\omega t + \pi) = -\,\text{Sin}\,\omega t$$

Fig. 4-20(b) suggests that if we reverse these two rays they should combine to produce the original ray reversed in direction. Ray ar [(identified by the single arrow in Fig. 4-2(b)] is reflected and refracted, producing the rays of amplitude ar^2 and art. Ray at (identified by the triple arrows) is also reflected and refracted, producing the rays of amplitudes att' and atr' as shown. Note that r is the amplitude reflection coefficient for a ray in medium 1 reflected from medium 2, and r' is the corresponding quantity for a ray in medium 2 reflected from medium 1. Similarly, t

---

* The principle is a consequence of time-reversal invariance which predicts that processes can run either way in time. Physicists have yet to find a process which they know cannot actually run backwards in time if offered the proper circumstances [see, for example A. Baker, *Modern Physics and Antiphysics* Chapter 3, Addison-Wesley (1970); *The Feynman Lectures on Physics*, Vol 1, Chapter 52, Addison-Wesley (1965)]

describes a ray that passes from medium 1 to medium 2; t' describes a ray that passes from medium 2 to medium 1.

The two rays in the upper left of Fig. 4-20(b) must be equivalent to the incident ray of Fig. 4-20(a), reversed; the two rays in the lower left of Fig. 4-20(b) must cancel each other. The former requirement leads to

$$r^2 + t\,t' = 1$$

or

$$t\,t' = 1 - r^2 \tag{4-29}$$

while the latter requirement leads to

$$rta + tr'a = 0$$

or

$$r' = -r \tag{4-30}$$

But we know from Lloyd's single mirror that there is a π change in phase when light in air is reflected at the boundary of a more dense medium and so the above result proves *that there is no abrupt phase change when light in a more dense medium is reflected at the boundary of air*, and this is indeed borne out by experiments. What happens to the electric and magnetic fields when there is a phase change of π by reflection at a denser medium? This can be understood by referring to Fig. 4-21 which shows a linearly polarized electromagnetic wave reflected by a denser medium. The electric vector is assumed to be parallel to the plane of incidence, as is shown in Fig. 4-21(a), but the results are the same whatever be the state of polarization of the incident beam. The horizontal lines are wavefronts spaced one-half a wavelength apart. The direction of the electric vector in the wavefront is shown by an arrow and the direction of the magnetic vector by dots and crosses. Dots implying that the direction of the magnetic field is coming out of the page (and perpendicular to E and to the direction of propagation); whereas the crosses imply that the magnetic field is in the downward direction. The lowest wavefront in the beam shown in Fig. 4-21(a) is just making contact with the reflecting surface, and the lowest wavefront in the beam shown in Fig. 4-21(b) is the reflected wavefront that originates at the reflecting surface at this instant.

Figs. 4-21(a) and (b) show that for the wave reflected by a denser surface, the direction of the electric vector is reversed relative to its direction in the incident wave, while that of the magnetic vector is not. The electric vector is said to be reflected with a *reversal of phase*.* On the other hand for reflection at a rarer medium, the magnetic vector is opposite in direction to that in the lowest wavefront while the electric vectors are in the same direction [Figs. 4-21(c) and (d)].

* That there is a reversal of phase of the electric vector and not of the magnetic vector can be shown from rigorous electromagnetic theory.

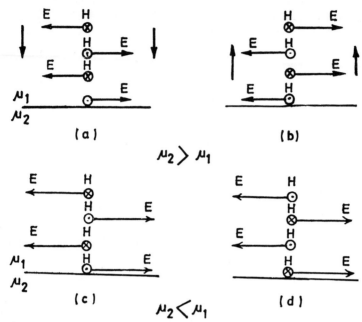

FIG. 4-21: Electric and magnetic vectors in a plane polarized train of waves (a) incident on and (b) reflected from the surface of an optically dense medium. The horizontal lines are wavefronts spaced one-half a wavelength apart. Electric and magnetic vectors in a plane polarized train of waves (c) incident on and (d) reflected from the surface of an optically rarer medium. The horizontal lines are wavefronts spaced one-half a wavelength apart.

4-13   Interference Involving Multiple Reflections

Everyone has seen the exquisite colours produced by a film of oil on the surface of a puddle on the road and also the colours produced by light reflected from a soap film. Hooke observed such colours in thin flakes of mica. Young felt that the key to all these phenomena was that interference occurs between light reflected from the top and bottom surfaces of the film. We will now study this in detail, dealing with some very simple cases which illustrate the principles concerned, and then lead on to the most commonly occurring case with thin films, concluding with Newton's Rings and Michelson's interferometer.

4-14   Reflection from a Plane Parallel Film

Let a ray of light from a source S be incident on the surface of such a film at A [Fig. 4-22(a)]. Part of this will be reflected as ray (a) and part refracted in the direction AC. Upon arrival at C, part of the latter will be reflected to B and part refracted to T. At B the ray CB will be again divided. A continuation of this process yields two sets of parallel rays, one

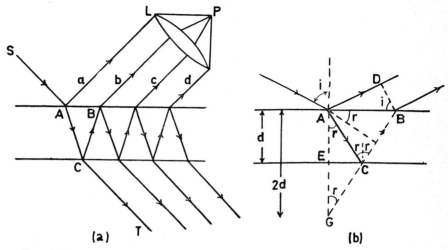

Fig. 4-22: (a) Multiple reflections in a plane parallel film.
(b) Optical path difference between two consecutive rays in multiple reflections.

on each side of the film. In each of these sets, the intensity decreases rapidly from one ray to the next. If the set of parallel reflected rays is now collected by a lens and focused at the point P, each ray will have travelled a different distance, and the phase relations may be such as to produce destructive or constructive interference at that point. It is such interference that produces the colours of thin films when they are viewed by the naked eye. In such a case L is the lens of the eye, and P lies on the retina.

In order to determine the phase difference between these rays, we must first evaluate the difference in the optical path traversed by a pair of successive rays, such as the rays (a) and (b). In Fig. 4-22(b) let d be the thickness of the film, μ its refractive index and i and r the angles of incidence and refraction. If BD is perpendicular to the ray (a), the optical paths from D and B to the focus of the lens will be equal. Starting at A the ray (b) has the path ACB in the film and the ray (a) has the path AD in air. The difference in these optical paths, Δ, is given by

$$\Delta = (AC + CB) \text{ of film} - AD \text{ of air}$$
$$= \mu (AC + CB) - AD$$

If BC is extended to intersect the perpendicular line AE at G, then

$$\angle ACF = 2r$$

and $\quad \angle CGA = \angle ACF -- \angle CAE = r$

$$\therefore \qquad AC = CG$$

Thus

$$\Delta = \mu \, (AC + CB) - AD$$
$$= \mu \, GB - AD$$

Now AF is drawn perpendicular to CB, therefore from triangles ADB and AFB

$$\text{Sin } i = \frac{AD}{AB}$$

and

$$\text{Sin } r = \frac{FB}{AB}$$

$\therefore$

$$\mu = \frac{\text{Sin } i}{\text{Sin } r} = \frac{AD}{FB}$$

or

$$AD = \mu FB$$

Thus

$$\Delta = \mu GB - \mu FB$$
$$= \mu GF$$

or

$$\Delta = 2\mu d \text{ Cos } r \qquad (4\text{-}31)$$

Since the ray (a) has undergone a phase change of π at reflection, while the ray (b) has not, so when

$$2\mu d \text{ Cos } r = m\lambda \qquad \text{(Minima)} \qquad (4\text{-}32)$$

we will have destructive interference (or minima).

Next we examine the phases of the remaining rays, (c), (d), (e), ... Since the geometry is the same, the path difference between the rays (c) and (b) will also be $2\mu d \text{ Cos } r$, but there are only internal reflections involved, so if Eq. 4-32 is satisfied, the ray (c) will be in the same phase as the ray (b). The same holds for all succeeding pairs, and so we conclude that under these conditions the rays (a) and (b) will be out of phase, but the rays (b), (c), (d), ... will be in phase with each other.

For the minima of intensity, the ray (b) is out of phase with the ray (a), but the ray (a) has a considerably greater amplitude han ray (b), so that these two will not completely annul each other. We can now prove that the addition of the rays (c), (d), (e) . . . ., which are all in phase with the ray (b), will give a net amplitude just sufficient to make up the difference and to produce *complete darkness* at the minima. Using a for the amplitude of the incident wave, r for the reflection coefficient, t and t' for the transmission coefficients (see Fig. 4-23) and adding the amplitudes of all the reflected rays, we obtain

$$A = - a \, [trt' + tr^3t' + tr^5t' + \, . \, . \, . \, .]$$
$$= - \frac{att'r}{(1 - r^2)}$$

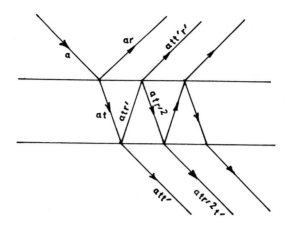

Fig. 4-23: Amplitudes of successive rays in multiple reflection.

But, from Stokes treatment (Eq. 4-29)

$$t\,t' = 1 - r^2$$

or

$$A = -\,ar \qquad (4\text{-}33)$$

This is just equal to the amplitude of the first reflected ray, so we conclude that where $2\mu d \cos r = m\lambda$, there will be complete destructive interference.

On the other hand, if conditions are such that

$$2\mu d \cos r = (m + \tfrac{1}{2})\lambda \qquad \text{(Maxima)} \qquad (4\text{-}34)$$

then, the ray (*b*) will be in phase with the ray (*a*), but the rays (*c*), (*e*), (*g*).... will be out of phase with the rays (*a*), (*d*), (*f*)...Since (*b*) is more intense than (*c*), (*d*) more intense than (*e*) etc., these pairs cannot cancel each other, and since the stronger series combines with (*a*), the strongest of all, there will be a maximum of intensity. We may now predict what will happen in four specially simple cases.

(1) In the first case the incident light is *parallel and monochromatic* and the *faces of the film are plane and parallel* to one another. Both *d* and *r* are constant, so Δ is constant for all points of the film. If it is an integral number of wave lengths, the film will be dark all over according to Eq. 4-32. If the path difference is an odd number of half waves, the illumination of the film is a maximum everywhere (Eq. 4-34), and, if it lies in between these possibilities, the illumination will be of constant intensity at every point. So in this case, *the illumination of the film seen by reflected light will be uniform.*

(2) In the second case, the incident light is *parallel and white, the film being of constant thickness as before.* Again d and r (and therefore Δ) are constants so those wavelengths for which the path differences is an integral number of waves will be absent from the reflected light. Hence the film will have a uniform colouration; it will be seen the same colour all over. We should mention here that, if a broad beam of strictly parallel light is incident on the film, only a small area AA' of the film will be visible by reflected light at any one position of the eye (Fig. 4-24); this is the area which can reflect light into the pupil of the eye and so will have about the same areas as the pupil of the eye itself. By moving the eye, different parts of the film can be scanned. Thus, with monochromatic light the intensity will remain the same and with white light the colour will remain the same.

(3) In the third case, monochromatic parallel light is incident on a film of varying thickness, so r is constant, while d varies; hence Δ will also vary. There will be minimum intensity in the reflected light for those thicknesses for which $\Delta = m\lambda$, and a maximum intensity for thicknesses for which $\Delta = (m + \frac{1}{2})\lambda$. We shall see a set of alternate dark and bright bands, any one band being a locus of constant path difference or, in other words, of constant thickness of the film.

(4) In the fourth case, if the incident light is white instead of monochromatic we shall see a set of coloured bands, any one band being a contour of constant thickness of the film. This case can be realised in the laboratory by directing a beam of parallel white light on to a vertical soap film formed by a circular wire frame and focussing an image of the soap film, formed by reflected light, on to a screen. As the soap solution drains to the bottom, making the film thinner at the top, a set of horizontal coloured bands form, the colours becoming more brilliant as the film gets thinner. Finally a black band forms at the top of the film itself when the film thickness is less than a wave-

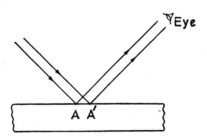

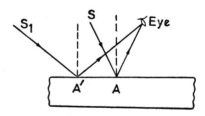

FIG. 4-24: If a broad beam of strictly parallel light is incident on the film, only a small area AA' of the film will be visible by reflected light at any one position of the eye.

FIG. 4-25: The point A of the film is seen by light which comes from the point S of the sky, while the point A' of the film is seen by light coming from another point S_1 of the sky.

length, the blackness being due to the change in phase of the wave reflected at the front surface of the film.

We can now consider the common case of the colours of a thin film of oil on a puddle of water seen by light reflected from the sky. It is clear from Fig. 4-25 that the point A of the film is seen by light which comes from the point S of the sky, while the point A' of the film is seen by light coming from another point S_1 of the sky.

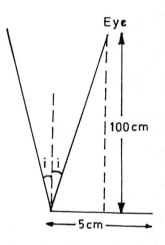

In general, each point of the film is seen by light from a different point of the sky, so that, although an extended source is used, the condition of the interfering wavetrains originating from a single point of the source is not violated. S_1 does send light on to A, but that light cannot enter the eye and contributes nothing to the illumination of the image of A formed at the eye so that those rays can be ignored for our purpose. In this case, both d and r vary in Fig. 4-25. But, if we consider a practical case of a film some 5 cm in linear dimensions viewed by an eye 100 cm vertically above the middle point of the film (Fig. 4-26), then

$$\text{Sin } i < \frac{2 \cdot 5}{100} = \frac{1}{40}$$

Assuming the refractive index to be 4/3, we obtain

$$\text{Sin } r < \frac{1}{40} \times \frac{3}{4} = \frac{3}{160}$$

or

Fig. 4-26: A film of 5 cm linear dimension is viewed by an eye 100 cm vertically above. The value of the sine of the angle of incidence lies between 0 and 2·5/100.

$$\text{Cos } r > \sqrt{1 - \left(\frac{3}{160}\right)^2} \approx 0 \cdot 9998$$

Thus the cosines of the angles lie between 1 and 0·9998, so the variation of path difference due to a variation in the angle of inclination of the light on the film can be ignored. So we will see a set of coloured bands, each band being a contour of equal thickness of the film. For this reason, these bands are often called fringes of equal thickness. The reader can easily verify for himself that this is what he does see, and he will notice that the colour of any particular region of the film alters, if the head is moved so as to change the angle of inclination of the light reaching his eye from that portion of the film. It may be noted that with an extended source, a finite area of the film can be seen for one position of the eye. This is the only effect of using an extended source compared to a point source.

In the above discussions we must have a 'thin' film, i.e. d should not be more than a few wavelengths of light. For very thick films ($d \approx 1$ cm), the path difference will be many wavelengths and therefore the phase difference at a given point on the film will change rapidly as we move even a small distance away from the point. For 'thin' films, however, the phase difference remains same for reasonably closer points. In other words, very small changes in the angle of incidence do not change the interference condition for 'thin' films but they do change it for 'thick' films.

Example
A water film ($\mu = 4/3$) in air is 3×10^{-5} cm thick. If it is illuminated with white light at normal incidence, what colour will it appear to be in the reflected light?

Solution
For normal incidence $\cos r = 1$. Thus the wavelengths for which we will have destructive interference will be given by

$$2\mu d = m\lambda$$

or

$$\lambda = \frac{2 \times \frac{4}{3} \times 3 \times 10^{-5}}{m} = \frac{8 \times 10^{-5}}{m} \text{ cm}$$

Thus, the possible values of λ for which we will have minima are 8×10^{-5} cm (red), 4×10^{-5} cm (violet), 2×10^{-5} cm (ultraviolet)..... Similarly the wavelengths which will correspond to constructive interference will be given by:

$$\lambda = \frac{8 \times 10^{-5}}{(m + \frac{1}{2})}$$

or $\lambda = 16 \times 10^{-5}$ cm, $5 \cdot 33 \times 10^{-5}$ (yellow), $3 \cdot 6 \times 10^{-5}$ (ultraviolet),....
If we restrict ourselves to the visible region ($4 \times 10^{-5} < \lambda < 8 \times 10^{-5}$ cm) we find that violet and red part will be absent whereas maximum intensity will occur around the yellow-green region. Thus, the colour will appear yellow-green in the reflected light.

4-15 Non-reflecting Films

The phenomenon of interference is utilized in the production of so called 'non-reflecting' glass. A thin layer or film of transparent material is deposited on the surface of the glass as shown in Fig. 4-27. The refractive index of this material is chosen at some value intermediate between that of air and the glass. (For example, the refractive index of magnesium flouride is 1·38 which is greater than the refractive index of air, but

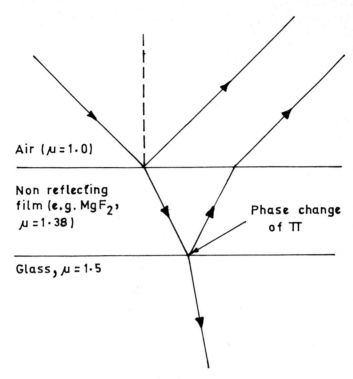

Air $(\mu = 1.0)$

Non reflecting
film (e.g. MgF$_2$,
$\mu = 1.38$)

Phase change
of π

Glass, $\mu = 1.5$

Fig. 4-27: Unwanted reflections from a glass surface can be reduced by coating the glass with a thin layer of MgF$_2$ (Phase change of π occurs at both the reflections).

smaller than the refractive index of a glass.) For such a case, in both the reflections, light is reflected from a medium more dense than that in which it is travelling and the same phase change occurs in each reflection. Thus the optical path difference for destructive interference is $(m+\tfrac{1}{2})\lambda$, or for $m = 0$, the condition is

$$2\mu d = \tfrac{1}{2}\lambda \qquad (4\text{-}35)$$

(We have assumed normal incidence.) Hence, the thickness of a coating needed to produce minimum reflection at the centre of the visible spectrum ($\lambda = 5.5 \times 10^{-5}$ cm) will be given by

$$\frac{5.5 \times 10^{-5}}{4\mu} \text{ cm}$$

If we use magnesium flouride as the coating material the thickness will be about 10^{-5} cm. Of course, some reflection does take place at both longer and shorter wavelengths and the reflected light has a purple colour. The overall reflection from a lens or prism surface can be reduced in this way from 4 to 5 per cent to a fraction of one per cent. The treatment is highly effective in reducing loss of light by reflection in instruments such as periscopes which have a large number of air-glass surfaces.

4-16   Newton's Rings*

If the convex surface of a lens is placed in contact with a plane glass plate, as in Fig. 4-28, a thin film of air is formed between the two surfaces. The

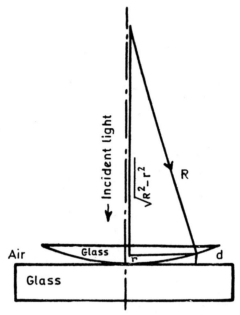

Fig. 4-28: When the convex surface of a lens is placed in contact with a plane glass plate, a thin film of air is formed between the two surfaces. The loci of points of equal thickness are circles concentric with the point of contact.

thickness of this film is very small at the point of contact, gradually increasing as one proceeds outward. The loci of points of equal thickness are circles concentric with the point of contact. When viewed by reflected light, the centre of the pattern is black, as with a thin soap film. However, here it is the ray from the bottom of the (air) film rather than from the

* Isaac Newton (1642-1727). The greatest genius in the history of physics. The ideas contained in his famous book *The Mathematical Foundations of Natural Science* written in Latin and published in 1687 were extended considerably only by Einstein. Newton devoted considerable time to the study of light and embodied the results in his famous *Optiks* the first edition of which was published in 1704. It seems strange that one of the most striking demonstrations of the interference of light, Newton's rings, should be credited to the chief proponent of the corpuscular theory of light.

top that undergoes a phase change π, for it is the one which is reflected from a medium of higher refractive index. The condition for a maximum remains unchanged, and is

$$2d = (m + \tfrac{1}{2})\lambda, \quad m = 0, 1, 2\ldots \tag{4-36}$$

(We have assumed normal incidence.) From Fig. 4-28, it is clear that if R is the radius of curvature of the lower surface of the lens then

$$d.\ (2R - d) = r^2 \tag{4-37}$$

However, if $d \ll 2R$ (which is indeed the case for typical experiments) we will have

$$d \simeq \frac{r^2}{2R} \tag{4-38}$$

Thus, the radius of the mth bright ring will be given by

$$r_m = \sqrt{(m + \tfrac{1}{2})\,\lambda R} \tag{4-39}$$

Similarly, the radii of the dark rings will be equal to $\sqrt{m\lambda R}$. (If white light is used, each spectrum component will produce its own set of circular fringes, with all the sets overlapping.)

Hence by measuring the radii of bright (or dark) rings the wavelength of the light producing them may be calculated or, conversely, if the wavelength is known, one can find the thickness of the air film. A typical experimental arrangement is shown in Fig. 4-29. (The observations are usually made at near normal incidence.) The glass plate G reflects the light down on to the film. After reflection, it is transmitted by G and observed in a microscope M. The radii of the rings (see Plate 4) can be measured by the travelling microscope. Often, because the lens surface is not clean, the central ring is not found to be dark. The wavelength can still be calculated by measuring the radii of two dark (or

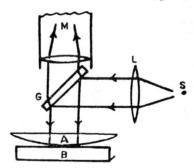

FIG. 4-29: An apparatus for observing Newton's rings.

bright) rings between which there occur p dark (or bright) rings. If these radii are r_m and r_{m+p} then

$$r_{m+p}{}^2 - r_m{}^2 = p\lambda R \quad \text{(independent of } p) \tag{4-40}$$

4-17 Application

The surface of an optical part which is being ground to some desired curvature may be compared with that of another surface, known to be correct, by bringing the two in contact and observing the interference fringes. For example, if a plane surface (an 'optical flat') is desired, a glass plate whose lower surface is accurately plane is placed over the surface to be tested. If both surfaces are accurately plane, the entire area of contact will be dark, or, if contact between them is made only at one edge, a series of straight interference fringes, parallel to the line of contact, will be observed (Problem 9). If the surface being ground is not plane, the interference fringes are curved (see Plate 5). By noting the shape and separation of the fringes, the departure of the surface from the desired form may be determined.

4-18   Michelson* Interferometer

The essential features of the formation of fringes in parallel-sided plates and wedges are the division of a beam of light by partial reflection at the the first surface and the subsequent superposition of the two disturbances after they have traversed unequal optical paths. In the Michelson interferometer the beam is divided (i.e. the amplitude is divided) into two beams of equal intensity by a half silvered plate. The essential optical parts of the Michelson interferometer (Fig. 4-30) consists of two highly

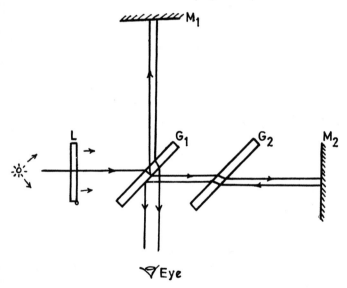

FIG. 4-30:   The Michelson interferometer.

polished plane mirrors, M_1 and M_2, and two plane parallel plates of glass, G_1 and G_2. The reverse of the plate G_1 is lightly silvered so that the light

---

* A. A. Michelson (1852-1931). American physicist of great genius. Michelson was awarded the Nobel Prize in 1907 for the invention of interferometer and spectroscopic and metrological investigations. During the latter part of his life he was Professor of Physics at the University of Chicago, where many of his famous experiments on the interference of light were done.

coming from the source S is divided into (1) a reflected and (2) a transmitted beam of equal intensity. The light reflected normally from the mirror M passes through G_1 a third time and reaches the eye. The light reflected from the mirror M_2 passes back through G_2 for the second time, is reflected from the surface of G_1 and reaches the eye.* The mirror M_1 can be moved along a well machined track by means of a screw. To obtain fringes, the mirrors M_1 and M_2 are made exactly perpendicular to each other by means of screws on mirror M_2. An extended source is used.

Circular fringes are produced with monochromatic light when the mirrors are exactly perpendicular to each other (see Plate 6). Their origin may be understood by referring to Fig. 4-31. Owing to serveral reflections

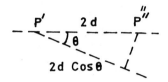

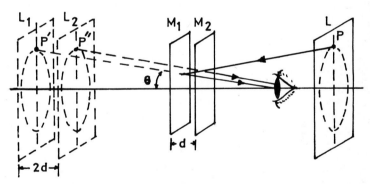

Fig. 4-31: Formation of circular fringes in the Michelson interferometer.

in the interferometer, we may now think of the extended source as being at L, behind the eye, and as forming two virtual images L_1 and L_2 in M_1 and M_2. These virtual sources are coherent in that the phases of corresponding points in the two are exactly the same at all instants. If d is the distance between M_1 and M_2 the virtual sources will be separated by $2d$. The path difference between the two rays coming to the eye from the corresponding points, P' and P'', is $2d \cos\theta$ as shown in Fig. 4-31. Hence when the eye is focused to receive parallel rays, the rays will reinforce each other to produce maxima for those angles θ which

---

* The purpose of the plate G_2, called the compensating plate, is to render the path in glass of the two rays equal. This is not essential for producing fringes in monochromatic light, but it is indispensable when white light is used. (Why?)

satisfy the relation:

$$2d \cos \theta = m\lambda \qquad (4\text{-}41)$$

4-19   Uses of the Michelson Interferometer

(1)   Determination of the Wavelength and Standardisation of the Metre

Suppose the interferometer has been adjusted* to give circular fringes with a nearly monochromatic source, for example a sodium flame. Then if the mirror M_1 is moved a small distance, x cm, then the distance $L_1 L_2$ increases by $2x$ cm. Now a displacement of the mirror by one half of a wavelength causes each fringe to move from its original position to that formerly occupied by the next adjacent fringe. (We have assumed $\cos\theta \approx 1$, which is indeed true for typical experimental arrangements.) Thus the slow displacement of M_1 will cause a continuous shift of the fringes and if the change from bright fringe to bright fringe takes place n times when the mirror is moved x cm, then

$$n\lambda = 2x \qquad (4\text{-}42)$$

Thus, if the distance x can be measured, by means of a micrometer screw, we can determine the wavelength λ.

Conversely (and herein we have an important application of interference) if a source emitting light of known wavelength is used, Eq. 4-42 provides us with an exact means of measuring distances.

In seeking for a permanent standard of length, so that the standard metre might be reproduced if it should ever be destroyed, it was decided that the wavelength of the light emitted by some chosen chemical element in an electrical discharge would be made a standard. After some search, one of the wavelengths emitted by cadmium vapour, lying in the red region of the spectrum, was chosen as the most suitable for this purpose. Making use of the interferometer, Michelson compared the standard metre with the wavelength of these cadmium waves. The value found was

$$1 \text{ metre} = 1{,}553{,}\ 164{\cdot}13 \text{ wavelengths}$$

or

$$\lambda = 6{\cdot}4384696 \times 10^{-7} \text{ m}$$

The precision of this result is better than one part in a million, and it represents one of the most precise physical measurements ever attempted.

(2)   Measurement of Small Differences in Wavelength

If a sodium flame is used as the source of light in a Michelson interferometer and the fringe pattern observed over a wide range of path differences, it is found that the fringes disappear and reappear periodically.

* The details of the adjustment are given by R. S. Longhurst in *Geometrical and Physical Optics* Section 8-9, Second Edition, Longmans (1967).

Between two successive disappearances the mirror has to be moved by 0·289 mm. This corresponds to a path difference of 0·578 mm. This is because a sodium flame emits two wavelengths λ_1 and λ_2 separated by a few Angstrom units. Whenever the

$$\text{Path Difference} = m\,\lambda_1 = (n + \tfrac{1}{2})\,\lambda_2 \qquad (4\text{-}43)$$

bright fringes for λ_1 occur at the same place as dark fringes for λ_2 and uniform illumination results. Now, whenever the path difference is changed by 0·578 mm, we go from one disappearance to the next. Thus, this path difference must contain exactly one more wavelength of λ_2 than of λ_1, or

$$\frac{0\cdot0578}{\lambda_2} - \frac{0\cdot0578}{\lambda_1} = 1$$

(λ measured in cm)

or

$$\lambda_1 - \lambda_2 = \Delta\lambda = \frac{\lambda^2}{0\cdot0578}$$

where

$$\lambda^2 = \lambda_1\lambda_2$$

Using $\lambda = 5\cdot89 \times 10^{-5}$ cm
we get

$$\Delta\lambda = 6\,\overset{\circ}{\text{A}}$$

This method of observing the change in the visibility of fringes with changing path difference was used by Michelson to analyze the spectral purity of other sources. For example, by this method he showed that the 6458 $\overset{\circ}{\text{A}}$ red line of cadmium is highly monochromatic (see also Chapter 5).

(3) Another direct use of Michelson interferometer comes from the fact that the path differences can be made arbitrarily large and the fringes remain visible over a very large path difference. The eventual limitation on the path difference is from coherence effects. This shall be discussed in greater detail in Chapter 5.

Before concluding this chapter it should be mentioned that in this chapter we have been dealing with phenomena which can only be explained in terms of wave theory. Moreover, these very phenomena provide us with various methods of measuring the actual wavelengths of light emitted by a source. Once wavelengths have been evaluated, instruments or optical arrangements by which interference fringes are obtained, enable us to make a great variety of measurements such as the accurate measurement of small distances, the flatness of surfaces and the analysis of the structure of a spectral line.

SUGGESTED READING

JENKINS, F. A., and H. E. WHITE, *Fundamentals of Optics*, III Edition, McGraw-Hill (1957). Chapters 13 and 14 recommended for further studies in interference.

LONGHURST, R. S., *Geometrical and Physical Optics*, II Edition, Longmans (1967). A number of interference phenomena and different kinds of interferometers are discussed in chapters 7, 8 and 9.

SLADKOVA, J., *Interference of Light*, Iliffe (1968). Numerous applications of the phenomenon of interference of light are discussed in the book.

PROBLEMS

1. Two point sources in Fig. 4-32 emit coherent waves. Show that the curves, for which the path differences for rays r_1 and r_2 are constant are hyperbolae.
   Extend the analysis to three dimensions. Show that the path difference cannot exceed d. What will be the shape of the corresponding curve?

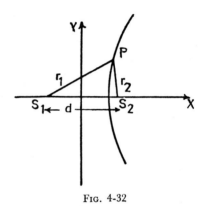

FIG. 4-32

2. In Fig. 4-33 S_1 and S_2 are two coherent point sources, which are at distance d apart. O' is the midpoint of $S_1 S_2$. O is a point equidistant from S_1 and S_2 and lies on the screen xx'. The point O is chosen as the origin and a line parallel to $S_1 S_2$ as the X-axis. The line $O'O$ lies on the Z-axis. The Y-axis is at right

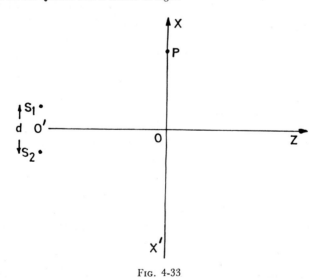

FIG. 4-33

angles to the X- and Z-axes and also perpendicular to the plane of the paper.

(a) Show that the locus of the point P on the screen, such that $S_2 P - S_1 P = \Delta$ is given by

$$x^2 = \left(\frac{d^2}{\Delta^2} - 1\right)^{-1} (y^2 + D^2) + \frac{\Delta^2}{4}$$

(b) For $D >> y$ find the shape of the fringes.

3. In the Fresnel's biprism experiment, a convex lens is put between the biprism and the eyepiece.

(a) Show that for $D > 4f$, there will be two positions of the lens when the image of the slits will be formed at the eyepiece, where D is the distance between the slit and the eyepiece and f, the focal length of the convex lens.

(b) If the distances between the images in the two cases are denoted by d_1 and d_2 then show that

$$d^2 = d_1 d_2$$

where d is the distance between the virtual images, S_1 and S_2.

(c) What will happen if $D < 4f$? Discuss.

4. In Young's double slit experiment, monochromatic light from a point source illuminates two parallel and narrow slits, the centres of the slits being 0·08 cm apart. In the interference pattern formed on the screen the distance between two adjacent bright fringes is 0·0304 cm. The screen is placed parallel to the plane of the slits and is at a distance of 50 cm. Calculate the wavelength of light.   ($\lambda = 4\cdot860 \times 10^{-5}$ cm)

5. In an experiment with Fresnel's biprism sodium light is used and bands 0·0196 cm in width are observed at a distance of 100 cm from the slit. A convex lens is then put between the observer and the prism, so as to give an image of the source at a distance 100 cm from the slit. The distance between the images is found to be 0·70 cm the lens being 30 cm from the slit. Calculate the wavelength of sodium light.   (5880 Å)

6. Fig. 4-34 represents the layout of Lloyd's mirror experiment. S is a point source of monochromatic light emitting waves of frequency 6×10^{14} vibrations/sec. A and B represent the two ends of a mirror placed horizontally. LOM represents the screen. The distances SP, PA, AB, and BO are 1 mm, 5 cm, 5 cm and 190 cm respectively.

(a) Determine the position of the region where the fringes will be visible and calculate the number of fringes.

(b) Calculate the thickness of a mica film ($\mu = 1\cdot5$) which should be introduced in the path of the direct ray so that the lowest fringe becomes the central fringe. The velocity of light is 3×10^{10} cm/sec.   [(a) 2 cm, 40 fringes. (b) 38 micron]

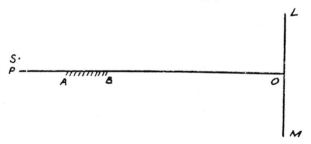

F$_{IG}$. 4-34

7. Fig. 4-35 shows Young's double slit arrangement which is illuminated by monochromatic light of wavelength 5460 Å. The slits S_1 and S_2 are 0·01 cm apart. What is the angular position, θ, of the first minimum and of the tenth maximum. (S is the mid point of $S_1 S_2$).   (0·16° and 3·8°)

8. In the Fresnel's biprism experiment (using monochromatic light of wavelength λ), a thin mica sheet of thickness e, is introduced in the path of one of the beams. Show that the spacing of the interference fringes remains unaffected due to the introduction of the plate;

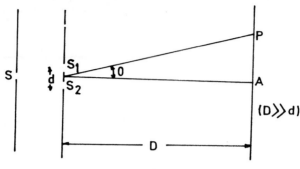

FIG. 4-35

and that the entire fringe system shifts laterally through a distance $D(\mu-1)e/d$ towards the side on which the plate is placed (μ is the refractive index of mica, d, the distance between the sources and D, the distance between the slit and the screen).

9. The two flat plates (shown in Fig. 4-36) are in contact at one edge and are separated at the other by a spacer which is 5×10^{-4} cm thick. The refractive indices for the upper and lower plates are 1·5 and 2·0 respectively. Fringes are observed in reflected light of wavelength 5×10^{-5} cm.

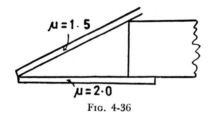

FIG. 4-36

(a) How many fringes are observed? What sort of fringe is seen at the contact point?

(b) If the whole apparatus is immersed in oil of refractive index 1·8, answer the same questions as in (a) (assume normal incidence).

[(a) 21 dark fringes and 20 bright ones (b) 36 dark fringes and 37 bright ones]

10. A plane wave of monochromatic light falls normally on a uniformly thin film of oil which covers a glass plate. The wavelength of the source can be varied continuously. Complete destructive interference of the reflected light is observed for wavelengths of 5×10^{-5} cm and 7×10^{-5} cm and for no wavelengths in between. If the refractive indices of oil and that of glass are 1·30 and 1·50 respectively, find the thickness of the film. $(6.74 \times 10^{-5}$ cm)

11. In the Fresnel's biprism experiment (using white light source) consider a point P on the screen such that

$$S_2P - S_1P = 5 \times 10^{-4} \text{ cm}$$

Calculate the wavelength of those colours in the visible region (i.e. 4×10^{-5} cm $< \lambda < 8 \times 10^{-5}$ cm) for which the point P will correspond to maximum intensity. Also calculate the wavelengths in the visible region for which the point P will have zero intensity. Hence discuss the colour of the light at the point P.

12. (a) Fig. 4-37 shows a source of light placed at a distance b from a Fresnel biprism. Assuming a very small refracting angle α, show that the distance between the two virtual (coherent) sources is approximately $2(\mu-1)b\alpha$ where μ is the refractive index of the material of the prism.

(b) If we use a sodium discharge lamp we have $\lambda \approx 6 \times 10^{-5}$ cm. Calculate the value of α for which the fringe width is 0·03 cm. Assume $c/b = 2$ and $\mu = 1.5$. $(\alpha \approx 20')$

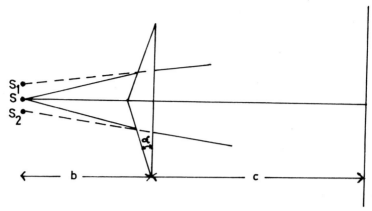

Fɪɢ. 4-37

(c) In the above experimental arrangement, the light intensity at the centre of the central fringe is four times the intensity that would result with a single slit. Why is this so?

13. When a thin sheet of glass (refractive index 1·5) is introduced into the path of one of the beams in a biprism arrangement, the position of the central bright fringe becomes that normally occupied by the fifth. If the wavelength used is 6000 Å, find the thickness of the glass sheet. (6 micron)

14. Two narrow pieces of glass plates are placed together so as to form a thin wedge when held in a vertical position and normally to the light from a yellow flame. Horizontal interference bands are seen. If the distance between the bands is 3 mm and the angle of the wedge is 1×10^{-4} radian, find the wavelength of light used. (6000 Å)

15. White light is incident on two parallel glass plates separated by an air film of 0·001 cm thickness, and the reflected light is examined with a spectroscope. Find the number of dark bands that can be seen in the spectrum between wavelengths 4×10^{-5} cm and 7×10^{-5} cm, when light is incident at an angle of 30° to the normal to the surface. (19)

16. The Newton's rings arrangement is used with a source emitting two wavelengths, $\lambda_1 = 6 \times 10^{-5}$ cm and $\lambda_2 = 4·5 \times 10^{-5}$ cm, and it is found that the nth dark ring due to λ_1 coincides with the $(n + 1)$th dark ring for λ_2. If the radius of curvature of the curved surface is 90 cm, find the diameter of the nth dark ring for λ_1. (2·538 mm)

17. The diameter of the tenth bright ring in a Newton's rings apparatus changes from 1·40 to 1·27 cm as a liquid is introduced between the lens and the plate. Find the index of refraction of the liquid. (1·21)

18. Newton's rings are observed between a plane surface and a lens supported at a variable distance above it. Sodium light is used and it is found that for certain distances the ring system disappears. Explain the phenomenon and calculate the distances of separation for which the fringes would be invisible. Wavelengths for sodium D_1 and D_2 lines are 5896 Å and 5890 Å. (0·0289 cm and 0·0578 cm)

19. Newton's rings are formed by reflection in the air film between a plane glass surface and a spherical surface of radius 50 cm and it is noticed that the centre of the system is bright. What do you conclude from the fact that the centre is bright? If the diameter of the 3rd bright ring is 0·181 cm and the diameter of 23rd bright ring is 0·501 cm, what is the wavelength of light used? (5460 Å)

20. (a) Assume a monochromatic beam of linearly polarized light incident normally on the surface of a very thin film (i.e. $2\mu d < < \lambda$). At a particular instant the electric and magnetic vectors are as shown in Fig. 4-21(a). What will be the directions of the electric and magnetic vectors (at that instant) after the wave has been reflected by

the lower surface of the film [i.e. try to construct a figure similar to Fig. 4-21(b)]?

(b) Consider the same problem when

$$t = \frac{1}{4}\frac{\lambda_0}{\mu}$$

21. Consider two coherent point sources S_1 and S_2 separated by a distance d. The shape of the interference pattern will depend on the relative positions of the coherent sources and the screen. Fig. 4-38 gives two different positions of the screen (C_1 and C_2). In the first case (C_1) the screen is placed at right angles to the perpendicular bisector of S_1S_2 and in the second case (C_2) the screen is placed at right angles to the line joining S_1S_2.

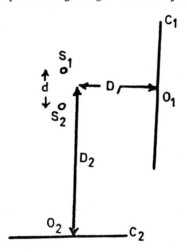

(a) Show that in the vicinity of the point O_1 the interference fringes have the form of equidistant straight lines parallel to one another. (This is what is observed in Young's experimental arrangement, see Problems 1 and 2).

(b) Show that in the vicinity of the point O_2 the interference patterns have the form of concentric circles with radii proportional to the square root of natural numbers. (Assume $\lambda < < d < < D_1,\ D_2$).

22. If the mirror M_2 in Michelson interferometer is moved through $2\cdot33 \times 10^{-2}$ cm, 792 fringes are counted. What is the wavelength of the light?   ($\lambda = 5\cdot88 \times 10^{-5}$ cm)

23. A Michelson interferometer is used with a sodium discharge tube as the light source. The yellow light of sodium lamp consists of two wavelengths, $5\cdot890 \times 10^{-5}$ cm and $5\cdot896 \times 10^{-5}$ cm. It is observed that the interference pattern disappears and reappears periodically as one moves the mirror M_2. Explain this phenomenon and calculate the change in path difference between two successive reappearances of the interference pattern.

24. In the example given in the text on nonreflecting films, we had determined the thickness of the film by assuming $m = 0$ (see Eq. 4-35). What would happen if the thickness corresponds to a large (integral) value of m?

Coherence

In *The War of the Worlds,* written before the turn of the century, H. G. Wells told a fanciful story how Martians invaded and almost conquered earth. Their weapon was a mysterious 'sword of heat', from which flickered 'a ghost of a beam of light'. It felled men in their tracks, made lead run like water and flashed anything combustible into masses of flame. Today Wells sword of heat comes close to reality in the laser....

— Thomas Meloy

5-1 Introduction

The coherence of a wave describes the accuracy with which it can be represented by a pure sine wave. Until now we have discussed optical effects in terms of waves whose wavelength λ and frequency ν can be exactly defined. In this chapter we will discuss the way in which uncertainties and small fluctuations in the frequency and wavelength can affect the observations in optical experiments. Waves which appear to be pure sine waves if they are observed only in a limited space or for a limited period of time are called *partially coherent waves*. We will define two different criteria of coherence. The first criterion expresses the correlation to be expected between a wave at a given time and a certain time later; the other between a given point and a certain distance away. The former leads to the concept of temporal coherence whereas the latter leads to the concept of spatial coherence.

Studies on coherence are particularly useful in interpreting interference experiments with lasers. This is due to the fact that the laser beam has certain features that are not present in other light sources. For example, in lasers, all atoms emit radiation simultaneously and the interval during which the phase of the emitted wave is constant is very much greater than in ordinary light sources. In radiations generated by lasers the phase remains constant during a period of the order of 1/100 or 1/10 sec, in some cases even longer. *Therefore, the interference of radiation emitted by two lasers may be observed even by means of very simple devices.* Indeed by using two different laser beams Magyar and Mandel* have succeeded in recording interference fringes. The two lasers are adjusted until the planes of polarization of the two beams are parallel and a linear polarizer is introduced as a further precaution. In a typical experiment, for $\lambda = 6943$ Å the measured fringe spacing was $2 \cdot 77 \pm 0 \cdot 03$ mm.

5-2 Temporal Coherence: Concept of Path Length

If the radiation field E, from a light source is an ideal sinusoidal function of time, then at any given position we could write

$$E = A \, \text{Cos} \, (\omega_0 t + \delta) \tag{5-1}$$

where A is the magnitude of the field, ω_0, the angular frequency and δ, the phase. As a function of time, the field appears as shown in Fig. 5-1.

FIG. 5-1: The ideal sinusoidal variation of the electric field with time.

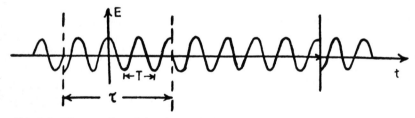

FIG. 5-2: The variation of the electric field with time where the phase undergoes abrupt changes after time intervals of the order of τ.

However, no emitted light produces a perfect sinusoidal variation of the field, and for any actual light source the magnitude A and the phase δ will vary with time. Therefore, a more realistic picture of the radiation from an emitting source might be as illustrated in Fig. 5-2. In Fig. 5-2 we have shown a time τ which represents the average time duration for which an ideal sinusoidal emission occurs. The period of the oscillation

* G. Magyar and L. Mandel, 'Interference Fringes Produced by Superposition of Two Independent Maser Light Beams', *Nature*, Vol. 198, p. 255, (1963).

is designated as T, where $\omega_0 T = 2\pi$. The spatial dimension, L, for which the light may be considered to be a perfect sinusoid is given by

$$L = \tau c = \frac{\tau}{T} \lambda \qquad (5\text{-}2)$$

where c is the velocity of light, and L is termed as the coherence length.

We may obtain a measure of the coherence length by means of the Michelson interferometer (see Chapter 4) shown in Fig. 5-3. Emission from a light source is split into two parts of approximately equal intensity, one part is reflected from a fixed mirror and the other from a movable mirror. The sum of these two reflections is then viewed. We know that two light waves produce a stationary interference pattern only if there is a definite amplitude and phase relationship between them.

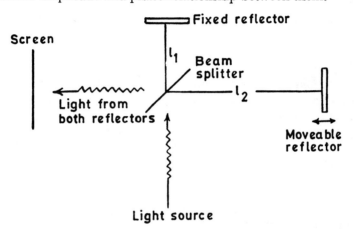

Fig. 5-3: Emission from a light source is split into two parts of approximately equal intensity, one part is reflected from a fixed mirror and the other from a moveable mirror.

If the path difference between these two beams is of the order of L, then there will not be any fixed phase relationship between the two interfering beams and the fringe pattern will disappear.* In other words, interference fringes will appear on the screen only if the difference in the two optical paths is less than the coherence length. It should be noted that the coherence length is not a very well defined quantity; its value is determined within a factor of 2 or so. Thus, when we start with equal path lengths interference fringes are seen and as we move one mirror the contrast of the fringes diminishes and eventually the fringes disappear. There is no definite cut off point at which the fringes suddenly lose distinctness.

The term τ is known as the *interval of coherence* such that phase correlation exists over time intervals $\Delta t \ll \tau$ but does not exist for $\Delta t \gg \tau$,

* In a typical Michelson interferometer experiment (Fig. 5-3), $L = 2l_1 \sim 2l_2$

the fading out of the circular fringes can be interpreted to mean that the time difference $\dfrac{L}{c}$ is comparable to τ.

The light from a sodium lamp has a coherence length of only 2-3 cm. Thus

$$\tau \sim \frac{L}{c} \sim 10^{-10} \text{ sec} \tag{5-3}$$

Although this is a short time interval in macroscopic terms, the number of oscillations for which the field remains coherent is

$$\frac{\tau}{T} = \frac{L/c}{\lambda/c} = \frac{L}{\lambda} \sim \frac{2\cdot 5 \text{ cm}}{5 \times 10^{-5} \text{ cm}} \sim 5 \times 10^4 \text{ cycles} \tag{5-4}$$

5-3 The Purity of a Spectral Line

The concept of coherence length is directly related to the *purity of a spectral line*. A perfectly sharp monochromatic line corresponds to a perfect sinusoid and therefore has an infinite value for τ (Fig. 5-1). However, as we have discussed, for every spectral line there is an upper limit on the distance of separation of mirrors beyond which the interference patterns cannot be obtained. This can be interpreted by supposing that even the best monochromatic sources emit a continuous distribution of wavelengths in some narrow interval between λ and $\lambda + \Delta\lambda$. When the path difference is small, the circular fringes for all the contributing wavelengths are effectively coincident. But as the path difference increases, the rate of expansion of the circles and the rate of production of new fringes at the centre is different for each wavelength between λ and $\lambda + \Delta\lambda$. We can adopt as *an order of magnitude criterion* of the optical path difference L, which causes the extremes λ and $\lambda + \Delta\lambda$ to produce fringe systems which are exactly out of step at the centre. If the centre is a black fringe for λ and a bright fringe for $\lambda + \Delta\lambda$, then (cf. Eq. 4-46)

$$L = m\lambda = (m - \tfrac{1}{2})(\lambda + \Delta\lambda) \tag{5-5}$$

or

$$L = \left(\frac{L}{\lambda} - \frac{1}{2}\right)(\lambda + \Delta\lambda)$$

or

$$\Delta\lambda = \frac{\lambda}{\left(\dfrac{2L}{\lambda} - 1\right)}$$

or

$$\Delta\lambda \sim \frac{\lambda^2}{2L} \tag{5-6}$$

where we have assumed $L \gg \lambda$ (see Eq. 5-4). Thus, if the fringes become noticeably indistinct when the optical path difference exceeds L, we can interpret this to mean that a spread of wavelengths is present having a line width given by Eq. 5-5.

Michelson found that the cadmium red line ($\lambda = 6438$ Å) is one of the most ideal monochromatic sources available, allowing fringes to be discerned to values of $L \sim 30$ cm. This corresponds to a wavelength spread

$$\Delta\lambda \sim \frac{\lambda^2}{2L} \sim \frac{(6 \cdot 4)^2 \times 10^{-10}}{60} \sim \cdot 007 \text{ Å}$$

This line was therefore used for the standardisation of the metre (see Chapter 4, Sec. 4-19).

We have thus given two alternative interpretations of the washing out of the interference pattern, viz. the concept of the temporal coherence and of the purity of the spectral line. However, these can be shown to be equivalent through the technique of Fourier analysis. The function shown in Fig. 5-2 can be regarded as a superposition of perfect sinusoidal waves having a variety of frequencies. The important contributions to this superposition come from frequencies close to the apparent frequency of the nearly sinusoidal wave. The frequency spread $\Delta\nu$ turns out to be inversely proportional to the coherence interval τ and obeys an order of magnitude relation

$$\tau \, \Delta \, \nu \sim 1 \tag{5-7}$$

Eq. 5-7 can also be derived from the qualitative arguments made above, if we assume the equivalence of the two interpretations. Since

$$\nu = \frac{c}{\lambda}$$

therefore (disregarding the sign)

$$\Delta \, \nu \sim c \, \frac{\Delta\lambda}{\lambda^2} \sim \frac{c}{2L}$$

or

$$\tau \, \Delta \, \nu \sim \left(\frac{L}{c}\right)\left(\frac{c}{2L}\right) \sim \frac{1}{2} \tag{5-8}$$

Eq. 5-8 implies that an infinitely sharp spectral line ($\Delta\nu = 0$) is associated with an infinite interval of coherence ($\tau = \infty$). That is, the ideal sinusoidal wave has no frequency spread and remains coherent over indefinitely long intervals of time. A decrease in the interval of coherence is associated with an increase in the breadth of a spectral line.

We have mentioned above that the light from a sodium lamp has a coherence length of only 2-3 cm. (This corresponds to $\Delta\lambda \sim 0 \cdot 06$ Å, implying a spectral purity of the order of one part in 10^6.) On the other hand a laser, in addition to its producing high power levels in a highly collimated beam, has an outstanding characteristic of having a degree of spectral purity much higher than that possible with a conventional light source. In a typical Michelson interferometer experiment using a helium-neon gas laser clear circular fringes could be obtained even for optical path differences equal to 9 metres, showing that the interval of coherence had not yet been exceeded. Experiments of a different type* show that the spectral purity is of the order of one part in 10^{14} for $\lambda = 11 \cdot 53 \times 10^{-5}$ cm. Thus, using Eq. 5-6

$$L \sim \left(\frac{\lambda}{\Delta\lambda}\right)\left(\frac{\lambda}{2}\right) \sim \left(\frac{10^{14}}{1}\right)\left(\frac{11 \cdot 53 \times 10^{-5}}{2}\right) \sim 6 \times 10^{9} \text{ cm} \qquad (5\text{-}9)$$

This corresponds to

$$\tau \sim \frac{L}{c} \sim \frac{1}{5} \text{ sec}$$

Example

In a Michelson interferometer experiment, the position of zero retardation is located by means of white light fringes. The white light source is replaced by a mercury arc, and a filter for the green line ($\lambda = 5461$ Å) is used. With increase in the optical path difference the visibility of fringes becomes poorer. After approximately 1000 fringes have passed, the pattern is difficult to perceive and does not improve with increasing path difference. Calculate the order of magnitude of the wavelength spread and the associated coherence interval.

Solution

$$L \sim 1000 \, \lambda \sim 5 \times 10^{-2} \text{ cm}$$

Therefore

$$\Delta\lambda \sim \frac{\lambda^{2}}{2L} \sim 2 \cdot 7 \text{ Å}$$

$$\tau \simeq \frac{L}{c} \sim 2 \times 10^{-12} \text{ sec}$$

and

$$\Delta\nu \sim \frac{1}{\tau} \sim 5 \times 10^{11} \text{ cycles/sec}$$

It is interesting to point out that in the case of an interference experiment with white light in which the eye is used to detect the interference fringes, we must consider the spectral sensitivity of the eye. This is

---

* See the discussion on optical beats (Section 5-7). The results are quoted from G. Magyar and L. Mandel 'Interference Fringes Produced by Superposition of Two Independent Maser Beams', *Nature*, Vol. 198, p. 255, (1963).

maximum at about 5500 Å and falls to zero at approximately 4000 Å and 7000 Å. Therefore, as far as the eye is concerned, the spectral "width" of a white light source is about 1500 Å and the corresponding coherence length is about 3 or 4 wavelengths. This is about the number of fringes that can be seen on either side of the zero fringe in the Michelson interferometer when using a source of white light such as a tungsten lamp.

5-4 Spatial Coherence

Until now we have considered the problem of coherence between two fields arriving at the same point in space, through different optical paths. We will now discuss the coherence between two fields at different points in space. This will be of importance in studying the coherence of the radiation fields of extended sources.

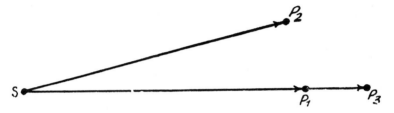

FIG. 5-4: Diagram to illustrate lateral and longitudinal coherence.

Suppose we have a single quasi-monochromatic source S (Fig. 5-4). Let the electric fields at the points P_1, P_2 and P_3 be E_1, E_2 and E_3 respectively. The two points P_1 and P_3 lie in the same direction from the source. They only differ in their distances from the source. The coherence between the fields E_1 and E_3 measures the *longitudinal spatial coherence* of the field. Obviously, the longitudinal coherence will merely depend on how large P_1P_3 is in comparison with the coherence length of the source, or equivalently, on the value of P_1P_3/c compared to the interval of coherence, τ. For whatever $E_1(t)$ is, $E_3(t)$ will always vary in the same way but at a time P_1P_3/c later. If $P_1P_3/c \ll \tau$, there will be high coherence between E_1 and E_3 whereas if $P_1P_3/c \gg \tau$ there will be little or no coherence.

Next, we consider the receiving point P_2 located at the same distance from S as P_1. In this case, the coherence between E_1 and E_2 measures the *lateral spatial coherence* of the field. Evidently, if S is a true point source, then the time dependence of the two fields E_1 and E_2 will be precisely the same, i.e. they will be mutually coherent. Incoherence between E_1 and E_2 will occur if the source has spatial extension rather than being a point. In order to understand this, let us go back to the Young's double slit experiment (Fig. 5-5). We had found that interference effects from ordinary light sources may be produced by putting a very narrow slit

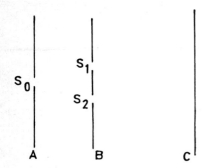

Fig. 5-5: Young's double slit experiment producing an interference pattern on C.

(S_0 in Fig 5-5) directly in front of the source. This ensures that the wavetrains that strike slits S_1 and S_2 (in screen B) originate from the same small region of the source. The beams emerging from S_1 and S_2 are coherent with respect to each other. If the phase of the light emitted from S_0 changes, this change is transmitted simultaneously to S_1 and S_2. Thus, at any point on screen C, a constant phase difference is maintained between the beams from these two slits and a stationary interference pattern occurs.

If the width of slit S_0 in Fig. 5-5 is gradually increased it will be observed experimentally that the maxima of the interference fringes become reduced in intensity and that the intensity in the fringe minima is no longer strictly zero. In other words the fringes do not remain very sharp and become less distinct. If S_0 is extremely wide, the lowering of the maximum intensity and the raising of the minimum intensity will be such that the fringes disappear, leaving only a uniform illumination. Under these conditions we say that *the beams from S_1 and S_2 pass continuously from a condition of complete coherence to one of complete incoherence.* Between these two limits, the beams are said to be partially coherent. What actually happens is this that if the source is so broad that one slit is illuminated mostly by one set of atoms and the other by another independent set, then the two slits are completely incoherent, i.e. their phases are uncorrelated. On the other hand, for a narrow source S_0, at a particular instant, the two slits are illuminated by the radiation emitted by one particular set of atoms.

We shall now derive a mathematical expression relating the coherence of the field to the size of the source. Since an extended source can be considered to be made up of many independent point sources, it will be convenient to study the case of two point sources which are mutually incoherent. Let S and S' be the positions of two incoherent sources (Fig. 5-6). We will calculate the minimum distance between S and S' for which the interference pattern on the screen will be washed out. Whenever the point S' is such that

$$S'S_2 - S'S_1 = \tfrac{1}{2}\lambda$$

the interference pattern due to S and S' will be out of step, i.e. the points at which S produces a bright fringe, S' will produce a dark fringe and vice versa.* From Fig. 5-6

---

* We are assuming that S and S' are incoherent sources and therefore on the screen, we will have to add up the intensities produced by each source (cf. Eq. 4-25).

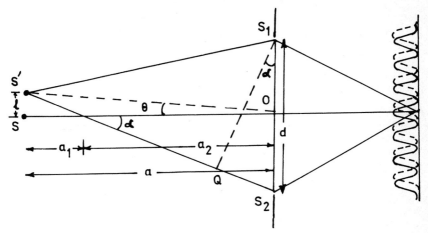

FIG. 5-6: Effect of two point sources on the double slit interference pattern.

$$S'S_2 - S'S_1 \approx S_2Q \approx \alpha d$$

But

$$\alpha \approx \frac{d/2}{a_2} = \frac{l}{a_1} \tag{5-10}$$

where l is the distance between the two sources S and S'. Thus

$$a = a_1 + a_2 \approx \left(l + \frac{d}{2}\right)\frac{1}{\alpha}$$

or $\tag{5-11}$

$$\alpha \approx \left(l + \frac{d}{2}\right)\frac{1}{a}$$

Therefore

$$S'S_2 - S'S_1 \approx \left(l + \frac{d}{2}\right)\frac{d}{a} \approx \frac{ld}{a}$$

where we have assumed $l \gg \frac{d}{2}$. Finally, for the interference pattern to be washed out

$$\frac{ld}{a} \approx \tfrac{1}{2}\lambda \tag{5-12}$$

Evidently, if we have an extended source, the spatial extension of which exceeds $\frac{\lambda a}{2d}$ then the interference pattern will not be observed on the screen.

We can rewrite Eq. 5-12 as

$$d \approx \tfrac{1}{2}\frac{\lambda}{l/a} \approx \frac{\lambda}{2\theta} \tag{5-13}$$

where θ is the angle that SS' subtends at 0. The quantity $\dfrac{\lambda}{\theta}$ is known as the lateral coherence width l_w. Evidently, if one were to perform the Young's double slit interference experiment then the distance between the two slits would have to be much less than the lateral coherence width in order to obtain distinct interference fringes.

Example

For a circular source the lateral coherence width is given by*

$$l_w = \frac{1 \cdot 22\lambda}{\theta} \tag{5-14}$$

Consider a pinhole of 1 mm diameter (Fig. 5-5). What should be the distance between S_1 and S_2 to obtain distinct interference fringes (Assume $SS_1 = SS_2 \approx 100$ cm and $\lambda \approx 5 \times 10^{-5}$ cm).

Solution

The angle θ will be given by

$$\theta \approx \frac{0 \cdot 1}{100} = 10^{-3} \text{ radians}$$

Therefore

$$l_w \approx \frac{1 \cdot 22 \times 5 \times 10^{-5}}{10^{-3}} \approx 0 \cdot 06 \text{ cm}$$

Thus the distance between $S_1 S_2$ should be small compared to $0 \cdot 6$ mm. (The corresponding distance for a $0 \cdot 1$ mm diameter pinhole would be about 6 mm.)

Thus if one were to perform an interference experiment in which a double slit aperture was used, as in Young's experiment, then the distance between the slits would have to be less than the lateral coherence width in order to obtain distinct interference fringes.

5-5 Michelson's Stellar Interferometer

A good application of the above concept lies in the determination of the angular diameter of a distant object such as a star. By using a variable double-slit interference arrangement as shown in Fig. 5-7, the lateral coherence width, l_w, can easily be found — it is merely the slit separation that results in the disappearance of interference fringes. The angular diameter of a star is then given by $1 \cdot 22\lambda / l_w$.

Due to their large distances, the angular diameters of stars are extremely small, of the order of a fraction of a second of arc. Hence the separation

---

* The relation is derived in M. Born and E. Wolf's *Principles of Optics*, Section 10.4.2, Fourth Edition, Pergamon Press (1970).

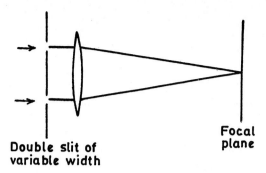

Double slit of variable width

Focal plane

FIG. 5-7: Interference fringes from a distant source.

of the holes may need to be very large before the disappearance of the fringes is osberved. This has two disadvantages: First, with the holes well apart the interference fringes become extremely close and secondly, a telescope of very large aperture is needed, although the greater part of the expensive lens is never used. These difficulties were overcome in an ingenious manner by Michelson by using a mirror system shown in Fig. 5-8. The coherence measured is clearly that between the light at

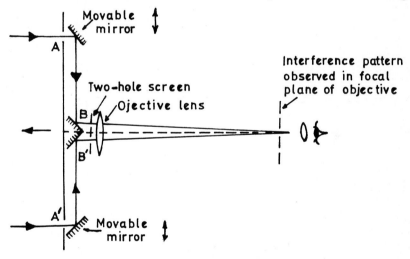

FIG. 5-8: Michelson's stellar interferometer.

apertures A and A', which can conveniently be mounted on racks to alter their separation. The interference pattern observed is that arising from the two apertures B and B', and therefore the fringe width can be made conveniently large by making the holes B and B' close together. Moreover the dimensions of the pattern do not change as A and A' are separated. For example in the case of the star Arctures, the first disappearance of the fringes occurred for $AA' = 24$ ft. Thus the angular diameter of the star would be given by

$$\alpha = \frac{1 \cdot 22 \times 5 \times 10^{-5}}{24 \times 12 \times 2 \cdot 54} \text{ radians}$$

$$= \frac{6 \cdot 1 \times 10^{-5}}{24 \times 12 \times 2 \cdot 54} \times \frac{180 \times 60 \times 60}{\pi} \text{ sec}$$

$$\approx 0 \cdot 02 \text{ sec}$$

From the known distance of Arctures, one then finds that its actual diameter is 27 times that of the sun.*

5-6 An Experiment with a Laser Beam

With a laser beam it is not necessary to use a pinhole to obtain interference fringes because the laser beam is essentially monochromatic and spatially coherent.

In order to carry out the experiment we place the slits in the path of the laser beam in such a way that the beam fills both the slits (Fig. 5-9).

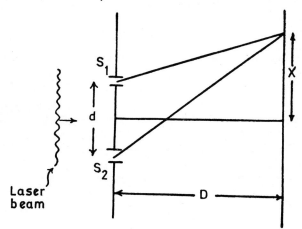

Fig. 5-9: A measurement of spatial coherence, as determined by the interference pattern between the light beams passing through slits S_1 and S_2.

In a dark room the screen should be placed several metres away from the slits, so that the fringes on the screen will be sufficiently far apart to measure the separation with reasonable accuracy.

With a given slit pair, cover one slit without disturbing the laser beam impinging on the other slit. The fringes will disappear, since the fringe pattern results from the interference of waves emerging from each slit. The pattern that appears on the screen is the single slit diffraction pattern and will be discussed in Chapter 6.

* For further details see A. A. Micnelson, *Studies in Optics*, Chapter 11, University of Chicago Press, (1927).

The spatial coherence of the laser light may be demonstrated by moving the laser so that the beam scans across the two slits. It will be observed that as long as the beam fills both slits, the fringe pattern remains fixed regardless of which portion of the beam is intercepted by the slits. *This demonstrates that there is always a fixed phase relationship between the beam of light that enters each slit, and that the laser light is spatially coherent for the given slit separation across the entire beam.*

5-7 Optical Beats

One can obtain interference between waves of different frequencies leading to the phenomenon of beats. Consider at a particular point we have two disturbances of same amplitude A_0 with frequencies ω_1 and ω_2 and with initial phases ϕ_1 and ϕ_2 respectively. Then, from the superposition principle, the resultant frequencies of the two will be given by

$$y(t) = A_0 \, \text{Sin} \, (\omega_1 t - \phi_1) + A_0 \, \text{Sin} \, (\omega_2 t - \phi_2) \tag{5-15}$$

$$= 2A_0 \, \text{Cos} \, [(\Delta \omega)t - \phi] \, \text{Sin} \, (\omega_0 t - \phi_0) \tag{5-16}$$

where

$$\Delta \omega = \frac{\omega_1 - \omega_2}{2}, \quad \omega_0 = \frac{\omega_1 + \omega_2}{2} \tag{5-17}$$

and

$$\phi = \frac{\phi_1 - \phi_2}{2}, \quad \phi_0 = \frac{\phi_1 + \phi_2}{2} \tag{5-18}$$

In Eq. 5-16, the quantity in sqare brackets represents the envelope of the superposed wave, and the second term represents a sinusoidal wave with the average frequency ω_0 (Fig. 5-10). For sound waves, this phenomenon is the familiar one of *beats*. When two tuning forks, one having a frequency of 440 cps and the other 442 cps are sounded together, the ear perceives the superposition of these waves as a sound with a frequency of 441 cps with an intensity which varies from zero to maximum and back with a frequency of 2 cps. We note that this is just twice the frequency $\Delta \nu = 1$ cps with which the amplitude factor varies, since the ear perceives only the magnitude of the amplitude, not its sign.

The interference between waves for different frequencies holds for optical phenomena as well. Suppose we have two plane polarized light waves producing electric fields E_1 and E_2. Let both the waves be polarized in the same direction, say along the z-axis. The total field at a particular point is the superposition of E_1 and E_2. Since both the waves are polarized in the same direction, we may drop the vector signs; and, therefore, to obtain the resultant field E we may just add the two fields

$$E = E_1 \, \text{Sin} \, (\omega_1 t - \phi_1) + E_2 \, \text{Sin} \, (\omega_2 t - \phi_2) \tag{5-19}$$

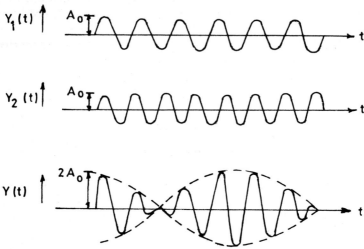

FIG. 5-10: Superposition of two sinusoidal waves with slightly different frequencies results in a wave that appears locally sinusoidal but whose amplitude varies from zero to twice the amplitude of each individual wave. The locally sinusoidal nature is described by the average frequency $(\omega_1 + \omega_2)/2$, while the envelope (shown by dotted curve) has a slower variation with time given by the 'beat frequency' $(\omega_1 \sim \omega_2)/2$.

$$= \frac{E_1 + E_2}{2} \left[\text{Sin} \ (\omega_1 t - \phi_1) + \text{Sin} \ (\omega_2 t - \phi_2) \right]$$

$$+ \frac{E_1 - E_2}{2} \left[\text{Sin} \ (\omega_1 t - \phi_1) - \text{Sin} \ (\omega_2 t - \phi_2) \right]$$

$$= (E_1 + E_2) \ \text{Cos} \ [(\Delta\omega) \ t - \phi] \ \text{Sin} \ (\omega_0 t - \phi_0)$$

$$+ (E_1 - E_2) \ \text{Sin} \ [(\Delta\omega) \ t - \phi] \ \text{Cos} \ (\omega_0 t - \phi_0) \quad (5\text{-}20)$$

where $\Delta\omega$, ω_0, ϕ, and ϕ_0 are same as defined in Eqs. 5-17 and 5-18. The terms Sin $(\omega_0 t - \phi_0)$ and Cos $(\omega_0 t - \phi_0)$ represent the 'fast' oscillations at the average frequency ω_0, whereas the terms Cos $[(\Delta\omega) t - \phi]$ and Sin $[(\Delta\omega) t - \phi]$ represent the 'slow' oscillations at the beat frequency. The energy density, which can be measured by a photocell (whose output current is proportional to the incident light intensity), is proportional to the average of $E^2(t)$ over one period $T \left(= \dfrac{2\pi}{\omega_0} \right)$ of the 'fast' oscillations at the average frequency ω_0:

$$<E^2 (t)> = (E_1 + E_2)^2 \ \text{Cos}^2 \ [(\Delta\omega)t - \phi] \ \tfrac{1}{2}$$

$$+ (E_1 - E_2)^2 \ \text{Sin}^2 \ [(\Delta\omega)t - \phi] \tfrac{1}{2} \quad (5\text{-}21)$$

where, in carrying out the average we have assumed $\text{Cos}^2 [(\Delta\omega)t - \phi]$ and $\text{Sin}^2 [(\Delta\omega)t - \phi]$ to remain constant* during the short time T, and have used

* This will be justified when $\Delta\omega \ll \omega_0$. (Why?)

$$\langle \text{Cos}^2 (\omega_0 t - \phi) \rangle = \langle \text{Sin}^2 (\omega_0 t - \phi) \rangle = \tfrac{1}{2} \qquad (5\text{-}22)$$

and
$$\langle \text{Sin} (\omega_0 t - \phi) \, \text{Cos} (\omega_0 t - \phi) \rangle = 0 \qquad (5\text{-}23)$$

Rearranging Eq. 5-19, we obtain

$$\langle E^2(t) \rangle = \tfrac{1}{2}[E_1^2 + E_2^2 + 2E_1 E_2 \, \text{Cos} \{2 \, (\Delta\omega)t - 2\phi\}]$$
$$= \tfrac{1}{2}[E_1^2 + E_2^2 + 2E_1 E_2 \, \text{Cos} \{(\omega_1 - \omega_2)t + (\phi_1 - \phi_2)\}] \qquad (5\text{-}24)$$

From Eq. 5-20 we can see that the superposition signal is modulated sinusoidally at the difference frequency $(\omega_1 - \omega_2)$. As long as the phases $\phi_1 (t)$ and $\phi_2(t)$ are constant the beat note remains steady. In the absence of other effects, the duration of a steady beat note is therefore, a measure of the coherence time (see also Problem 5).

The first demonstration of beats resulting from the superposition of incoherent beams was given by Forrester *et al.* in 1955.* Later, with development of the laser beams such beating experiments became easier to perform. Javan *et al.*† and Lipsett and Mandel‡ determined the coherence times of the helium-neon laser and the ruby laser respectively, by superposition of beams from two independent sources.

The layout of the experimental arrangement of Lipsett and Mandel is shown in Fig. 5-11. The two light beams are sampled individually at photocells 1 and 2 and are then combined by a beam splitter that

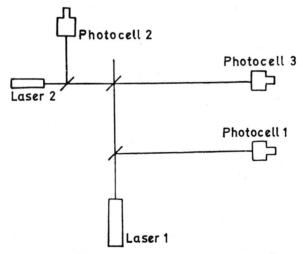

Fig. 5-11: The layout of the experimental arrangement of Lipsett and Mandel for observing optical beats.

* A. T. Forrester, R. A. Gudmundsen and P. O. Johnson 'Photoelectric Mixing of Incoherent Light', *Physical Review*, Vol. 99, p. 691 (1955).

† A. Javan, E. A. Ballik and W. L. Bond, 'Frequency characteristics of a Continuous-Wave He-Ne Optical Maser', *Journal of the Optical Society of America*, Vol. 52, p. 96 (1962).

‡ M. S. Lipsett and L. Mandel, 'Coherence Time Measurements of Light from Ruby Optical Masers', *Nature*, Vol. 199, p. 553 (1963).

directs the superposed beams to photocell 3. The actual apparatus is reminiscent of a Michelson interferometer. (In the experiment, the beams are superposed to within 0·03 mm and aligned within 0·1 milli-radians.)

The maximum frequency $|\nu_1 - \nu_2|$ that the system can detect is more than 60 megacycle/sec and is limited mainly by the response of the oscilloscope. Frequencies below about 10 megacycle/sec tend to be obscured by the spike envelopes (Fig. 5-12).*

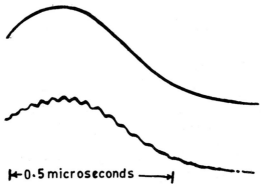

FIG. 5-12: Oscilloscope trace of the sum of the intensities of the laser beams (upper) and the intensity of the superposed laser beam (lower). (From M. S. Lipsett and L. Mandel, *Nature*, **199**, p. 553, 1963).

A typical beat note of Lipsett and Mandel's experiment is reproduced in Fig. 5-12 and shows a modulation on the lower trace. The beatnote can be seen to persist for the full duration of the spike, drifting in frequency from 33 megacycle/sec to about 21 megacycle/sec in the course of 0·7 microsecond. No systematic trend for the beat frequency to increase or decrease was found from other similar traces. The coherence time (Fig. 5-12) is of the order of $\frac{1}{2}$ microsecond, which is also the half width of the spike.

We may conclude that two different light beams do interfere but we do not readily see any interference pattern using ordinary light sources because of their short coherence time.

Soon, no doubt, with the availability of laser sources, someone will be able to demostrate two sources shining on a wall, in which the beats are so slow that one can see the wall get bright and dark!

---

* For the laser pulse used by Lipsett and Mandel amplitudes E_1 and E_2 (Eq. 5-19) remained constant within a time shorter than the coherence time of the light. (This is typical for a solid laser.) Therefore, the oscilloscope trace of the sum of the intensities of the laser beams changes with time having a half width of about $\frac{1}{2}$ microseconds (Fig. 5-12).

SUGGESTED READING

BORN, M. and E. WOLF, *Principles of Optics,* Fourth Edition, Pergamon Press (1970). Chapter X gives a detailed but highly mathematical account on interference and diffraction with partially coherent light.

FOWLES, G. R. *Introduction to Modern Optics,* Holt, Rinehart and Winston (1968). Chapter 3 on Coherence and Interference is directly related to the concepts developed in this chapter. The treatment is mathematical but on a level which is simpler than that given by Born and Wolf.

LIPSON, S. G. and H. LIPSON, *Optical Physics,* Cambridge University Press (1969), Chapter 8 gives a good account on coherence. At the end of the chapter the statistics of photon emission is also discussed.

MANDEL, L. and E. WOLF, 'Coherence Properties of Optical Fields', *Reviews of Modern Physics,* Vol. 37, p. 231 (1965). Section 7 gives a brief review of all the interference experiments performed using laser beams. The detailed mathematical interpretations of the experiments have also been given.

PANTELL, R. H. *Experiments in Optics with a Laser Source,* Optics Technology Inc. (1969). This booklet is intended as a supplement to an undergraduate lecture and laboratory course in optics. Elegantly written, the booklet discusses some interesting experiments in optics performed by using a continuous helium-neon gas laser.

SCHAWLOW A. L. 'Optical Masers', *Scientific American,* June, 1961. The article gives a good description of the structure, properties and some important scientific applications of lasers.

SCHAWLOW, A. L. (Editor) *Lasers and Light,* W. H. Freeman and Co. (1969). This is a collection of 32 articles from *Scientific American* on various aspects of light.

TOWNE, D. H. *Wave Phenomena,* Addison-Wesley (1967). Chapter 11 on 'Interference pattern from a pair of point sources' is particularly relevant for this chapter.

PROBLEMS

1. It was shown that the coherence length is given by $L = \tau c$. Further, the coherence time τ is related to $\Delta \nu$ by the relation

$$\tau \Delta \nu \sim 1$$

   Using this relation prove that $L \sim \lambda^2 / \Delta \lambda$.

2. Suppose we have a filter that transmits wavelengths 5000 ± 0.5 Å. If this filter is placed in front of a source of white light what is the coherence length of the transmitted light?

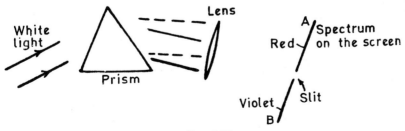

FIG. 5-13

3. Let a continuous spectrum be formed on the screen AB (Fig. 5-13) with linear dispersion of 20 Å/mm, i.e. in traversing 1 mm of the screen the wavelength changes by 20 Å. (This spectrum may be formed by a prism on the focal plane of a lens.) By making a small

slit on the plane of the screen one can obtain an almost monochromatic beam. **For an** exit slit of width 0·02 cm, what is the coherence time and the coherence length of **the** light with mean wavelength of 5×10^{-5} cm.   ($L = 0.0625$ cm; $\tau = 2.08 \times 10^{-12}$ sec)

4. The largest star measured by Michelson with his stellar interferometer was the red giant Betelgeuse. Complete disappearance of the fringes first occurred when the mirrors were 25 inches apart. Compute the angular diameter in seconds of the star disk (Assume $\lambda = 5.7 \times 10^{-5}$ cm).

5. Show that, in order to observe beats, we should have the following requirements of the coherence time

$$\tau_1 > \frac{1}{|\omega_1 - \omega_2|}$$

and

$$\tau_2 > \frac{1}{|\omega_1 - \omega_2|}$$

i.e. we require both coherence times to be long compared with the beat period.

6. Why is it easy to observe beats with two different tuning forks vibrating with slightly different frequencies? What is their coherence time?

7. If we have two pinholes which are separated by less than 2×10^{-3} cm and these are placed directly in front of the sun show that the interference pattern will be observed on the screen (assume $\lambda = 6000$ Å and the angular diameter of the sun = 0·5°).

8. What is the line width in Angstroms of the light from a laser source whose coherence length is 10 Km? The mean wavelength is 6328 Å. What is the spread in frequency?

9. What is the coherence length for a good mercury arc that emits 5461 Å with a line width of 0·006 Å. How does this compare with the laser whose coherence length is 100 cm?

10. A pinhole of diameter 0·5 mm is used as a source for the Young's interference experiment (Fig. 4-8 of Chapter 4) using a sodium lamp ($\lambda = 5.89 \times 10^{-5}$ cm). If the distance from the pinhole to slits is 100 cm what is the maximum spacing such that the interference fringes are just visible.   (0·07 cm)

11. Calculate approximately the width of the slit (in the Fresnel's biprism experiment) for which the fringe pattern will remain distinct.

12. Suppose we consider a He-Ne laser ($\lambda = 6328$ Å) that is pulsed on, by means of a fast shutter, for 0·01 nanosecond. If a filter is placed in front of this light what must be the minimum transmission band width for the filter if most of the light is to pass through?

Diffraction

But, soft! What light through yonder window breaks....

— *Romeo and Juliet*

6-1 Introduction

This chapter is continuation of the chapter on interference. According to Feynman, no one has ever been able to define the difference between interference and diffraction satisfactorily. It is a question of usage, and their is no specific, important physical difference between them. The best we can do, is to say that when there are only a few interfering sources, say two, then the result is usually called interference, but if there are a number of interfering sources, the word diffraction is commonly used. So we will not worry about whether it is interference or diffraction, but continue from the chapter on interference.

We will, therefore, discuss the situation where there are n equally spaced sources, all of equal amplitude but different from one another in phase, either because they are driven differently in phase, or because they are viewed from an angle such that there is a difference in time delay. For one reason or another in order to obtain the resultant disturbance we have to add something like this:

$$R = A\,[\text{Cos}\ \omega t + \text{Cos}\ (\omega t + \phi) + \text{Cos}\ (\omega t + 2\phi)$$
$$+ \ \ldots\ldots\ + \text{Cos}\ \{\omega t + (n-1)\phi\}] \qquad (6\text{-}1)$$

where ϕ is the phase difference between one source and the next rource as seen in a particular direction. The above summation has been carried out in Chapter 1 (see the discussion following Eq. 1-50). The result is

$$R = A_R \, Cos \left[\omega t + \tfrac{1}{2} (n - 1)\phi \right] \tag{6-2}$$

where the amplitude A_R is given by

$$A_R = A \frac{Sin \; n\phi/2}{Sin \; \phi/2} \tag{6-3}$$

The resultant intensity, I, is thus of the form

$$I \sim A_R{}^2 = A^2 \frac{Sin^2 \; n\phi/2}{Sin^2 \; \phi/2} \tag{6-4}$$

We may also write

$$I = I_0 \frac{Sin^2 \; n\phi/2}{Sin^2 \; \phi/2} \tag{6-5}$$

Eq. 6-5 can be checked for $n = 2$. Writing

$$Sin \; \phi = 2 \, Sin \; \phi/2 \; Cos \; \phi/2$$

we find that

$$A_R = 2A \, Cos \; \phi/2$$

and

$$I = 4I_0 \, Cos^2 \; \phi/2 \tag{6-6}$$

which is of the same form as Eq. 4-21.

We will now discuss the diffraction pattern produced by a narrow slit, i.e. we will try to find out how a beam of parallel light (when passing through a narrow slit) spreads out into the region of geometrical shadow. However, before we go into the details of calculating the diffraction pattern, it should be mentioned that diffraction phenomena are usually divided into the following two general categories:

(1) Those in which the source of light and the screen on which the pattern is observed are effectively at infinite distances from the aperture causing the diffraction, and

(2) those in which either the source or the screen, or both are at finite distances from the aperture.

The phenomena coming under category (1) are called Fraunhofer diffraction and those coming under category (2) are called Fresnel diffraction. Fraunhofer diffraction is much simpler to treat theoretically. The experimental arrangement for observing Fraunhofer diffraction is also simple and is shown in Fig. 6-1. The light from a source is rendered parallel by a lens and the diffracted beam is focused on a screen by another lens

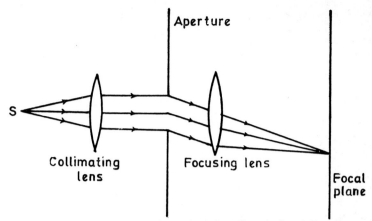

FIG. 6-1: Arrangement for observing Fraunhofer diffraction.

placed behind the aperture. This arrangement effectively removes the source and screen to infinity. In the observation of Fresnel diffraction, on the other hand, no lenses are necessary (Fig. 6-2) but in this case the wavefronts are not plane, and the theoretical treatment is consequently more complex.* We will discuss only the Fraunhofer class of diffraction.

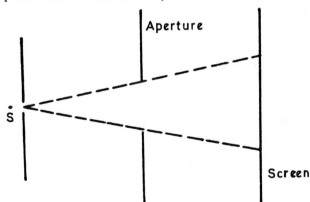

FIG. 6-2: A typical arrangement for observing Fresnel diffraction.

6-2 Calculation of Single Slit Diffraction Pattern

We would now calculate the diffraction pattern produced when a plane wave (emitted, say, by a distant point source) is incident on a slit. (A slit is a rectangular aperture whose length is large compared to its breadth.) The study of the diffraction pattern is based on the superposition of the Huygens' secondary wavelets which can be thought of as being sent out from every point on the wavefront at the instant that occupies the plane of the slit.

* See however, Chapter 2, Section 2-6.

Fig. 6-3 represents a section of a slit illuminated by parallel light. Since we have a continuous distribution of point sources, we would

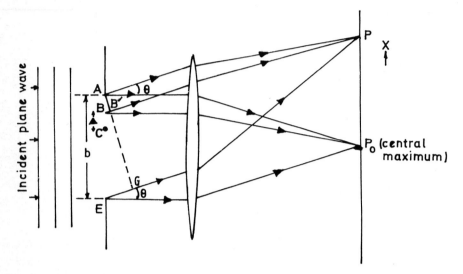

FIG. 6-3: The Fraunhofer diffraction pattern by a single slit. (The width of the slit is greatly exaggerated.)

perform an integration (a superposition) over the contributions from infinitesimal elements of the slit. Instead of an integration over a continuous distribution of sources, we can (and will) consider a discrete sum over contributions from n identical equally spaced 'sources'.* In the limit of n going to infinity, we will have a continuous distribution of radiating sources. (The advantage of using n discrete sources rather than a continuous distribution is that we thereby obtain at the same time the solution for the radiation pattern produced by n narrow slits for arbitrary n from $n = 2$ to infinity.)

If the width of the slit is equal to b, then b is the width of the region that contains linear array of n Huygens' secondary wavelet sources. Let the separation between adjacent sources be Δ. Then we have

$$b = (n - 1)\Delta \qquad (6\text{-}7)$$

Suppose the plane wave is incident in the $+ z$ direction and the n sources are parallel to the x-axis as shown in Fig. 6-3. (These n sources are shown as A,B,C,......E.)

At a distant field point P, each source gives a contribution that has the same amplitude A, because P is distant enough (as compared to b) so that *in the dependence of amplitude on distance* we can assume the distance

---

* The integration has been worked out in *Fundamentals of Optics*, by F. A. Jenkins and H. A. White, p. 290-292, (McGraw-Hill). The final result is same.

is very nearly the same for all the sources. However, because of (even slightly) different path lengths to the point *P*, the field produced by one source will, in general, differ in phase from that produced by another source. This phase difference can easily be calculated by drawing a perpendicular, *AG*, to the parallel diffracted rays under consideration. It is obvious that all the secondary wavelets start in the same phase from different points in *AE*. Now the optical paths of all the rays from *AG* to the focal point *P* are equal. Therefore, the path difference between two consecutive rays is equal to *BB'* ($= \Delta \sin \theta$). This corresponds to a phase difference of $\frac{2\pi}{\lambda} \Delta \sin\theta$. Thus, the resultant electric field, *E*, at the point *P* is equal to the resultant of *n* sources of equal amplitudes with phases increasing in arithmetical progression:

$$E = A \left[\cos\omega t + \cos(\omega t + \phi) + \ldots + \cos\{\omega t + (n-1)\phi\} \right]$$

where

$$\phi = \frac{2\pi}{\lambda} \Delta \sin\theta \tag{6-8}$$

The result for the resultant amplitude for such a problem given by Eq. 6-2 is

$$E_\theta = A \frac{\sin n\phi/2}{\sin\phi/2} \tag{6-9}$$

We now let the spacing Δ go to zero and let *n* go to infinity in such a way that $(n-1)\Delta$ remains constant (equal to *b*). The total phase shift Φ between the contributions of the first and the nth source at *P* is exactly $(n-1)\phi$. This is approximately $n\phi$ for large values of *n*. Thus,

$$\Phi = (n-1)\phi = (n-1)\frac{2\pi}{\lambda}\Delta\sin\theta \tag{6-10}$$

or

$$\Phi = \frac{2\pi}{\lambda} b \sin\theta \tag{6-11}$$

and

$$\Phi \approx n\phi \tag{6-12}$$

Thus, the resultant amplitude becomes

$$E_\theta = A \frac{\sin\frac{1}{2}n\phi}{\sin\frac{1}{2}\phi} \approx A \frac{\sin\frac{1}{2}\Phi}{\sin\frac{1}{2}[\Phi/n]} \tag{6-13}$$

For large values of *n*, we can neglect* all terms except the first one in the Taylor series expansion of $\sin\frac{1}{2}\frac{\Phi}{n}$ in Eq. 6-13. Thus,

---

* This is rigorously correct because we will let *n* approach infinity.

$$\text{Sin} \left(\tfrac{1}{2} \frac{\Phi}{n} \right) \approx \frac{\Phi}{2n}$$

and

$$E_\theta \approx nA \, \frac{\text{Sin} \tfrac{1}{2} \Phi}{\tfrac{1}{2} \Phi}$$

We can make one further simplification. As n approaches infinity, we must let A approch zero in such a way that nA is constant $(= E_0)$. We finally obtain

$$E_\theta = E_0 \, \frac{\text{Sin} \, \beta}{\beta} \tag{6-14}$$

where

$$\beta = \frac{\pi \, b \, \text{Sin} \, \theta}{\lambda} \tag{6-15}$$

The quantity β is a convenient variable, which signifies one half of the phase difference between the contributions coming from opposite edges of the slit. The intensity on the screen is then* of the form

$$I \sim E_\theta{}^2 = E_0{}^2 \, \frac{\text{Sin}^2 \, \beta}{\beta^2}$$

or

$$I = I_0 \, \frac{\text{Sin}^2 \beta}{\beta^2} \tag{6-16}†$$

6-3   The Diffraction Pattern

Fig. 6-4 shows plots of Eq. 6-14 for the amplitude (dotted curve) and of Eq. 6-16 for the intensity, taking the values of the constants E_0 and I_0 equal to 1. The maximum intensity of the strong central band comes at the point P_0 of Fig. 6-3, where evidently all the secondary wavelets will arrive in phase because the path difference Δ is zero. For this point β is zero, and although the quotient $(\text{Sin} \, \beta/\beta)$ becomes indeterminate, we know that

$$\underset{\beta \to 0}{\text{Lt}} \, \frac{\text{Sin} \, \beta}{\beta} = 1$$

The point P_0 is referred to as the position of the *principal maximum*. From this principal maximum the intensity falls to zero at $\beta = \pm \pi$, then passes through several *secondary maxima*, with equally spaced points of

---

* See Section 3-7.

† Alternatively we could have derived Eq. 6-16 by using the graphical method (see Fig. 1-30); in the limit of A going to zero each component would have been on the arc of a circle [See, for example, D. Halliday and R. Resnick, *Physics*, Part II, Sec. 44-4, Second Edition, John Wiley (1965)].

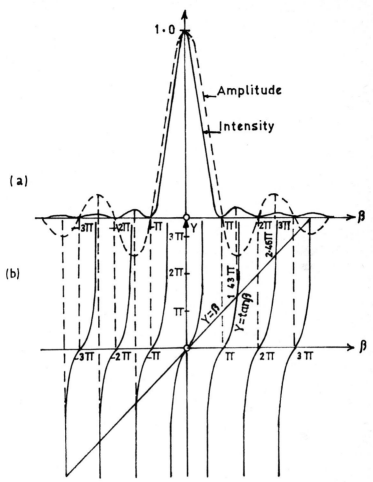

Fɪɢ. 6-4: (a) The variation of the amplitude and intensity as a function of β.
(b) The roots of the equation tan β = β

zero intensity at $\beta = \pm\,\pi,\, \pm\,2\pi,\, \pm\,3\pi,\, \ldots\ldots$, or in general at $\beta = \pm\,m\pi$. The secondary maxima do not fall halfway between these points, but are displaced toward the centre of the pattern by an amount which decreases with increasing m. The exact values of β for these maxima can be found by differentiating Eq. 6-14 with respect to β and equating to zero. This yields the condition

$$\tan \beta = \beta \qquad\qquad (6\text{-}17)$$

The values of β satisfying Eq. 6-17 are easily found graphically as the intersections of the curve $y = \tan \beta$ and the straight line $y = \beta$. In Fig. 6-4(b) these points of intersection lie directly below the corresponding secondary maxima.

The intensities of the secondary maxima may be calculated to a very close approximation by finding the values of $\text{Sin}^2\beta/\beta^2$ at the halfway positions, i.e. where $\beta = 3\pi/2, 5\pi/2, 7\pi/2, \ldots$ Using these values of β, we find that the intensities of the secondary maxima are approximately $4/9\pi^2, 4/25\pi^2, 4/49\pi^2, \ldots$ or $1/22\cdot2, 1/61\cdot7, 1/121\ldots$ of the intensity of the principal maximum. Table 1 which gives exact values of the maxima shows that the approximation is good and gets better for larger values of m.

TABLE I

RELATIVE INTENSITIES OF THE MAXIMA OF DIFFRACTION
PATTERNS OF RECTANGULAR AND CIRCULAR APERTURES*

	Rectangular Aperture	Circular Aperture
Central Maxima	1	1
Ist Maxima	0·0496	0·0174
2nd Maxima	0·0168	0·0042
3rd Maxima	0·0083	0·0016

* After G. R. Fowles *Introduction to Modern Optics*, p. 119, Holt, Rinehart & Winston, (1967).

It is interesting to point out that the positions of zero intensity can easily be obtained without making any detailed mathematical calculations. To show this, consider diffraction at an angle θ such that

$$b \text{ Sin } \theta = \lambda \qquad (6\text{-}18)$$

We divide the slit into two halves (Fig. 6-5). The secondary wavelet from the point in the slit adjacent to the upper edge will travel $\lambda/2$ farther than that from the point at the centre* and so these two will produce fields with a phase difference of π and will give a resultant field of zero amplitude at P_1. Similarly the wavelet from the next point below the upper edge will cancel that from the next point below the centre, and we may continue this pairing off to include all points in the wavefront, so that the resultant effect at P_1 is zero.

Next, for an angle of diffraction given by

$$b \text{ Sin } \theta = 2\lambda \qquad (6\text{-}19)$$

we divide the slit into four parts, the pairing of points again gives zero intensity, since the parts cancel in pairs. Thus, we obtain zero intensity for angles of diffraction given by

$$b \text{ Sin } \theta = m\lambda; \quad m = \pm 1, \pm 2, \pm 3 \ldots\ldots \qquad (6\text{-}20)$$

Example

A lens whose focal length is 40 cm forms a Fraunhofer diffraction

* These two points are shown as the 'first pair' in Fig. 6-5.

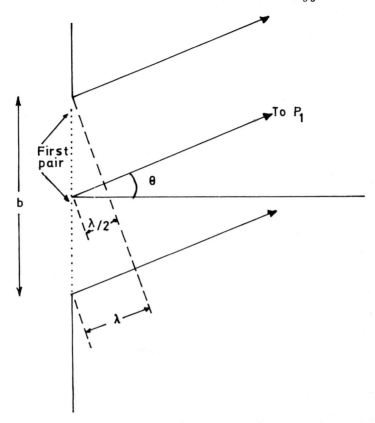

FIG. 6-5: When $b \sin\theta = \lambda$, waves from the sources cancel each other.

pattern of a slit whose width is 0·03 cm. Calculate the distance of the first dark band and of the next bright band from the axis (wavelength of light used is 6×10^{-5} cm). The lens is placed close to the slit.

Solution

In the single slit diffraction pattern, the angles of diffraction corresponding to zero intensity are given by

$$b \sin\theta = m\lambda, \text{ where } m = 1,2,3,\ldots$$

For the first dark band $m = 1$, hence

$$\sin\theta = \frac{\lambda}{b} = \frac{6 \times 10^{-5}}{3 \times 10^{-2}} = 2 \times 10^{-3}$$

or

$$\theta \approx 2 \times 10^{-3} \text{ radians}$$

If the distance of first dark band from the axis is x cm then the angle of diffraction θ is given by

$$\theta \simeq \frac{x}{f} = \frac{x}{40} \text{ radians}$$

or

$$x = 2 \times 10^{-3} \times 40 \text{ cm} = 0\cdot8 \text{ mm}$$

The angle of diffraction corresponding to first bright band on either side of the central band is approximately given by

$$b \text{ Sin } \theta' = \frac{3\lambda}{2}$$

or

$$\theta' \approx \frac{3\lambda}{2b} = 3 \times 10^{-3} \text{ radians}$$

Hence, the distance x' of the first bright band from the axis is

$$x' = \theta'f$$
$$= 3 \times 10^{-3} \times 40 \text{ cm} = 1\cdot20 \text{ mm}$$

It should be remembered that for given values of b and λ, β cannot take arbitrarily large values. In fact, the maximum value of $|\beta|$ is given by

$$|\beta_{max}| = \frac{\pi b}{\lambda} \tag{6-21}$$

corresponding to $\theta = \pi/2$. Thus, if the width of the slit, b, equals λ we will barely see the first diffraction maximum (which will occur at $\theta = \pm \frac{\pi}{2}$). On the other hand, if the slit is wide enough (say $b = 10\lambda$) we will see 10 diffraction minima on either side of the principle maximum. Indeed, the intensity distribution in the diffraction pattern is quite sensitive to the width of the slit. In Fig. 6-6 are plotted the relative intensities in single slit diffraction pattern for $b/\lambda = 1$, 5 and 10.

Thus, we find that when light passes through a narrow slit, it spreads out to a certain extent into the region of the geometrical shadow. This spreading out is small if the dimension of the aperture is large compared to the wavelength of the light. On the other hand, if the dimension of the aperture is comparable to the wavelength of light used, then one obtains measurable intensity of light inside the 'geometrical shadow'. This is one of the simplest examples of diffraction.*

6-4   A Possible Home Experiment

One of the simplest ways to see a single slit diffraction pattern is as follows: Take two cards which have sharp edges (something like two post

---

* The phenomenon of diffraction is therefore often interpreted as 'the failure of light to travel in straight lines'.

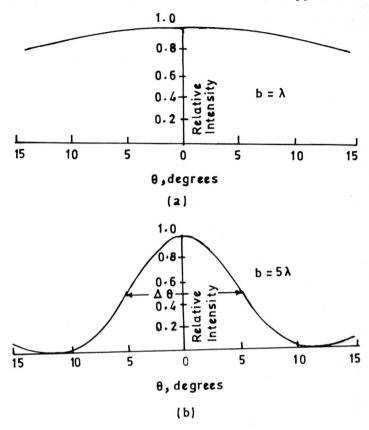

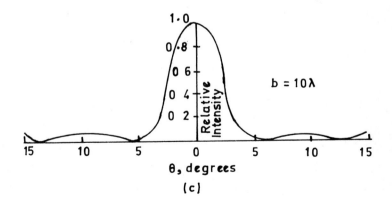

FIG. 6-6: The relative intensity for three values of b/λ. The arrow in (*b*) shows the half width of the central maximum [After D. Halliday and R. Resnick, *Physics*, Part II, p. 1110, John Willey, (1965).]

cards). Hold the two cards very close to one another in front of your eyes and look at a line source (or if a line source is not available then look at the filament of a bulb placed far away). The two edges of the card form a slit of variable width. Vary the slit width from 'zero' to 'infinity', where 'zero' is zero and 'infinity' is about 0·1 cm. If you are looking towards an incandescant lamp you will see beautifully coloured diffraction pattern, each wavelength giving its own diffraction pattern.

6-5   Angular Width of a Diffraction Limited Beam

Referring to Figs. 6-4 and 6-6, we notice that the main feature of the intensity plot is that the intensity is large only in an angular band roughly between $\theta = - \frac{1}{2}\lambda/b$ and $\theta = + \frac{1}{2}\lambda/b$; i.e.

$$\Delta\theta \approx \frac{\lambda}{b} \tag{6-22}$$

where θ is measured in radians. Thus, a beam of width b has angular width $\approx \lambda/b$, and spreads by an amount $W \approx \dfrac{\lambda L}{b}$ in travelling a distance L. Fig. 6-7(a) shows the spreading of a beam made by a plane wave

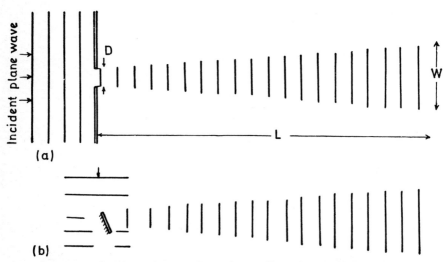

Fig. 6-7: Beam of width D has angular width $\approx \lambda/D$. and 'spreads' by an amount $W \approx L\lambda/D$ in travelling a distance L.
(a) Beam made by a plane wave incident on a hole in an opaque screen.
(b) Beam made by a plane wave incident on a plane mirror [After F. S. Crawford, *Waves*, p. 474, McGraw-Hill, (1968).]

incident on a slit in an opaque screen and Fig. 6-7(b), shows the spreading of a beam made by a plane wave incident on a plane mirror. Each one of the sketches can be taken to represent either water waves, sound waves, or electromagnetic waves. (It may be visible light of wavelength 5×10^{-5} cm, or may be microwaves of wavelength 10 cm.)

6-6 Application: Laser Beam versus Flashlight Beam

Suppose you have a diffraction limited laser beam* of $b = 0.2$ cm, with wavelength 5×10^{-5} cm. How much does the beam diameter increase after a distance of 1000 m? The angular spread of the beam is

$$\Delta\theta \approx \frac{\lambda}{b} \cong \frac{5 \times 10^{-5}\,\text{cm}}{2 \times 10^{-1}\,\text{cm}} = 2.5 \times 10^{-4}\ \text{radians}$$

The spatial spread, W, is given by

$$W \approx L\,\Delta\theta \approx \frac{L\lambda}{b}$$

$$= 1000 \times 2.5 \times 10^{-4}\ \text{meters} \approx 25\ \text{cm}$$

Thus, the spread is of the order of a quarter of a metre at a distance of 1000 m. On the other hand for a flash light with a beam of diameter 0·2 cm formed by a 'point' filament at the focus of a lens, how small would the filament have to be for the flashlight beam to be diffraction limited? If the filament is not a 'point', then different parts of the filament give 'independent' beams (see the lower figure on the jacket). The angular spread due to the finite size of the filament turns out to be approximately the width of the filament divided by the focal length f:

$$\Delta\theta \approx \frac{\Delta x}{f} \tag{6-23}$$

If we want to obtain a diffraction limited (rather than filament-size limited) flashlight beam which has to start with, has a width of 0·2 cm, then we want $\Delta\theta$ due to the filament to be less than the diffraction width, which is about 2.5×10^{-4} radians according to the above calculations. For a typical flashlight the filament is about 0·5 cm from the lens, i.e. $f \approx 0.5$ cm. Thus the filament must have the dimension Δx given by

$$\Delta x < f\,\Delta\theta \approx 0.5 \times 2.5 \times 10^{-4} = 1.25 \times 10^{-4}\ \text{cm}$$

Such a small filament is difficult to make. (This is the reason why a laser beam was used in finding the distance of the moon.)

6-7 Calculation of the Half Width of the Principal Maximum in the Single Slit Diffraction Pattern

The half width, $\Delta\theta$, of the principal maximum is defined in Fig. 6-6(b).

---

* A laser beam is almost perfectly parallel, the spreading out is due to diffraction and therefore the beam is called diffraction limited. Because of the lack of spreading of a laser beam it can be brought to a very sharp focus giving rise to an extremely high intensity (see Example in Chapter 3, page 107 bottom). See also Appendix D and the upper figure on the jacket.

(For an angle of diffraction equal to $\Delta\theta/2$, the intensity of the principal maximum becomes one half of its peak value.) Obviously $\Delta\theta$ should be such that

$$\frac{\text{Sin}^2\,\beta}{\beta^2} = \tfrac{1}{2}$$

where

$$\beta = \frac{\pi b}{\lambda}\,\text{Sin}\left(\frac{\Delta\theta}{2}\right)$$

We know that when

$$\beta = 0, \quad \frac{\text{Sin}^2\,\beta}{\beta^2} = 1$$

and when

$$\beta = \pi, \quad \frac{\text{Sin}^2\,\beta}{\beta^2} = 0$$

Thus, the value of β is approximately $\pi/2$ for which $\dfrac{\text{Sin}^2\beta}{\beta^2} = \tfrac{1}{2}$.

Solving this, we obtain

$$\frac{\text{Sin}^2\,\pi/2}{(\pi/2)^2} = \frac{4}{\pi^2} = 0.4$$

This value is less than 0·5, so the value of β should be less than $\pi/2$. After a few more trials we find that for $\beta = 1.40$ radians ($= 80°$), $\text{Sin}^2\,\beta/\beta^2$ value is very nearly equal to $\tfrac{1}{2}$. The corresponding value of $\Delta\theta$ will be given by

$$\Delta\theta = 2\,\text{Sin}^{-1}\left[\frac{1.40\lambda}{\pi b}\right]$$

Let us now take the specific case of $b = 5\lambda$.
For this, $\Delta\theta$, will be given by

$$\frac{\pi b\,\text{Sin}\,\dfrac{\Delta\theta}{2}}{\lambda} = 1.40$$

or

$$\Delta\theta = 2\,\text{Sin}^{-1}\left[\frac{1.40\,\lambda}{\pi\,5\lambda}\right]$$
$$\approx 2\,\text{Sin}^{-1}\,(0.088)$$
$$= 10.2°$$

which is in agreement with Fig. 6-6(b).

6-8   Some General Observations

The linear width of the diffraction pattern on the screen (Fig. 6-4) will be proportional to the slit-screen distance, which is the focal length f, of a lens placed close to the slit. The linear distance d, for which the intensity is appreciable is approximately given by

$$d \approx \frac{f\lambda}{b} \tag{6-24}$$

The width of the pattern increases in proportion to the wavelength, so that for red light it is roughly twice as wide as for violet light (the slit width etc., being the same). If white light is used, the central maximum is white in the middle, but is reddish on its outer edge, shading into purple and impure colours farther out (Problem 23).

The angular width of the pattern for a given wavelength is inversely proportional to the slit width b, so that as b value is made larger, the pattern shrinks rapidly to a smaller scale. Thus, for $f \approx 20$ cm, $\lambda \approx 5 \times 10^{-5}$ cm and $b \approx 1$ cm; $d \approx 10^{-3}$ cm. This fact, that when the width of the aperture is large compared to a wavelength the diffraction is practically negligible, led the early investigators to conclude that light travels in straight lines and that it could not be a wave motion. Sound waves, whose wavelengths are of the order of metres, will evidently be diffracted through large angles in passing through an aperture of an ordinary size, such as an open window.

It is important to note that in what has been discussed above (and also in Chapter 4) we have only used Huygens' principle and had nothing to do with the electromagnetic character of light. Indeed the treatment will be valid for any 'wave-phenomena'.

6-9   Diffraction by a Rectangular Aperture*

In the preceding sections the intensity function for a slit was derived by summing up the effects of the secondary wavelets originating from a linear section of the wavefront. In Fig. 6-3, the length of the slit is at right angles to the plane of the paper and what we have considered is just a section of the wavefront. Indeed, an experimental arrangement for studying the diffraction pattern is similar to that shown in Fig. 6-8.

In order to calculate the contribution from parts of the wavefront out of the plane of paper (Fig. 6-3) we must divide the slit into a large (say m) number of sections, each section consisting of n sources. The field produced by each section can be calculated using an approach similar to the one given earlier, and then contribution from each section is added up.† The mathematics is similar but more cumbersome. The result of the calculation gives a product of two single slit diffraction patterns:

---

* May be skipped in the first reading.

† This summation will essentially be a sum of nn_1 terms.

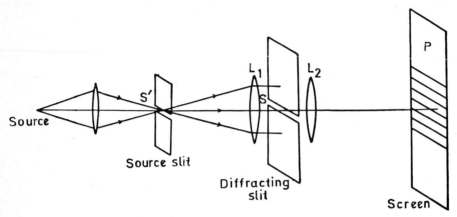

FIG. 6-8: An outline of the experimental arrangement for obtaining the single slit diffraction pattern.

$$I \sim b^2 l^2 \frac{\text{Sin}^2\beta}{\beta^2} \frac{\text{Sin}^2\gamma}{\gamma^2} \qquad (6\text{-}25)$$

where b and l are the width and length of the slit respectively, and

$$\beta = \frac{\pi b \text{ Sin } \theta}{\lambda}$$

$$\gamma = \frac{\pi l \text{ Sin}\Omega}{\lambda}$$

The angles θ and Ω are measured from the normal to the aperture at its centre, in planes through the normal parallel to the sides b and l respectively. These angles can be precisely defined as follows: In Fig. 6-9(a) $ABCD$ is a rectangular aperture and O the centre. The normal, ON, is perpendicular to the plane of the paper. Fig. 6-9(b) contains the normal and the side EF. From any arbitrary point, if we drop a perpendicular on this plane which meets the plane at Q' then $\angle Q'ON$ defines θ. Similarly Ω is defined in Fig. 6-9(c). The diffraction pattern is shown in Fig. 6-10 (see also Plate 7). For $l \gg b$, the maxima and minima in the y-direction (Fig. 6-10) are very close to one another and get smeared up. Thus, for a long (and narrow) rectangular slit we get a diffraction pattern as predicted by Eq. 6-5.

6-10 Diffraction by a Circular Aperture

The diffraction pattern formed by plane waves passing through a circular aperture [Fig. 6-11(a)] is of considerable importance because of its application in studying the resolving powers of optical instruments. Unfortunately it is a problem of considerable difficulty. The intensity distribution for this problem was first calculated by Airy in 1835. The

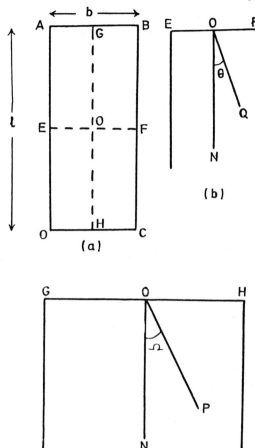

Fig. 6-9: The definition of the angles θ and Ω.
(*a* shows a rectangular aperture). ON is normal
to the plane of the aperture at the point O.

intensity distribution is shown in Fig. 6-11(*b*). (It is obvious that because of the symmetry of the problem, the intensity pattern on the screen will be the same on the circumference of a circle.) The first diffraction minima occurs for

$$\rho = 3 \cdot 832 \tag{6-26}$$

where

$$\rho = \frac{2\pi}{\lambda} R \, \mathrm{Sin}\theta \tag{6-27}$$

R being the radius of the aperture. The angular radius of the first dark ring is thus given by

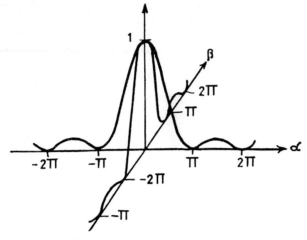

FIG. 6-10: Diffraction pattern of a rectangular aperture.

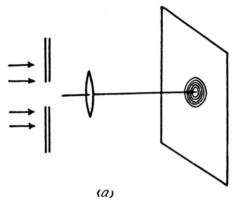

(a)

FIG. 6-11: (a) Fraunhoffer diffraction of a
circular aperture.

$$\text{Sin}\,\theta = \frac{3\cdot832\lambda}{2\pi R} = \frac{1\cdot22\lambda}{D} \qquad (6\text{-}28)$$

where D ($= 2R$) is the diameter of the aperture. The bright central area is known as the *Airy disk* (See Plate 8.) It extends to $\theta = \text{Sin}^{-1}$ ($1\cdot22$ λ/D.) This disk is surrounded by a number of fainter rings. Neither the disk nor the rings are sharply limited, but shade off gradually at the edges, being separated circles of zero intensity. In Table I are listed the values of the relative intensities of the first few maxima of the diffraction patterns of circular and rectangular apertures.

An interesting application of the above phenomenon arises when a directional beam of sound is desised. To a person who is unacquainted with diffraction phenomenon it might seem that the most efficient way

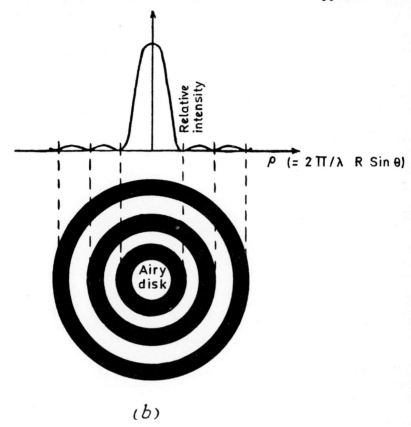

(b)

Fig. 6-11: (b) The corresponding intensity distribution.

to direct sound at a distant listener would be to shout into a cylindrical tube pointed at the listener, thus 'starting the sound off in the right direction' [Fig. 6-12(a)]. Instead, of course, a megaphone of large terminal diameter is more efficient, since then at least the shorter wavelength sounds produce a narow beam [Fig. 6-12(b)].

6-11 Diffraction Limited Optics

In Chapter 2 we have assumed the existence of perfect lenses. We should now take note of the fact that even if a lens has perfect surfaces, the very fact of its finite size limits the sharpness of the image. As an example of such a 'diffraction limited' lens, let us consider how small a point we may focus a laser on to. We suppose the laser emits a perfectly parallel beam (i.e. a plane wave) of diameter d_1, which is brought to focus by a lens of focal length f, and diameter d_2 [Fig. 6-13(a)]. The smaller of the two diameters defines the aperture in question. We can see this by imagining an actual aperture 'stopping down' a larger beam

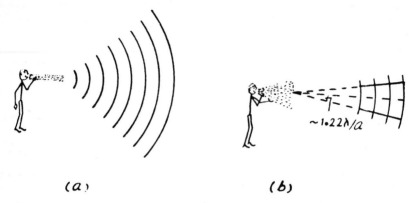

FIG. 6-12: The directionality of a sound beam is greatly enhanced by using a mega-phone of large diameter compared with wavelength [After D. H. Towne *Wave Phenomena*, p. 282, Addison-Wesley (1967).]

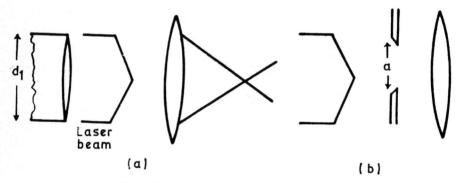

FIG. 6-13: A diffraction-limited laser beam.

or a lens, as in Fig. 6-13(b), where we have used a as the aperture size. The 'image' at the focal point of the lens in not a true point, but rather the diffraction pattern of the aperture.* This has a central maximum of angular width $\Delta\theta \approx \dfrac{\lambda}{a}$ and therefore of diameter $d \approx f\dfrac{\lambda}{a}$. In order to reduce the value of d, we can use a short focus lens (for example a microscope objective) and can keep a as large as possible. But inevitably the spot is of finite size because even under optimum conditions $f/a \approx 1$ (why?) leaving $d \simeq \lambda$.

---

* It should be pointed out that in actual lenses the extent of the image will be more than what is predicted by the diffraction pattern. This is caused by the various lens 'defects' mentioned in Chapter 2. Here what we are trying to mention is this that even if all of these 'defects' could be eliminated by suitable shaping of the lens surfaces or by introducing correcting lenses, the diffraction pattern would still remain. It is an inherent property of the lens aperture, and of the wavelength of light used.

6-12   Limit of Resolution

While considering single slit diffraction pattern, we found that when light passes through a narrow slit, it spreads out to a certain extent into the region of the geometrical shadow. The same is true when light passes through a circular aperture. In general, the spread out is more if the aperture is small. This diffraction caused by a small hole is of importance in designing optical instruments. It determines the ultimate limit of their magnification.

In order to understand how the diffraction phenomenon limits the useful magnification, let us consider what happens when two distant point sources, which are close together, send light through a pinhole. In Fig. 6-14(a), I_1 and I_2 are the images of the sources S_1 and S_2 res-

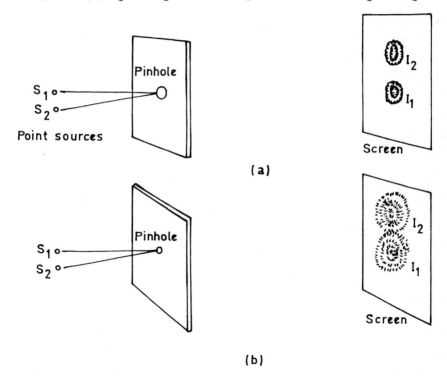

FIG. 6-14: (a) When light from two point sources S_1 and S_2 passes through a pinhole, the diffraction patterns do not overlap and the images are resolved.
(b) When the hole is very small, the diffraction patterns overlap and the images may not be resolved.

pectively. These images are diffraction patterns, and for a smaller hole size [Fig. 6-14(b)] the images become larger and more fuzzy. When the images overlap considerably, it is difficult to decide from looking at the screen, whether the pattern is that of two separate sources or a single odd shaped source. Under these circumstances, i.e. when the images cannot be distinguished, we say that the sources are 'unresolved'. When we can separate them, we say that they are resolved. The *resolution* of an optical instrument is a measure of its ability to give separated images of objects that are close together.

Instead of a pinhole, if we have a lens then we can focus the light from two point sources and produce what appear to be sharp images. However, if these images are carefully examined, we would find that a lens can not eliminate the spreading of light by diffraction because light in passing through a lens is passing through a hole of limited size.

6-13   Limit of Resolution for Optical Instruments

In Chapter 2 we found that the magnifying power of a telescope (or of a microscope) depends (except for certain numerical factors) only on the focal lengths of the lenses making up the optical system of the instrument. It appears as though any desired magnification might be attained by a proper choice of focal lengths. Unless the instrument is properly designed, while the magnifying power becomes larger the image does not gain in detail, even though all lens aberrations have been corrected. This limit to the useful magnification is set by the diffraction of light waves, which essentially implies that the laws of geometrical optics do not hold strictly for a wavefront of limited extent. *Physically, the image of a point source is not the intersection of rays from the source but the diffraction pattern of those waves from the source that pass through the lens system.**

In Section 6-12 we have shown that the light from a distant point source, diffracted by a circular opening, is focussed by a lens not as a geometrical point but as a disk of finite radius surrounded by dark and bright rings. An optical system is said to be able to *resolve* two point sources if the corresponding diffraction patterns are sufficiently small or sufficiently separated to be distinguished. Fig. 6-15 shows the intensity distribution of the diffraction patterns (formed at the focal plane of a lens) for two distant point objects with small angular separations. In Fig. 6-15(*a*) the objects are not resolved, i.e. they cannot be distinguished from a single point object. In Fig. 6-15(*b*) they are barely resolved and in Fig. 6-15(*c*) they are fully resolved. We 'adopt' the following criterion

---

* As we have seen, the extent of this diffraction pattern is of the order of $f\lambda/a$ and hence geometrical optics is a valid approximation as long as the dimension of the lens systems are such that even if the image extends over a distance of the order of $f\lambda/a$, we may still consider it as a 'point image'. However, the diffraction pattern will set a limit to the resolution of two images. This implies that there is no point in removing the various lens aberrations to the extent where geometrical optics approximation will itself breakdown. Thus, for example, we should not try to remove the spherical aberration of a lens to such an extent that the paraxial and marginal rays get focussed within a distance less than $f\lambda/a$.

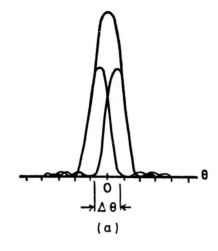

(a)

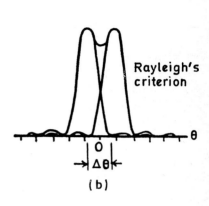

Rayleigh's criterion

(b)

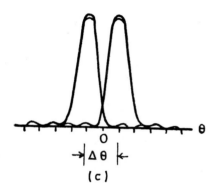

(c)

Fig. 6-15: The diffraction pattern of two distant objects formed on the focal plane of a converging lens. In (a) the angular separation of the objects is so small that the images are not resolved. In (b) the objects are barely resolved (according to Rayleigh's criterion). In (c) the objects are fully resolved.

as the definition of 'resolution' of two images: If the angular separation of two point sources is such that the maximum of the diffraction pattern of one source falls on the first minimum of the diffraction pattern of the other [Fig. 6-15(b)] then the two images are said to be just resolved. This is called *Rayleigh's criterion* (see Plate 9.) This criterion, though useful, is arbitrary; often other criteria are used for deciding whether two objects are resolved or not. Thus, according to the Rayleigh's criterion, the minimum angular resolution of a telescope is (see Eq. 6-28):

$$\theta = \mathrm{Sin}^{-1}\frac{1{\cdot}22\lambda}{D} \tag{6-29}$$

where D is the diameter of the circular aperture which limits the beam from the primary image, or usually that of the objective. Since the angles involved are rather small, we can replace $\mathrm{Sin}\,\theta$ by θ, then

$$\theta \approx \frac{1{\cdot}22\,\lambda}{D} \tag{6-30}$$

If the angular separation between the objects is greater than θ, we can resolve the two objects; if it is less, then they remain unresolved.

A Numerical Example

To get an idea of the linear size of the above diffraction pattern, let us calculate the radius of the first dark ring in the image formed at the focal plane of an ordinary field glass. The diameter of the objective is 3 cm and focal length 25 cm. White light has an effective wavelength of 5.5×10^{-5} cm. Thus, the angular radius of this ring is

$$\theta = \frac{1.22 \times 5.6 \times 10^{-5}}{3}$$

$$= 2.38 \times 10^{-5} \text{ radians}$$

The linear radius $= \theta \times$ focal length

$$\approx 6 \times 10^{-4} \text{ cm}$$

The central disk for this telescope is thus about 0·01 mm in diameter when the object is a point source such as a star. Applying the Rayleigh's criterion, the smallest angular separation of a double star which could be theoretically resolved by this telescope is $\approx 2.38 \times 10^{-5}$ radians (≈ 4.54 second).

The above considerations show that the minimum angle that can be resolved, by a telescope is inversely proportional to D, the diameter of the objective. Indeed, in one of the largest telescopes in existence, an objective of diameter of 40 inches is used. (This corresponds to $\theta \approx 0.14$ second.)

With a given objective in a telescope, the angular size of the image as seen by the eye is determined by the magnification of the eyepiece. However, increasing the size of the image by increasing the power of the eyepiece does not enhance the detail that can be seen. This is because it is impossible by magnification to bring out detail which is not originally present in the primary image. Each point in an object becomes a small circular diffraction pattern in the image, so that if an eyepiece of very high power is used, the image appears blurred and no greater detail is seen, thus diffraction by the objective limits the resolving power of a telescope.

To reduce diffraction effects in microscopes* we often use ultraviolet light, which, because of its shorter wavelength, permits finer details to be examined than would be possible with visible light. In an electron microscope, the electron beam may have an effective wavelength of the order of 5×10^{-10} cm which is shorter by almost a factor of 10^5 than visible light ($\lambda \approx 5 \times 10^{-5}$ cm). (An electron beam, under some circumstances, behaves like waves. This will be discussed in Chapter 9.)

---

* The resolving power of a microscope is discussed in Appendix F.

This permits the detailed examination of tiny objects like viruses. If a virus is examined with an optical microscope, its structure would be concealed by diffraction.

6-14 Angular Resolution of the Human Eye

Take a millimetre scale or make equidistant marks on a piece of paper and find the distance at which the lines become blurred and are thus unresolved. Typically, you will find that 1 mm can be barely resolved at 2 metres and not at all at 4 metres. Thus, for the human eye at the center of field of view (i.e. looking directly at the lines) we find an angular resolution limit, $\Delta\theta$, given by

$$\Delta\theta \approx \frac{1\,\text{mm}}{2\,\text{m}} = \frac{1}{2000} \text{ radians}$$

$$\approx 1\cdot7 \text{ min}$$

Now the diameter of the pupil of the eye is about 2 mm (one can approximately measure this by using a ruler held near the eye and looking through a plane mirror). Thus, the angular full width $\Delta\theta$, of the image of a distant point is given by

$\Delta\theta$ (diffraction limited)

$$\approx \frac{\lambda}{D} \approx \frac{5 \times 10^{-4} \text{ mm}}{2 \text{ mm}} = \frac{1}{4000}$$

Thus, the eye likes to have the points separated by an angular separation of about twice the diffraction width before it sees them resolved.

In order to verify that the (rough) agreement between eye resolution and diffraction width is not accidental, repeat the above experiment looking through a pinhole in a piece of paper. The pinhole should have diameter of about 1 mm (assuming your pupil is about 2 mm). You will (most probably) find that the angular resolution will get worse roughly by a factor of 2.

6-15 Diffraction Pattern Produced by a Double Slit

The interference of light from two narrow slits which are close together was first demonstrated by Young, and it has already been discussed in Chapter 4 as a simple example of the interference of two light beams. In discussion of the double slit experiment we have assumed that the slits are arbitrarily narrow which implies that the central part of the diffusing screen was uniformly illuminated by the diffracted waves from each slit. It is important to understand the modifications of the interference pattern which occur when the width of the individual slit is made greater, until it becomes comparable with the distance between them. This corresponds more nearly to the actual conditions under which the experiment is usually performed. We shall, therefore, discuss the Fraunhofer diffraction pattern formed by a double slit.

6-16   Derivation of the Equation for Intensity

Let us consider two parallel slits AB and CD, each of width b separated by an opaque space BC (Fig. 6-16). The distance between the two slits

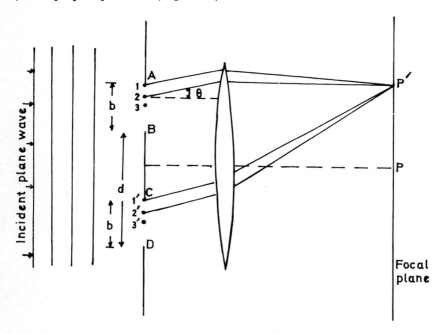

FIG. 6-16: Fraunhofer diffraction by a double slit. The point P' on the screen receives light which is diffracted by an angle θ.

is d. Let a plane wave front of light (of wavelength λ) be incident normally on the plane of the slits. As before the study of the diffraction pattern will be based on the superposition of the Huygens' secondary wavelets which can be thought of as being sent out from every point on the wavefront at the instant that it occupies the plane of the slits. Further, we will consider a discrete sum over contributions from n identical equally spaced 'sources' from one slit and an equal number from the other slit. Thus, we will have to make a sum over $2n$ terms. Now if we are considering diffraction at an angle θ, i.e. trying to find out the intensity at the point P' (Fig. 6-16), then the path difference between two consecutive rays is $\Delta \sin\theta$, the corresponding phase difference being $\dfrac{2\pi}{\lambda} \Delta \sin\theta$ (see the discussion on the single slit). However, two corresponding points on the two slits (like the points 1 and 1' or 2 and

2') will differ in phase by $\frac{2\pi}{\lambda} d \operatorname{Sin}\theta$. Thus the resultant field at the point P' will be obtained by summing up the following $2n$ terms (see Eq. 6-1).

$$E = A \left[\operatorname{Cos} \omega t + \operatorname{Cos} (\omega t + \phi) + \ldots + \operatorname{Cos} \{\omega t + (n-1)\phi\} \right]$$
$$+ A \left[\operatorname{Cos} (\omega t + \Phi_1) + \operatorname{Cos} (\omega t + \Phi_1 + \phi) + \right.$$
$$\left. \ldots + \operatorname{Cos} \{\omega t + \Phi_1 + (n-1)\phi\} \right] \qquad (6\text{-}31)$$

where

$$\phi = \frac{2\pi}{\lambda} \Delta \operatorname{Sin}\theta$$

and

$$\Phi_1 = \frac{2\pi}{\lambda} d \operatorname{Sin}\theta$$

Each series can easily be summed up (see Chapter 1, Sec. 1-17) and the result is:

$$E_\theta = A \left[\frac{\operatorname{Sin} n\phi/2}{\operatorname{Sin} \phi/2} \operatorname{Cos} (\omega t + \delta) \right.$$
$$\left. + \frac{\operatorname{Sin} n\phi/2}{\operatorname{Sin} \phi/2} \operatorname{Cos} (\omega t + \delta + \Phi_1) \right] \qquad (6\text{-}32)$$

where

$$\delta = \tfrac{1}{2} (n-1) \phi$$

Rearranging, we obtain

$$E_\theta = 2A \frac{\operatorname{Sin} n\phi/2}{\operatorname{Sin} \phi/2} \operatorname{Cos} \frac{1}{2} \Phi_1 \operatorname{Cos} \left(\omega t + \delta + \frac{\Phi_1}{2} \right) \qquad (6\text{-}33)$$

In the limit of n going to infinity and Δ going to zero the resultant intensity distribution will be given by

$$I \sim 4E_0{}^2 \frac{\operatorname{Sin}^2\beta}{\beta^2} \operatorname{Cos}^2\gamma \qquad (6\text{-}34)$$

or

$$I = I_0 \frac{\operatorname{Sin}^2\beta}{\beta^2} \operatorname{Cos}^2\gamma \qquad (6\text{-}35)$$

where, as before,

$$\beta = \frac{\pi b \operatorname{Sin}\theta}{\lambda}$$

and

$$\gamma = \frac{\pi}{\lambda} d \operatorname{Sin}\theta$$

The factor $\operatorname{Sin}^2\beta/\beta^2$ in Eqs. 6-34 and 6-35 is the intensity distribution for the single slit (of width b) diffraction pattern. The factor $\text{Cos}^2\gamma$ (in Eqs. 6-34 and 6-35) is characteristic of the interference pattern produced by two beams of equal intensity and phase difference 2γ (see Eq. 4-21). The resultant intensity will be zero when either of the two factors is zero. For the first factor this will occur when $\beta = \pi,\ 2\pi,\ 3\pi,\ \ldots$ and for the second factor when $\gamma = \pi/2,\ 3\pi/2,\ 5\pi/2,\ \ldots$.

The intensity distributions for a two slit diffraction pattern is shown in Fig. 6-17 for different slit widths. The curves can be explained by saying

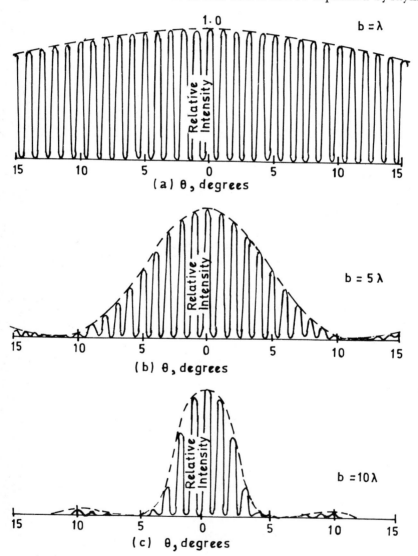

Fig. 6-17: The double slit diffraction pattern for $d = 50\lambda$ [After D. Halliday and R. Resnick, *Physics*, Part II, p. 1116, John Wiley, (1965).]

that the light from the two slits undergoes interference to produce fringes of the type obtained with two beams, but that the intensities of these fringes are limited by the amount of light arriving at the given point on the screen by virtue of the diffraction occurring at each slit. The relative intensities in the resultant pattern as given by Eqs. 6-34 and 6-35 are those obtained by multiplying the intensity function for the interference pattern from two infinitely narrow slits of separation d by the intensity function from a single slit of width b. Thus, the result may be regarded as due to the joint action of interference between the rays coming from corresponding points in the two slits, and, of diffraction, which determines the amount of light emerging from either slit at a given angle (Fig. 6-18).

6-17 Position of the Maxima and Minima

From Eqs. 6-34 and 6-35 it is seen that the intensity will be zero when either

$$\gamma = \frac{\pi}{2}, \frac{3\pi}{2}, \frac{5\pi}{2}, \ldots \tag{6-36}$$

or, when

$$\beta = \pi, 2\pi, 3\pi, \ldots \tag{6-37}$$

Eq. 6-36 gives the minima for the interference pattern and since

$$\gamma = \frac{\pi}{\lambda} d \sin\theta$$

they occur at angles θ such that

$$d \sin\theta = \frac{\lambda}{2}, \frac{3\lambda}{2}, \ldots = (m + \tfrac{1}{2}) \lambda \quad \text{(Minima)} \tag{6-38}$$

where $m = 0, 1, 2, 3, \ldots$.
Next, Eq. 6-37 gives the minima for the diffraction pattern, and since

$$\beta = \frac{\pi}{\lambda} b \sin\theta$$

they occur at such angles θ, such that

$$b \sin\theta = \lambda, 2\lambda, \ldots = p\lambda \quad \text{(Minima)} \tag{6-39}$$

where $p = 1, 2, 3, \ldots$.
  The exact positions of the maxima are not given by any simple relation, but their approximate positions may be found by neglecting the variation of the factor $\dfrac{\sin^2\beta}{\beta^2}$; an assumption justified only when the slits are very narrow and when the maxima near the centre of the pattern are considered. The positions of the maxima will then be determined solely by the $\cos^2\gamma$ factor, which has maxima for

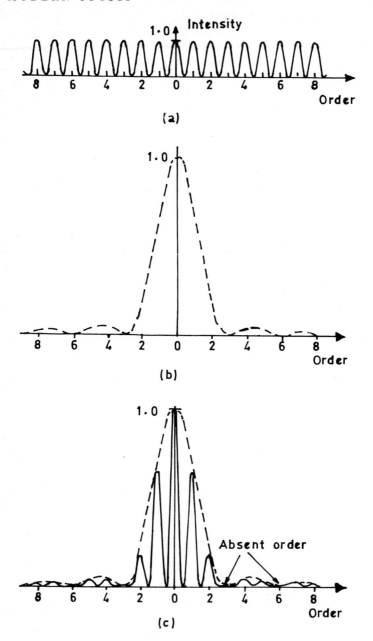

FIG. 6-18: (a) The interference pattern by two point sources.
(b) The single slit diffraction pattern.
(c) The double slit pattern.

$$\gamma = 0, \pi, \ldots$$

i.e. for

$$d \sin\theta = 0, \lambda, 2\lambda, \ldots = m\lambda \quad \text{(Approximate Maxima)}$$

where $m = 0, 1, 2, 3, \ldots$.

The number m represents the order of interference. Now if the slit width b, is kept constant and the separation of the slits d, is varied, the scale of interference pattern varies, but that of the diffraction pattern remains the same. However, a particular order interference maxima may be absent if the diffraction angle corresponds to a minima given by Eq. 6-39. This is known as a missing order. Thus, a particular missing order occurs when there exists a θ for which the two following equations are simultaneously satisfied:

$$d \sin\theta = m\lambda, \text{ where } m = 1, 2, 3, \ldots \tag{6-40}$$

$$b \sin\theta = p\lambda, \text{ where } p = 1, 2, 3, \ldots \tag{6-41}$$

Using Eqs. 6-40 and 6-41 we obtain:

$$\frac{d}{b} = \frac{m}{p} \tag{6-42}$$

Since m and p are both integers, d/b must be in the ratio of two integers to have missing orders (If the ratio d/b is not exactly equal to a ratio of two integers, then the intensity of a particular order will not be zero but would be very small.)

Example

In a double slit diffraction pattern the screen is 40 cm away from the slits. Calculate the wavelength of light if the fringe spacing is 0·25 cm. What is the linear distance from the central maximum to the first minimum of the fringe envelope? (Assume the slit width $b = 0·002$ cm and the separation of the two slits $d = 0·01$ cm.)

Solution

Fringe width $= \dfrac{\lambda D}{d}$

$$\therefore \quad \lambda = \frac{0·25 \times 0·01}{40}$$

$$= 6·25 \times 10^{-5} \text{ cm}$$

The first diffraction minima will occur when

$$\frac{\sin\beta}{\beta} = 0$$

or when

$$\beta = \pi$$

i.e.

$$b \, \mathrm{Sin}\theta = \lambda$$

$$\therefore \quad \mathrm{Sin}\theta = \frac{\lambda}{b} = \frac{6 \cdot 25 \times 10^{-5}}{2 \times 10^{-3}} \approx 3 \cdot 13 \times 10^{-2}$$

or

$$\mathrm{Sin}\theta \approx \theta \approx 0 \cdot 031 \text{ radians}$$

The required linear distance is

$$D \tan 0 \cong D\theta \approx 40 \times \cdot 031$$

$$= 1 \cdot 24 \text{ cm}$$

Thus, the number of fringes $\approx 2 \times \dfrac{1 \cdot 24}{0 \cdot 25}$

$$= 9 \cdot 92$$

There are about ten fringes in the central peak of the fringe envelope.

Example
What requirements must be met for the central maximum of the envelope of the double slit diffraction pattern to contain exactly thirteen fringes?

Solution
For this, the seventh minimum of the interference factor $\mathrm{Cos}^2 \gamma$ must correspond to the same angle of diffraction for which the first minima (of the diffraction pattern) occurs. Thus, for the seventh minimum

$$\gamma \left(= \frac{\pi}{\lambda} d \, \mathrm{Sin}\theta \right) = (6 + \tfrac{1}{2}) \pi$$

$$\therefore \quad \mathrm{Sin}\theta = \frac{13 \, \lambda}{2 \, d}$$

Next, for the same value of θ the first diffraction minima must occur, i.e.

$$\beta \left(= \frac{\pi b \, \mathrm{Sin}\theta}{\lambda} \right) = \pi$$

or

$$\mathrm{Sin}\theta = \frac{\lambda}{b}$$

Thus, the required condition will be met if

$$\frac{13 \, \lambda}{2 \, d} = \frac{\lambda}{b}$$

or

$$\frac{d}{b} = \frac{13}{2}$$

6-18   The Intensity Distribution from N Parallel Slits

Let us consider N parallel slits each of width, b, and each separated from the other by a distance, d.

Let a plane wavefront of light of wavelength λ be incident normally on the plane of the slits (Fig. 6-19). As before, the study of the diffraction

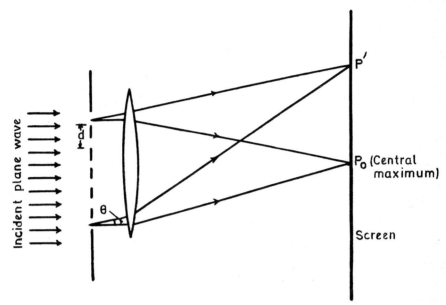

Fig. 6-19: A diffraction grating.

pattern will be based on the superposition of the Huygens' secondary wavelets which can be thought of as being sent out from every point on the wavefront at the instant that it occupies the plane of the slits. Thus, we will have to consider a discrete sum of contributions from n identical equally spaced 'sources' from each of the N slits. We will, therefore, be making a sum over nN terms. Making the same series of arguments as in the case of a double slit, we arrive at the following expression for the resultant field at the point P' (corresponding to the angle of diffraction θ):

$$
\begin{aligned}
E_\theta = A\,[&\text{Cos } \omega t + \text{Cos }(\omega t + \phi) + \ldots + \text{Cos }\{\omega t + (n-1)\phi\} \\
&+ \text{Cos }(\omega t + \phi_1) + \text{Cos }(\omega t + \phi_1 + \phi) \\
&+ \ldots + \text{Cos }\{\omega t + \phi_1 + (n-1)\phi\} + \ldots \\
&+ \text{Cos }\{\omega t + (N-1)\phi_1\} + \text{Cos }\{\omega t + (N-1)\phi_1 + \phi\} \\
&+ \ldots + \text{Cos }\{\omega t + (N-1)\phi_1 + (n-1)\phi\}]
\end{aligned}
\tag{6-43}
$$

or

$$E_\theta = A \left[\frac{\mathrm{Sin}\,\frac{n\phi}{2}}{\mathrm{Sin}\,\frac{\phi}{2}}\, \mathrm{Cos}\,(\omega t + \delta) + \frac{\mathrm{Sin}\,\frac{n\phi}{2}}{\mathrm{Sin}\,\frac{\phi}{2}}\, \mathrm{Cos}\,(\omega t + \delta + \phi_1) \right.$$

$$\left. + \ldots + \frac{\mathrm{Sin}\,\frac{n\phi}{2}}{\mathrm{Sin}\,\frac{\phi}{2}}\, \mathrm{Cos}\,\{\omega t + \delta + (N-1)\phi_1\} \right]$$

or

$$E_\theta = A \left[\frac{\mathrm{Sin}\,\frac{n\phi}{2}\,\mathrm{Sin}\,\frac{N\phi_1}{2}}{\mathrm{Sin}\,\frac{\phi}{2}\,\mathrm{Sin}\,\frac{\phi_1}{2}}\, \mathrm{Cos}\,(\omega t + \delta + \tfrac{1}{2}(N-1)\,\phi_1) \right] \tag{6-44}$$

where

$$\phi_1 = \frac{2\pi}{\lambda}\,d\,\mathrm{Sin}\theta \tag{6-45}$$

and

$$\delta = \tfrac{1}{2}(n-1)\,\phi \tag{6-46}$$

In the limit of n going to infinity the intensity distribution becomes

$$I \sim E_0{}^2\,\frac{\mathrm{Sin}^2\,\beta}{\beta^2}\,\frac{\mathrm{Sin}^2\,N\gamma}{\mathrm{Sin}^2\,\gamma} \tag{6-47}$$

which may be written as

$$I = I_0\,\frac{\mathrm{Sin}^2\,\beta}{\beta^2}\,\frac{\mathrm{Sin}^2\,N\gamma}{\mathrm{Sin}^2\,\gamma} \tag{6-48}$$

where

$$\gamma = \frac{\pi d\,\mathrm{Sin}\,\theta}{\lambda} \tag{6-49}$$

and

$$\beta = \frac{\pi\,b\,\mathrm{Sin}\theta}{\lambda} \tag{6-50}$$

For $N = 1$, we obtain the single slit pattern (Eq. 6-16) and for $N = 2$, we obtain the double slit pattern (Eqs. 6-34 and 6-35).

6-19 Principal Maxima

The new factor $\mathrm{Sin}^2 N\gamma/\mathrm{Sin}^2\gamma$ may be said to represent the *interference* term for N slits. It possesses maximum values equal to N^2 for $\gamma = 0$, π, 2π, ... Although the expression for the above values of γ becomes indeterminate, the result may be obtained by noting that

$$\lim_{\gamma \to m\pi} \frac{\text{Sin } \mathcal{N}\gamma}{\text{Sin } \gamma} = \lim_{\gamma \to m\pi} \frac{\mathcal{N} \text{ Cos } \mathcal{N}\gamma}{\text{Cos } \gamma} = \pm \mathcal{N} \qquad (6\text{-}51)$$

The negative sign occurs when either m or $\mathcal{N}$ is an odd integer. Thus, for $\gamma = m\pi$, we obtain a large value for the intensity:

$$I = \mathcal{N}^2 I_0 \text{ Sin}^2 \beta/\beta^2 \qquad (6\text{-}52)$$

and we obtain, what are known as the *principal maxima*. The corresponding angle of diffraction will be given by

$$d \text{ Sin}\theta = 0, \lambda, \ 2\lambda, \dots = m\lambda \ \text{(Principal Maxima)} \qquad (6\text{-}53)$$

where $m = 0, 1, 2, \dots$

The relative intensities of the different order principal maxima are governed by the single slit diffraction envelope $\sin^2 \beta/\beta^2$.

6-20   Minima and Secondary Maxima

To find the minima of the function $\text{Sin}^2\mathcal{N}\gamma/\text{Sin}^2\gamma$, we note that the numerator becomes zero more often than the denominator, and this occurs at the values $\mathcal{N}\gamma = 0, \pi, 2\pi, \dots$ or, in general, $p\pi$. In the special case when $p = 0, \mathcal{N}, 2\mathcal{N}, \dots$, γ will also be a multiple of π and the denominator will also vanish. Indeed, for these values of γ we obtain the principal maxima. However, for

$$\gamma = \frac{p\pi}{\mathcal{N}} \qquad (6\text{-}54)$$

(where $p \neq 0, \mathcal{N}, 2\mathcal{N}, \dots$), the numerator of $\text{Sin}^2 \mathcal{N}\gamma/\text{Sin}^2\gamma$ is zero but not the denominator. Thus, for these directions, the intensity will be zero and we will obtain minima. The corresponding angles of diffraction will be given by:

$$d \text{ Sin}\theta = \frac{\lambda}{\mathcal{N}}, \ \frac{2\lambda}{\mathcal{N}} \dots, \ \frac{(\mathcal{N}-1)\,\lambda}{\mathcal{N}}, \ \frac{(\mathcal{N}+1)\,\lambda}{\mathcal{N}}, \ \frac{(\mathcal{N}+2)\,\lambda}{\mathcal{N}},$$

$$\dots, \ \frac{(2\mathcal{N}-1)\,\lambda}{\mathcal{N}}, \ \frac{(2\mathcal{N}+1)\,\lambda}{\mathcal{N}}, \ \frac{(2\mathcal{N}+2)\lambda}{\mathcal{N}}, \dots \text{(Minima)} \quad (6\text{-}55)$$

where we have omitted the values $0, \mathcal{N}\lambda/\mathcal{N}, 2\mathcal{N}\lambda/\mathcal{N}, \dots$ which correspond to principal maxima. Between two adjacent principal maxima there will hence be $(\mathcal{N} - 1)$ points of zero intensity. The two minima on either side of a principal maximum are separated by twice the distance of the others (show this). Between the other minima the intensity rises again, and gives rise to what are known as secondary maxima (Fig. 6-20). The intensity of these secondary maxima are much smaller than the intensity of the principal maxima (Plate 10). Indeed, (assuming the term $\dfrac{\text{Sin}^2 \beta}{\beta^2}$ not to vary too much with θ), the intensity of the secondary maxima adjacent to the principal maximum is about 4 per cent of the intensity

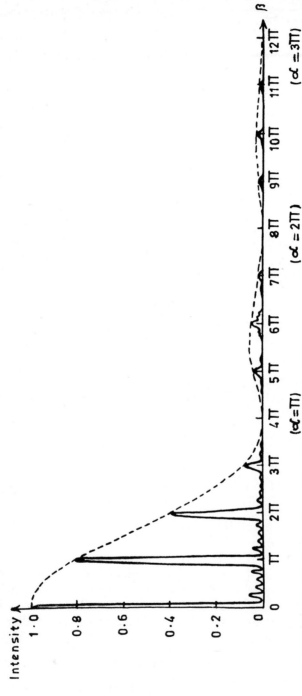

FIG. 6-20: Intensity distribution for grating with six slits ($b = 1/4d$). The dotted line gives the single slit diffraction pattern. [After R. S. Longhurst, *Geometrical and Physical Optics*, p. 252, Longmans, (1967).]

of the adjacent principal maximum. (This is true only for large values of N, see Problem 18.) The variation of $\text{Sin}^2 N\gamma / \text{Sin}^2 \gamma$, for $N = 6$, is shown in Fig. 6-20.

It should be noted that we can not have an arbitrarily large number of principal maxima. In fact, since $\text{Sin}\,\theta$ cannot exceed unity, the value of m cannot be greater than d/λ.

6-21   The Diffraction Grating

Until now we have been studying the diffraction pattern by a large number of parallel equidistant slits each separated from the adjacent one by an opaque strip of constant width. Any arrangement which is equivalent in its action to a number of parallel equidistant slits described above is called a diffraction grating. Gratings are often used to measure wavelengths and to study the structure and intensity of spectral lines. Few devices have contributed more to our knowledge of modern physics.

A typical grating for use with transmitted light is a glass plate upon which are ruled a number of equally spaced lines or grooves, usually several thousand per centimetre. The chief requirement for a good grating is that the lines should be as nearly equally spaced as possible over the whole ruled surface, which in different gratings varies from 1 to 10 inches in width. This is a difficult requirement to fulfill, and there are few places in the world where ruling machines of precision adequate for the production of fine gratings have been constructed. Fine grooves are ruled with a diamond point. After each groove has been ruled, the machine lifts the diamond point and moves the grating forward by a small rotation of the screw which drives the carriage carrying it. To have the spacing of the rulings constant, the screw must be of constant pitch, and it was not until the manufacture of a nearly perfect screw had been achieved by Rowland, in 1882, that the problem of successfully large gratings was solved. Rowland's arrangement gave 14438 lines per inch, corresponding to $d = 2 \cdot 54/14438 = 1 \cdot 693 \times 10^{-4}$ cm. This value of d is approximately three wavelengths of yellow light ($\lambda \approx 5 \cdot 5 \times 10^{-5}$ cm), and the third order ($m = 3$) is the highest that can be observed in this colour with normal incidence (why?). Correspondingly higher orders can be observed for shorter wavelengths.

Fig. 6-21 shows a simple grating spectroscope, used for viewing the spectrum of a light source, assumed to emit a number of discrete wavelengths, or *spectral lines*. The light from source S is focused by lens L_1 on a slit S_1 placed in the focal plane of lens L_2. The parallel beam of light emerging from collimator C falls on the grating. The rays, diffracted at an angle θ, are focused on the focal plane of the lens L_3. The image formed in this plane is examined using an eyepiece E. A symmetrical diffraction pattern is formed on the other side of the central position,

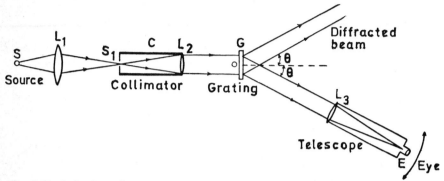

FIG. 6-21: A simple grating spectrometer used to analyze the wavelengths of the light emitted by a source. The collimator sends out a parallel beam of light and the diffracted beam is received by the telescope which can rotate about a vertical axis passing through O.

as shown by the dotted lines. The entire spectrum can be viewed by rotating the telescope T through various angles. The spectrum can be recorded on a photographic plate placed in the focal plane of L_3. (What would the diffraction pattern look like if the slit S_1 is replaced by a point source?)

6-22   Dispersion

The positions of the mth order principal maxima are given by
$$d \operatorname{Sin}\theta = m\lambda, \quad m = 0, 1, 2, \ldots \tag{6-56}$$
The difference in the angles of diffraction increases with the order number. To express this separation, the quantity frequently used is called the *angular dispersion*, which is defined as the rate of change of the angle of diffraction with change of wavelength. An expression for this quantity is obtained by differentiating the above equation with respect to λ and then substituting the ratio of finite increments for the derivative. The result is

$$\frac{\Delta\theta}{\Delta\lambda} = \frac{m}{d \operatorname{Cos}\theta} \quad \left(\begin{array}{l}\text{Angular}\\ \text{Dispersion}\end{array}\right) \tag{6-57}$$

From the above equation we can deduce the following:

(1) For a given small wavelength difference, $\Delta\lambda$, the angle is directly proportional to the order m. Hence the second order spectrum is approximately twice as wide as the first order, the third order is approximately thrice as wide as the first order. We have written 'approximately' because for a higher order spectrum the factor $\operatorname{Cos}\theta$ will decrease. This will result in a greater dispersion. However, if θ does not become large, $\operatorname{Cos}\theta$ will not differ much from unity, and this factor will not be of much importance. Indeed, if we neglect the influence of $\operatorname{Cos}\theta$, the different spectral lines in one order will differ

in angle by amounts which are directly proportional to their differences in wavelength. Such a spectrum is called a *normal spectrum*, and one of the chief advantages of gratings over prism instruments is this simple linear scale for wavelengths in their spectra.

(2) Secondly, $\Delta\theta$ is inversely proportional to the slit separation, *d*, which is usually referred to as the *grating space*. The smaller the graing space, the more widely spread will be the spectra.

Now for $d = 1.693 \times 10^{-4}$ cm (this value corresponds to 14438 lines per inch)

$$\frac{\Delta\theta}{\Delta\lambda} = \frac{m}{\text{Cos } \theta} \times \frac{1}{1.693 \times 10^{-4}}$$

$$\approx m \ (6 \times 10^3) \text{ radians/cm}$$

where we have neglected the influence of Cos θ. Thus, in the second order spectrum

$$\frac{\Delta\theta}{\Delta\lambda} \approx 1.2 \times 10^4 \text{ radians/cm}$$

On the other hand, for prism made of flint glass (with normal dimensions), the angular dispersion is approximately given by:

$$\frac{\Delta\theta}{\Delta\lambda} \approx - 5 \times 10^2 \text{ radians/cm}$$

where θ, in the case of a prism, denotes the angle of emergence from the prism. (The negative sign indicates that the angle of emergence decreases with λ.)

Example

Find the angular breadth of the first order visible spectrum produced by a plane diffraction grating having 15,000 lines per inch when white light is incident normally on the grating. Assume that the wavelength of white light extends from 4000 to 7000 Å. Prove that the violet of the third order visible spectrum overlaps the red of the second order.

Solution

The grating element $d = \dfrac{2.54}{15000} = 1.69 \times 10^{-4}$ cm

The angle of diffraction of the violet in the first order is given by

$$d \text{ Sin} \theta_v = \lambda_v$$

or $\text{ Sin } \theta_v = \dfrac{4 \times 10^{-5}}{1.69 \times 10^{-4}} = 0.237$

or $\theta_v = 13° \ 14'$ for the first order violet spectrum.

Similarly $\theta_r = 24°\ 30'$ for the first order red spectrum.

Hence the angular breadth* of the first order visible spectrum is

$$24°\ 30' - 13°\ 40' = 10°\ 50'$$

The diffraction angle of the *third order violet* is given by

$$\text{Sin } \theta_v = \frac{3 \times 4 \times 10^{-5}}{d}$$

and of the *second order red* is given by

$$\text{Sin } \theta_r = \frac{2 \times 7 \times 10^{-5}}{d}$$

The second angle is greater than the first, whatever be the value of d. Therefore, the violet of the third order will always overlap with the red of the second order provided, of course, the third order exists.

Example

Sodium light is incident normally upon a plane transmission diffraction grating having 10,000 lines per inch. Calculate the angular separation of the D_1 and D_2 lines in the first order spectrum as observed in a telescope. The focal length of the objective and the eyepiece of the telescope being 24 cm and 2 cm respectively. (The wavelengths of the D_1 and D_2 lines of sodium light are 5890 and 5896 Å.)

Solution

Grating element, $d = \dfrac{2 \cdot 54}{10000}$

$$= 2 \cdot 54 \times 10^{-4} \text{ cm}$$

In the first order $m = 1$, and the grating equation becomes

$$d \text{ Sin} \theta = \lambda$$

Since the wavelength difference is small, we expect the angular dispersion to be small and therefore Eq. 6-57 can be used to find the value of $\Delta\theta$. Thus,

$$\frac{\Delta\theta}{\Delta\lambda} = \frac{1}{d} = \frac{1}{2 \cdot 54 \times 10^{-4}} \text{ cm}^{-1}$$

$$\therefore \quad \Delta\theta = \frac{6 \times 10^{-8}}{2 \cdot 54 \times 10^{-4}} = 2 \cdot 36 \times 10^{-4} \text{ radians}$$

($\Delta\theta$ is indeed very small and so we are justified in using Eq. 6-57.)

The angular magnifying power of the telescope is given by

* It should be noted that Eq. 6-57 is to be used only when $\Delta\theta$ (measured in radians) is $<<<1$.

$$\frac{\Delta\theta'}{\Delta\theta} = -\frac{f_o}{f_e}$$

where f_o and f_e are the focal lengths of the objective and eyepiece respectively. Substituting the given values, we obtain

$$\Delta\theta' = -\frac{24}{2} \times 2\cdot36 \times 10^{-4} \text{ radians}$$

$$= -2\cdot83 \times 10^{-3} \text{ radians}$$

$$\approx 0\cdot16°$$

The angular separation of sodium lines as seen by the telescope is $0\cdot16°$.

6-23 Widths of the Principal Maxima

We have shown that the principal maxima occur when

$$d\,\text{Sin}\theta = 0, \lambda, 2\lambda, \ldots = m\lambda$$

and the minima occur when

$$d\,\text{Sin}\theta = \frac{\lambda}{N}, \frac{2\lambda}{N}, \ldots, \frac{(N-1)\lambda}{N}, \frac{(N+1)\lambda}{N}, \frac{(N+2)\lambda}{N}, \ldots$$

(see Eqs. 6-53 and 6-55). Thus if we denote θ_m as the angle of diffraction for the mth order principal maximum and $\theta_m \pm \Delta\theta$ as the angles of diffraction for the first minima on either side, then

$$d\,\text{Sin }\theta_m = m\lambda$$

and

$$d\,\text{Sin }(\theta_m + \Delta\theta) = m\lambda \pm \frac{\lambda}{N}$$

Assuming $\Delta\theta$ to be small, we obtain

$$d\,\text{Sin }(\theta_m \pm \Delta\theta) = d\,[\text{Sin}\theta_m\,\text{Cos }\Delta\theta \pm \text{Cos }\theta_m\,\text{Sin}\Delta\theta]$$

$$\approx d\,[\text{Sin}\theta_m \pm \Delta\theta\,\text{Cos}\theta_m]$$

$$= m\lambda \pm \Delta\theta\,d\,\text{Cos}\theta_m \qquad (6\text{-}58)$$

where $\Delta\theta$ is measured in radians. Thus, Eq. 6-58 gives:

$$\Delta\theta d\,\text{Cos}\theta_m = \frac{\lambda}{N}$$

or the angular half width of the mth order principal maximum is

$$\Delta\theta = \frac{\lambda}{Nd\,\text{Cos}\theta_m} \qquad (6\text{-}59)$$

6-24 Resolving Power of a Grating

To distinguish light waves whose wavelengths are close together, the principal maxima of these wavelengths formed by the grating should be as narrow as possible. In other words, the grating should have a high resolving power, R, which is defined by the relation:

$$R = \frac{\lambda}{\Delta\lambda} \qquad (6\text{-}60)$$

where λ is the mean wavelength of two spectral lines that can barely be recognized as separate and $\Delta\lambda$, the wavelength difference between them. The smaller the value of $\Delta\lambda$, is the closer the lines can be and still be resolved; hence greater will be the resolving power of the grating.

The resolving power of a grating is usually determined by the same consideration (the Rayleigh's criterion) as that is used to determine the resolving power of a lens. If two principal maxima are to be barely resolved, they must, according to this criterion, have an angular separation $\Delta\theta$, such that the maximum of one line coincides with the first minimum of the other (Fig. 6-22.) Thus, if two wavelengths λ and $\lambda + \Delta\lambda$

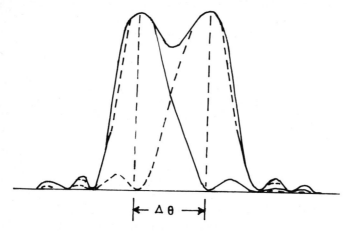

Fig. 6-22: Angular separation of two spectral lines which are resolved by a diffraction grating.

are to be just resolved in the mth order then the principal maximum of order m of $\lambda + \Delta\lambda$ must form at the same angle as the first minimum of wavelength λ, following the principal maxima of the same order. If we denote this common angle by θ, then, for the two lines to be just resolved we must have

$$d\,\mathrm{Sin}\theta = m\,(\lambda + \Delta\lambda) \qquad (6\text{-}61)$$

and

$$d\,\mathrm{Sin}\theta = m\lambda + \frac{\lambda}{N} \qquad (6\text{-}62)$$

$$\therefore \quad m\Delta\lambda = \frac{\lambda}{N} \qquad (6\text{-}63)$$

or

$$R = \frac{\lambda}{\Delta\lambda} = mN \qquad (6\text{-}64)$$

Thus, for a grating to barely resolve the two sodium lines ($\lambda_1 \approx 5890$ Å and $\lambda_2 \approx 5896$ Å) in the first order, the number of lines needed is

$$N = \frac{5893}{(5896 - 5890)} \approx 1000$$

Thus, according to Eq. 6-64 the resolving power of a grating is directly proportional to the order m, and in a given order it is proportional to the total number of lines effective in the formation of the diffraction pattern. It can be seen that the resolving power is independent of the spacing of the lines. Of course all the lines will be effective, only if the diameters of the collimating and telescope lenses are great enough so as to fill the entire ruled surface of the grating of the incident beam. Moreover the apparent non-dependence of resolving power on d is illusory, for it can easily be shown that the resolving power of a grating cannot exceed dN/λ (Problem 19).

6-25 X-ray Diffraction*

While discussing the electromagnetic waves we had pointed out that X-rays are electromagnetic radiation with wavelengths of the order of 1 Å, i.e. X-rays are the same type of waves as the visible light except for the difference in wavelength We would, therefore, expect X-rays to produce diffraction patterns similar to those produced by visible light. However, since the wavelengths are of the order of 1 Å, in order to obtain an observable diffraction pattern, the width of the slit (in a diffraction grating) must also be of the same order. Early investigators attempted to observe diffraction of X-ray beams passing through extremely narrow slits, a few thousandths of a millimetre wide. The results obtained were marginal and not very convincing. It occurred to Laue, in 1913, that if the atoms in a crystal were arranged in a regular way (with interatomic spacings of the order of 1 Å), a crystal might serve as a three dimensional diffraction grating for X-rays. The experiment was performed by Friedrich and Knipping and it succeeded, thus establishing that X-rays could be diffracted and that the atoms in a crystal were arranged in a regular manner.

Figs. 6-23 (*i*), (*ii*) and (*iii*) show the periodic arrangement of atoms in a simple cubic lattice, a face centred cubic lattice and a body centred cubic lattice respectively. In a simple cubic lattice [Fig. 6-23(*i*)]

---

* An excellent account of X-ray diffraction is given in 'X-ray Crystallography' by Sir Lawrence Bragg, *Scientific American*, July 1968.

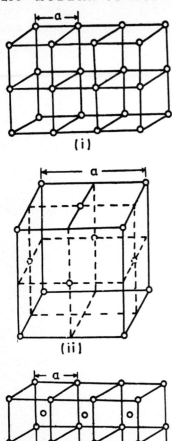

FIG. 6-23: Periodic array of atoms.
(i) Simple cubic lattice.
(ii) Face centered cubic lattice.
(iii) Body centered cubic lattice.

the atoms lie on the corners of a cube. If in addition, we have atoms lying at the centres of the cube as well, we obtain the body centred cubic (abbreviated as *bcc*) lattice [Fig. 6-23(*iii*)]. On the other hand, if atoms lie at the centres of the faces of the cube we obtain the face centred cubic (abbreviated as *fcc*) lattice [Fig. 6-23(*ii*)]. The noble metals (like copper, silver and gold) crystallize in the *fcc* form while the alkali metals (lithium, sodium and potassium) crystallize in the *bcc* form. There exists many other types of periodic structures but the main point which should be emphasized here is the fact that *an ideal crystal is composed of periodic arrangement of atoms in space.*

In order to understand the diffraction of X-rays by such a crystal, let us consider a plane electromagnetic wave incident upon a square array of atoms as shown in Fig. 6-24. As the incident wave passes through, the atoms become sources of secondary waves. The total diffraction pattern is then the result of the superposition of these secondary waves. Now, for the row of atoms shown in Fig. 6-24, the path length from source to observer is the same for all atoms in the row, if the lines to the source and the observer are at equal angles, as is shown in Fig. 6-24. Next, we examine the condition under which the scattered radiation from adjacent rows is also in phase. As shown in Fig. 6-24(*b*), the path difference between the waves scattered in the same column but adjacent rows is $2d \sin\theta$. For constructive interference to occur, this must be an integral multiple of λ. Thus, the condition that must be satisfied in order for the radiation from the *entire array* to arrive at the observer in phase is

$$2d \sin\theta = m\lambda \qquad (6\text{-}65)^*$$

where $m = 1, 2, 3, \ldots$. When this condition is satisfied, a strong maximum in the diffraction pattern is observed. If the condition is not

---

* The values of m are limited by the condition that $\sin\theta$ must be less than unity.

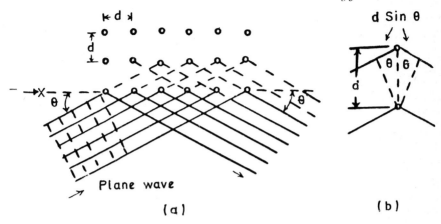

FIG. 6-24: Scattering of radiation from a square array of atoms. When the angle of incidence is equal to the angle of reflection, we obtain constructive interference between radiation scattered from different atoms in a row. When the condition $2d \sin\theta = m\lambda$ is also satisfied, interference between radiation scattered from different atoms in adjacent rows is also constructive.

satisfied, no strong maximum is observed, since there is then no position of the observer for which all the radiation will arrive in phase from all directions. Eq. 6-65 is called the *Bragg condition*, in honour of Sir William Bragg, one of the pioneers in X-ray diffraction analysis.

It is convenient to describe this interference effect in terms of reflections of the wave from the lines parallel to the *x*-axis in Fig. 6-24. When strong interference maxima occur, they are always at angles such that the angles of incidence and reflection are equal, and there is the additional condition of Eq. 6-65, which must be satisfied for a strong reflection to occur.

The entire discussion can be extended to a three-dimensional array of atoms. Again the concept of reflection planes is useful; if the array has regular structure, it is possible to construct a set of parallel planes, passing through all the atoms (see Fig. 6-25.) The periodic arrangement of atoms shown in Fig. 6-25 is the simple cubic structure. (The argument is however valid for any crystal structure.) The scattered waves will combine to give a relatively intense wavetrain in that particular direction in which the incident wave would be reflected as if each crystal plane were

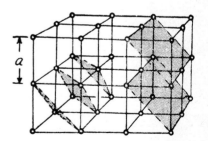

FIG. 6-25: Two different families of planes for a simple cubic crystal. The spacing of the planes on the left is $a/\sqrt{3}$; that of the planes on the right is $a/\sqrt{2}$. There are many other sets of crystal planes.

a mirror; furthermore the distance between the adjacent planes, d, and the angle θ should be such that Eq. 6-65 is satisfied. The set of parallel planes is in general not unique; it is always possible to construct several different sets of parallel planes. Correspondingly, there are various sets of angles corresponding to the directions of maximum intensity. Fig. 6-25 illustrates two sets of planes for the simple case of a cubic array of atoms.

Let us discuss the case of bcc structure in some detail. The x, y and z directions are mutually orthogonal and the distances between the nearest two atoms in these directions are equal [Fig 6-26(i)]. Consider an atomic

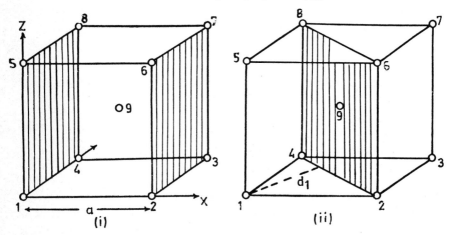

FIG. 6-26: (i) (100) planes and (ii) (110) planes in a bcc lattice.

plane passing through the atoms 1, 4, 8 and 5. Along the x-axis, the atom nearest to 1 is atom number 2 and therefore a plane passing through the atoms 2, 3, 7 and 6 is parallel to the one passing through the origin.* If an X-ray beam diffracted from the plane defined by 1, 4, 8 and 5 has a path difference of λ with respect to the beam diffracted from the plane defined by 2, 3, 6 and 7, then we expect a diffraction maximum in the direction specified by

$$2a \sin\theta = \lambda \qquad (6\text{-}66)$$

where a is the distance between the planes, i.e. between the atoms 1 and 2. However, in this particular structure, there is an atomic plane parallel to the above planes which lies half way between the two planes passing through 9. If Eq. 6-66 is satisfied then the path difference between the

* The orientation of a family of planes is usually specified by Miller indices (see Appendix G). The reflection by the family of planes shown in Fig. 6-26(i) is designated as (100) reflection where (100) denote the Miller indices of the family of planes. A second order reflection from these planes is designated as (200) reflection. The plane shown in Fig. 6-26(ii) is designated as (110) plane.

X-ray beams diffracted by the plane through 1, 4, 5 and 8 and the one through 9 is $\lambda/2$ (and since the density of atoms in all these planes are equal*), we will get destructive interference.

If however, the path difference between the diffracted beams from 1, 4, 5, 8 and 2, 3, 6, 7 is 2λ, then the path difference between the beams diffracted from 1, 4, 5 and 8 and from the plane passing through 9 will be λ giving rise to constructive interference. We will, therefore, get a maximum when

$$2a\,Sin\theta = 2m\lambda \qquad (6\text{-}67)^\dagger$$

Consider another type of plane passing through the atoms 4, 2, 6, 8 and 9 [Fig. 6-26(ii)]. The plane parallel to this and passing through the origin and containing the atom number 5 has not been shown in Fig. 6-26(ii). If the path difference between the diffracted beams from these two planes is λ, there will be constructive interference when

$$2d_1\,Sin\theta_1 = m\lambda \qquad (6\text{-}68)^\ddagger$$

where d_1 $(= a/\sqrt{2})$ is the interplanar spacing for this family of planes. The shaded plane shown in Fig. 6-26(ii) is nearest to the origin and there is no other plane in between the two and parallel to these. In a similar manner, we can draw different families of planes which will give rise to constructive interference.

Now if a collimated beam of non-monochromatic X-rays (i.e. continuously distributed in wavelength), is allowed to fall on a single crystal and if a photographic plate is interposed in the path of the scattered rays (Fig. 6-27), then a regular pattern, which is characteristic of the crystal structure, appears. It is called a *Laue pattern* after the German Physicist Max Von Laue who in 1912 had suggested that if the wavelengths of X-rays are of the same order of magnitude as the atomic spacing in crystals, a crystal could be used as a 3-dimensional grating. Each dot in the pattern corresponds to the direction of scattering from a family of planes. The atomic arrangements in the crystal can be deduced from a careful study of the positions and intensities of the Laue spots.

A more straight forward procedure for the analysis of the X-ray diffraction pattern is the so-called 'powder method' developed independently in 1916 by Peter J. W. Debye and Paul Scherrer in Switzerland and by Albert W. Hull in the U.S.A. In this method, the scatterer, instead of being a single crystal, is a powder containing a large number of small (single) crystals, all randomly oriented. A monochromatic beam of X-rays is aimed at the powder and usually it is convenient to employ a cylindrical

* This can easily be shown if we consider the infinite lattice.

† These reflections are designated as (200), (400) in Fig. 6-29(a) corresponding to $m = 1, 2$ respectively (m cannot be greater than 2).

‡ These reflections are designated as (110), (220) in Fig. 6-29(a). (For a *bcc* lattice, it can be shown that Bragg reflections occur only when the sum of Miller indices is an even number.)

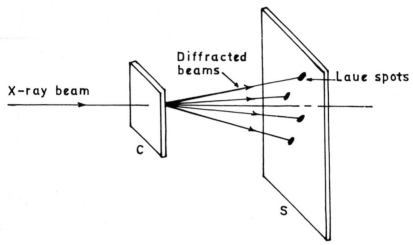

FIG. 6-27: A non-monochromatic X-ray beam is allowed to fall on crystal *C*. Strong diffracted beams appear in different directions, forming the so-called Laue pattern on a screen.

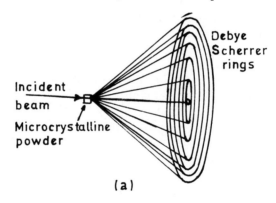

(a)

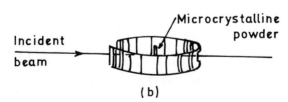

(b)

FIG. 6-28: Powder photographs are made by aiming monochromatic X-rays at micro-crystalline powder which consists of small single crystals oriented randomly. The diffracted beams corresponding to a particular set of planes will then form a cone. If recorded on a photographic plate perpendicular to the incident beam, as shown in (a), each diffraction order will appear as a ring surrounding the central spot. It is usually more convenient to employ a cylindrical photographic film whose axis is perpendicular to the incident radiation as shown in (b). When the film is unrolled one obtains a photograph similar to the one shown in Fig. 6-29.

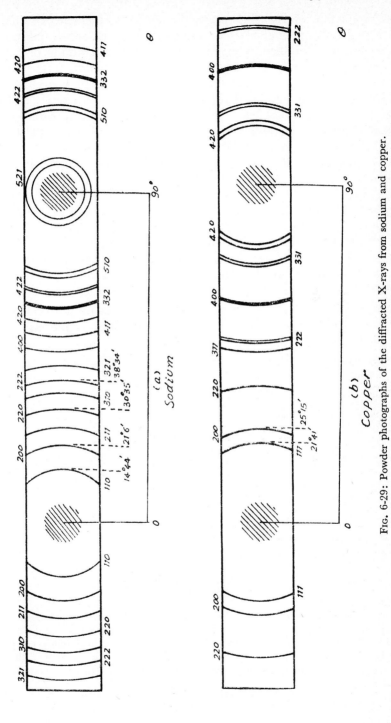

FIG. 6-29: Powder photographs of the diffracted X-rays from sodium and copper.

photographic film whose axis is perpendicular to the incident X-radiation (Fig. 6-28.) Since small crystals are randomly oriented all possible angles of incidence are obtained. Therefore, for a particular family of planes, different single crystals diffract the X-rays on a conical surface whose semi-vertical angle is twice the Bragg angle, θ. If the photographic film is placed as shown in Fig. 6-28(a) then one obtains a series of rings (known as the Debye-Scherrer rings), each ring corresponding to a particular order Bragg reflection. On the other hand, when the film is placed as shown in Fig. 6-28(b) one obtains arcs of circles (Fig. 6-29), each arc corresponding to a particular order Bragg reflection. From these arcs one can calculate* the Bragg angle θ. If the wavelength is known then using Eq. 6-65 one can obtain the interplanar spacing d, and once the interplanar spacings are known one can find the crystal structure.

In Fig. 6-29 are shown the positions of Bragg maxima for powdered sodium and a copper wire. From the film itself one can find the Bragg angle and the result (for sodium) is that intense maxima occur for $\theta = 14° 44'$, $21° 6'$, $26° 9'$, $30° 35'$, etc. The X-ray beam used corresponded to two wavelengths, $1 \cdot 540$ Å and $1 \cdot 544$ Å. Because of the presence of two wavelengths one obtains double lines at higher scattering angles for each family of planes. The lattice constant for sodium is $4 \cdot 28$ Å and one can easily show that the Bragg angles will indeed be as given above. One can carry out a similar analysis for the *fcc* structure (see Problem 25).

SUGGESTED READING

BRAGG, L. 'X-ray Crystallography', *Scientific American*, July 1968. In this article Bragg has discussed, how X-ray crystallography had originated and how it has evolved from the study of simple crystals like rocksalt to the anlysis of the molecular structures of enormously complex biological molecules.

JENKINS, F. A. and H. E. WHITE, *Fundamentals of Optics*, Third Edition, McGraw-Hill (1957). Recommended for a more detailed treatment on diffraction.

WEBB, R. *Elementary Wave Optics*, Academic Press (1967). The book gives a good introduction to more modern concepts in optics.

PROBLEMS

1. A parallel beam of monochromatic light of wavelength 5461 Å is normally incident on a slit which is 0·045 cm wide. A lens of focal length 40 cm is placed behind the slit. Find the distance from the principal maximum to the first minimum in the diffraction pattern formed in the focal plane of the lens. ($4 \cdot 85 \times 10^{-2}$ cm)

2. A single slit is illuminated by light whose wavelengths are λ_a and λ_b, so chosen that the

---

* It can easily be shown that the distance of the arcs from the point 0 is directly proportional to θ.

first diffraction minimum of λ_a coincides with the second minimum of λ_b. What relationship exists between the two wavelengths? $\hspace{2cm}$ ($\lambda_a = 2\lambda_b$)

3. In a single slit diffraction pattern the distance between the first minimum on the right and the first minimum on the left is 5·2 mm. The screen on which the pattern is displayed is 80 cm from the slit and wavelength is 5460 Å. Calculate the slit width. $\hspace{1cm}$ (0·17 mm)

4. Find the half-angular breadth of the central bright band in the Fraunhoffer diffraction pattern of a slit 14×10^{-5} cm wide, when the slit is illuminated by a parallel beam of monochromatic light of wavelengths (a) 400 m μ and (b) 700 m μ. $\hspace{0.5cm}$ [(a) 16·6° (b) 30°]

5. (a) Diffraction images of two slit sources are said to be just resolved when the angular separation, α, of the two separation is such that the maximum of one pattern falls exactly on tne first minimum of the other. Show that for such a case

$$\alpha \simeq \theta_1 \simeq \frac{\lambda}{b}$$

where θ_1 is the angle of diffraction for the first minimum.

(b) Considering the criterion for the resolution of two diffraction patterns of unequal intensity to be that the drop in intensity between the maxima shall be 20% of the weaker one, find the angular separation required when tne intensities are in the ratio 5:1. The results are to be expressed in terms of θ_1 which is the angular separation when the intensities of the two sources are equal. (Use graphical method to solve this problem.) $\hspace{0.5cm}$ (1·13 θ_1)

6. Find the diameter of the first bright ring in the focal plane of a 36 in lens whose focal length is 56 ft. (Assume $\lambda = 5500$ Å) $\hspace{2cm}$ (3.36×10^{-3} cm)

7. The two headlights of a car are 4 ft apart. Assuming that the resolving power of the eye is determined only by diffraction effects at the circular pupil aperture, calculate at what maximum distance will the eye resolve them? (Assume pupil diameter $= \frac{1}{2}$ cm and $\lambda = 5500$ Å) $\hspace{2cm}$ (9100 metres)

8. The intensity distribution for Fraunhofer diffraction of a circular aperture is of the form:*

$$I = I_0 \left[\frac{2J_1\,(\rho)}{\rho} \right]^2$$

where $\rho = \dfrac{2\pi}{\lambda}\, R\,Sin\theta$, and $J_1\,(x)$ is the Bessel function of the first kind and is defined by the equation

$$J_1(x) = \sum_{r=0}^{\infty} \frac{(-1)r}{r!(r+1)!} \left(\frac{x}{2} \right)^{2r+1}$$

(The Bessel functions are often tabulated in handbooks.) Using the tables or the above series expansion, plot the intensity distribution function as a function of θ for $\lambda = 6 \times 10^{-5}$ cm and $\lambda = 10^{-4}$ cm

9. Assume that light is incident on a grating at an angle i as is shown in Fig. 6-30. Show that the intensity distribution function is still of the form of Eq. 6-47 with β and γ defined by the following equations:

$$\beta = \frac{\pi b\,(Sin\,\theta + Sin\,i)}{\lambda}$$

and

$$\gamma = \frac{\pi d\,(Sin\,\theta + Sin\,i)}{\lambda}$$

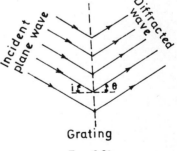

Grating

Fɪɢ. 6-30

* A straight forward derivation of the intensity distribution has been given in *Introduction to Modern Optics* by G. R. Fowles, p. 118.

(Hint: When the angle of diffraction is θ, the phase difference between the disturbances from the corresponding points in two successive slits is

$$\frac{2\pi}{\lambda} \, d \, (\text{Sin } \theta + \text{Sin } i)$$

Only the special case $i = 0$ has been treated in this chapter.)

10. The separation of sodium lines (mean $\lambda = 5893$ Å) in the second order spectrum of a transmission grating containing 5000 lines per cm is 2.5′ for normal incidence. What is the difference in wavelength between the two yellow lines? (5·9 Å)

11. A wire grating made of 200 wires per cm placed at equal distances apart is illuminated with light of wavelength 6×10^{-5} cm. If the diameter of each wire is 0·025 mm calculate the angle of diffraction for the third order spectrum and also find absent spectra, if any. (2° 4′; 2nd order missing)

12. A grating with 8000 rulings per inch is illuminated with white light at perpendicular incidence. (a) Show by calculation that only the first order spectrum is isolated but the second and third order spectra overlap if we assume that the wavelength of light extends from 4000 to 7000 Å. (b) What is the expected dispersion in the third order in the vicinity of the intense green line $(\lambda = 5460 \text{ Å})$?

13. An astronomical telescope has an objective of diameter 100 in. Assuming the mean wavelength of light to be 5.5×10^{-5} cm estimate the smallest angular separation of two stars which can be resolved by it. (0·0544 sec)

14. Calculate the useful magnifying power of a telescope of one inch objective assuming the resolving power of the eye to be equal to 2 min and the wavelength of light 6×10^{-5} cm. (20·2)

15. What is the minimum number of lines a grating must have so that if it can just resolve the sodium doublet $(\lambda_1 = 5890 \text{ Å} \text{ and } \lambda_2 = 5896 \text{ Å})$ in the third order spectrum? (328)

16. Light is incident normally on a grating 0·5 cm wide with 2500 lines. Find the angular separation of the two sodium lines, 5890 Å and 5896 Å, in the first order spectrum. Can they be seen distinctly? (1′5″; yes)

17. Calculate the number of lines per inch on a plane transmission grating which when used on a spectrometer with telescope and collimator of aperture 3/4 in will just resolve the two D lines (5890 Å and 5896 Å) in the second order spectrum. (799)

18. Show that for grating with large number of lines, the intensity of the first secondary maximum is about 4% of the principal maximum.

19. Show that the resolving power of a grating cannot exceed $\dfrac{dN}{\lambda}$.

20. What is the angular dispersion of the visible spectrum in the diffraction pattern produced by a transmission grating with 14438 lines per inch in the first order? (14°)

21. Show that in a grating if the opaque and the transparent strips are of equal width (i.e. $2b = d$) then all the even orders (except for $m = 0$) are absent.

22. Show that in a flash light the angular spread due to the finite size of a filament of linear dimension Δx is approximately $\Delta x/f$.

23. Show that in the single slit diffraction pattern if we use white light, the central maximum is white in the middle, but is reddish on its outer edge, shading into purple colours farther out.

24. A parallel beam of monochromatic light of wavelength 5.55×10^{-5} cm is incident on two round pinholes each of 0·4 mm in diameter. How close together would they have to be placed for the two Airy discs (central circular maxima) to overlap by one-half the radius of each when observed 1 metre behind the pinholes? (2·538 mm)

25. The average diameter of the pupil of the eye is 2·5 mm in daylight. Assuming the resolution to be limited by diffraction, calculate the distance at which two small green coloured objects $(\lambda \approx 5000 \text{ Å})$ 40 cm apart be barely resolved by the naked eye. (2·04 Km)

26. The powder photographs for sodium (which is a *bcc* lattice with lattice constant $= 4·28$

Å) and for copper (which is a *fcc* lattice with lattice constant $= 3.61$Å) are shown in Fig. 6-29. For sodium intense maxima are observed for $\theta = 14° 44', 21° 6', 26° 9', 30° 35',$ $34° 41', 38° 34', 42° 19', 46° 21', 49° 46', 53° 34', 57° 34', (61° 50', 62° 6'), (66° 34',$ $66° 54')$ and $(80° 12', 81°12')$. (The angles in parentheses correspond to two different wavelengths.) Show that these maxima correspond to Bragg reflections by (110), (200), (211), (220), (310), (222), (321), (400), (411), (420), (332), (422), (431), or (510) and (521) planes respectively.

Similarly show that in case of copper, the θ values of $21° 41', 25° 15', 37° 6', 45° 2', 47°$ $39', (58° 32', 58° 47'), (68° 24', 68° 48'), (72° 32', 73° 0')$ correspond to Bragg reflections by (111), (200), (220), (311), (222), (400), (331) and (420) planes respectively. (The X-ray beam consists of wavelengths 1.540 Å and 1.544 Å; these are known as $CuK\alpha_1$ and $CuK\alpha_2$ lines.)

))) **7**

Photography by Coherent Light—Holography

7-1 Introduction

Although there have been great refinements in photographic techniques and in the invention of new photographic materials, the basic principles of photography have not changed over the past one hundred years. Essentially, the photographic process consists of recording an illuminated three-dimensional scene on a two-dimensional light sensitive surface. Thus, the light reflected from the objects in the scene is focussed on the light sensitive surface by some kind of an image forming device, which can be a complex series of lenses (as in an expensive camera) or may simply be a pinhole in an opaque screen. Fig. 7-1 shows the basic physics involved in ordinary photography. For simplicity a 'pinhole camera' has been shown which enables the image to form on the surface of a photographic film. When ordinary (incoherent) light falls on the photographic plate (Fig. 7-1), the eye sees only a two-dimensional image of the original object.

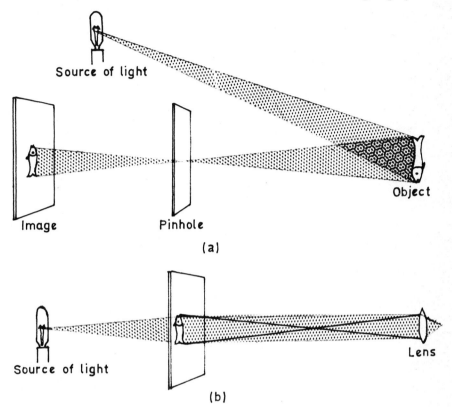

FIG. 7-1 (a): Ordinary photography consists of recording an illuminated three-dimensional object on a two-dimensional light sensitive surface. The light reflected from the object is focused on the surface by some kind of image-forming device, which may be simply a pinhole in an opaque screen.

(b) When ordinary incoherent light is shone through the photographic transparency, the eye sees only a static two-dimensional image of the original object.

In this chapter we will discuss a radically different concept in photographic optics. Invented in 1947 by Dennis Gabor of the Imperial College of Science and Technology, London, this process, which can be called photography by wavefront reconstruction, does not record an image of the object being photographed but rather records the reflected light waves themselves. The photographic record is called a *hologram*; it bears no resemblance to the original object but nevertheless contains in a kind of optical code all the information about the object that would be contained in an ordinary photograph and much additional information that cannot be recorded by any other photographic process.

The creation of an intelligible image from the hologram is known as the reconstruction process. In this stage the captured waves are in effect released from the hologram record, whereupon they proceed onward, oblivious to the time lapse in their history. The reconstructed waves are indistinguishable from the original waves and are capable of all the phenomena that characterize the original waves. For example, they can be passed through a lens and brought to a focus, thereby forming an image of the original object, even though the object has long since been removed. If the reconstructed waves are intercepted by the eye of an observer, the effect is exactly as if the original waves had been observed— the observer sees essentially the original object itself in full three-dimensional form, complete with parallax (the apparent displacement of an object when seen from different directions) and many other effects that occur in the normal 'seeing' process.

The above mentioned photographic process is known as *holography*. Although holography was known as early as 1947, it was not well known to many scientists and was considered (for the following 15 years) as a subject of 'academic interest'. The early efforts were hampered by the lack of an adequate source of coherent light. The invention of the laser in 1960 opened the way to major advances in wavefront reconstruction photography. In 1962, E. N. Leith and J. Upatnieks using a laser could obtain high quality holograms.*

7-2 Principle

Light waves reflected by an object are characterised by their amplitude and phase. In the case of a point scatterer of light [Fig. 7-2(a)] the reflected wavefronts are spherical and concentric around the point of origin. If the reflecting object is not a single point but a complex object, it can then be regarded as a collection of a number of points, and the resulting wave pattern reflected from the surface of the object can be regarded as the sum of many such sets of spherical waves, each set concentric about its point of origin [Fig. 7-2(b)]. The exact form of the wave pattern reflected from an extended and irregular object is, in general, highly complex. The main problem of wavefront reconstruction photography is to record this complex wave pattern at a given plane at some instant of time. Such a record can be thought of as a 'freezing' of the wave pattern; the pattern remains frozen until such time as one chooses to reactivate the process, whereupon the waves are 'read out' of the recording medium. To capture the wave pattern completely both the amplitude and the phase of the waves must be recorded at each point on the recording surface.

For recording the wave pattern, it is not sufficient to place a photographic plate in the path of the wave, because the photographic plate is sensitive to the intensity, i.e. to the square of the amplitude and cannot

* The original work appeared in the paper entitled 'Reconstructed Wavefronts and Communication Theory' by E. N. Leith and J. Upatnieks in the *Journal of the Optical Society* of *America*, Vol. 52, No. 10, pp. 1123-1130 (1962).

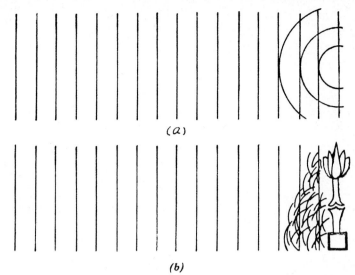

FIG. 7-2: Light waves are reflected from a point scatterer (a) in a series of ever expanding spherical shells, called wavefronts, that are concentric about the point of origin. If the reflecting object is complex (b) it can then be regarded as a collection of a large number of points, and the resulting wave pattern reflected from the surface of the object can be regarded as the sum of many such sets of spherical waves, each set concentric about its point of origin. The central problem of wavefront reconstruction photography is to record this pattern at a given plane at some instant of time.

be used to record the phase information. The problem is, therefore, to assemble some appropriate apparatus that can convert these phase relations into intensity variations which can be recorded photographically. Gabor solved this problem by superposing to this wave a second known wave. The interference produced between the two waves increases the intensity at points where phases are in concordance and decreases the intensity at points where phases are in opposition.

The experimental arrangement (Fig. 7-3) consists of illuminating the object by a laser beam.* Each point on the object reflects light to the entire photographic plate; conversely, each point of the photographic plate receives light from the entire object. A part of the laser beam is reflected by a plane mirror and produces, what is known as, the reference beam. This reference beam interferes with the light reflected from the object and produces an interference pattern (known as hologram) on the photographic plate.

In the reproduction stage (Fig. 7-4) the hologram is illuminated with a collimated laser beam and two images are produced by the diffracted waves emerging from the hologram. One of them, when projected back

* Why we use a laser beam will be explained shortly.

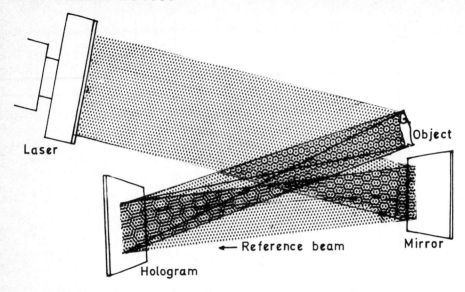

Laser

Object

Reference beam

Mirror

Hologram

FIG. 7-3: In the recording stage of wavefront reconstruction photography no lens or other image forming device is used and consequently no image is formed. Instead each point on the object reflects light to the entire hologram; conversely, each point of the hologram receives light from the entire object. The reference beam produces by means of interference effects, a visible display of the wave pattern of the light impinging on the hologram from the object.

toward the illuminating source, seem to emanate from an apparent object located at the position where the original object was located. These waves are said to produce a virtual image. The other wave produces a real image, which can be photographed directly, without the need for a lens, by simply placing a photographic plate at the position of the image (Fig. 7-4).

7-3 Theory

Let us try to work out this mathematically.*

We will first consider the recording of the hologram. The disturbance produced (on the plane of the photographic plate) by the wave reflected from the object can be expressed by the expression:

$$E_s = a_s \ (x, y) \ \text{Cos} \ [\omega t + \phi_s(x, y)] \tag{7-1}$$

where the x-y plane is defined by the plane of the photographic plate; $a_s(x, y)$ and $\phi_s(x, y)$ represent the amplitude and phase of the disturbance respectively, and ω is the angular frequency.

* This treatment is limited to two-dimensional hologram and its reconstruction. Three-dimensional effects become important when the emulsion thickness is 15-20 times the wavelength of the laser beam.

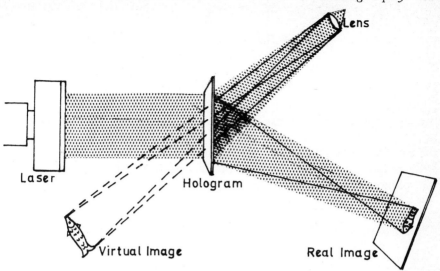

Fig. 7-4: In the reproduction stage the hologram is illuminated with a collimated beam of monochromatic light and two images are produced by waves emerging from the hologram interference grating. One of the waves, when projected back toward the illuminating source, seems to emenate from an apparent object located at the position where the original object was located. These waves are said to produce a virtual image. The other first order diffracted waves have conjugate, or reversed, curvature. These waves produce a real image, which can be photographed directly, without the need for lens, by simply placing a photograpnic plate at the position of the image.

Thus, the intensity, I, is given by

$$I \sim \langle E_s{}^2 \rangle = \langle a_s{}^2(x,y) \operatorname{Cos}^2(\omega t + \phi_s) \rangle$$
$$= \tfrac{1}{2} a_s{}^2(x,y) \tag{7-2}$$

where the sign $\langle \ldots \rangle$ implies the time average.* Eq. 7-2 does not contain any more the phase. Since a photographic plate measures the intensity, it does not give any information of the phase.

Next, let us consider the 'reference beam'. The disturbance produced by this beam on the plane of the photographic plate can be represented by:

$$E_r = a_r \operatorname{Cos}(\omega t - \alpha x) \tag{7-3}$$

where a_r denotes the amplitude and αx represents a linear variation of the phase of the wave in the plane of the photographic plate in the x-direction (one obtains this variation of phase by inclining the beam with respect to the normal of the plate).

The two beams E_s and E_r interfere. The resultant disturbance and intensity are given by

$$E_s + E_r = a_s(x,y) \operatorname{Cos}[\omega t + \phi_s(x,y)] + a_r \operatorname{Cos}(\omega t - \alpha x) \tag{7-4}$$

* The time average has to be carried out in the manner described in Chapter 3, Sec. 3-7.

$$I <(E_s + E_r)^2>$$

$$= <a_s^2 \, \text{Cos}^2 \, (\omega t + \phi_s) + a_r^2 \, \text{Cos}^2 \, (\omega t - \alpha x)$$

$$+ \, 2 \, a_s a_r \, \text{Cos} \, (\omega t + \phi_s) \, \text{Cos} \, (\omega t - \alpha x)>$$

$$= \tfrac{1}{2} \, a_s^2 + \tfrac{1}{2} \, a_r^2 + a_s a_r \, \text{Cos} \, [\phi_s(x,y) + \alpha x] \tag{7-5}$$

Thus the blackening of the photographic plate depends on three terms, the last term being the most important as it gives the effect of the functions $a_s \, (x_s, y)$ and $\phi_s(x,y)$. When a_r is large compared to a_s (as is usually the case) we may neglect the first term.

By making a judicious choice of the exposure, the transparency of the photographic plate is given by:

$$T_A = T_0 + 2B \, a_r a_s \, \text{Cos} \, [\phi_s(x,y) + \alpha x] \tag{7-6}$$

where T_0 is a constant which depends on a_r, and B a constant.

7-4   Reconstruction

If we illuminate the hologram by a beam (Fig. 7-4) which is identical to the reference beam $a_r \, \text{Cos} \, (\omega t - \alpha x)$, the transmitted wave will be given by the expression:

$$E_t = T_A \, a_r \, \text{Cos} \, (\omega t - \alpha x) = T_0 \, a_r \, \text{Cos} \, (\omega t - \alpha x)$$

$$+ \, 2B \, a_r^2 \, a_s \, (x,y) \, \text{Cos} \, (\omega t - \alpha x) \, \text{Cos} \, [\phi_s(x,y) + \alpha x]$$

$$= T_0 \, a_r \, \text{Cos} \, (\omega t - \alpha x) + B \, a_r^2 \, a_s \, (x,y) \, \text{Cos} \, [\omega t + \phi_s(x,y)]$$

$$+ \, B a_r^2 \, a_s \, (x,y) \, \text{Cos} \, [\omega t - 2\alpha x - \phi_s(x,y)] \tag{7-7}$$

We can distinguish the following three terms on the right hand side of Eq. 7-7:

1.   The term $T_0 \, a_r \, \text{Cos} \, (\omega t - \alpha x)$ represents the attenuated incident wave.

2.   The term $B a_r^2 \, a_s \, (x,y) \, \text{Cos} \, [\omega t + \phi_s \, (x,y]$, within a constant factor $B a_r^2$, represents the wave surface identical to the wave surface emitted by the object. Thus, this wave surface when projected back toward the illuminating source, seems to emanate from an apparent object located from a place where the original object was located. We say that these waves produce a virtual image, and an observer who looks through the plate would see the object in three dimensions. This virtual image can be photographed by a camera (Fig. 7-4).

3.   The term $B a_r^2 \, a_s \, (x,y) \, \text{Cos} \, [\omega t - 2\alpha x - \phi_s \, (x,y)]$ represents a wave surface which is also a replica of the original waves, except that they have conjugate, or reversed, curvature; originally diverging spherical waves from an object point are converted into converging spherical waves. [This is because of the negative sign in front of $\phi_s(x,y)$.] These waves, therefore, produce a real image, which can be photographed

directly without a lens, by placing a photographic plate at the image position. The term $2\alpha x$ represents an angular displacement of this image with respect to the virtual image.*

7-5 Holograms

Holograms and the images they produce have many curious and fascinating properties. The pertinent information recorded on the holographic film can be seen under magnification and consists of highly irregular fringes that bear no apparent relation to the subject. It is quite unlikely that one could learn to interpret a hologram visually without actually reconstructing the image (see Plate 11).

When the hologram is placed in a beam of coherent light, however, the images embodied in it are suddenly revealed. The identity between the reconstructed waves and the original waves that impinged on the plate when the hologram was made implies that the image produced by the hologram should be indistinguishable in appearance from the original object. This identity is in fact realized. The virtual image, for instance, which is seen by looking through the hologram as if it were a window, appears in complete three-dimensional form, and this three-dimensional effect is achieved entirely without the use of stereo pairs of photographs and without the need for such devices as stereo viewers.

The image has additional features of realism that do not occur in conventional stereo-photographic imaging. For example, as the observer changes his viewing position the perspective of the picture changes, just as it would if the observer were viewing the original scene. Parallax effects are evident between near and far objects in the scene; if an object in the foreground lies in front of something else, the observer can move his head and look around the obstructing object, thereby seeing the previously hidden object. Moreover, one must refocus one's eyes when the observation is changed from a near to a more distant object in the scene. In short, the reconstruction has all the visual properties of the original scene, and we know of no visual test one can make to distinguish the two.

Similarly, the real image can be viewed by an observer, who will find it suspended in space between himself and the plate.

A hologram made in the manner just described has several interesting properties in addition to those having to do with the three-dimensional nature of its reconstruction. For example, each part of the hologram, no matter how small, can reproduce the entire image; thus the hologram can be broken into small fragments, each of which can be used to construct a complete image. As the pieces become smaller, resolution is lost,

* In the reconstruction process can we use a laser beam emitting light whose wavelength is different from that used in the recording process?

since resolution is a function of the aperture of the imaging system. This curious property is explained on the basis of an observation made— each point on the hologram receives light from all parts of the subject and therefore contains, in an encoded form, the entire image.

Another curious property of the wavefront reconstruction process is that it does not produce negatives. The hologram itself would normally be regarded as a negative, but the image it produces is a positive. If the hologram were copied by contact printing, the hologram would be reversed in the sense that opaque areas would now become transparent and vice versa. The image reconstructed from the copy, however, would remain a positive and would be indistinguishable from the image produced by the original except for the small degradation in quality that normally occurs in photographic copying. This property arises because the information on the hologram is embodied in the fringe contrast and in the fringe spacings; neither of these is altered by the reversal of polarity.

7-6 Requirements

Wavefront reconstruction photography, although appearing to offer exciting possibilities, has in the past been confined to the laboratory. The major reason for this is the strict coherence requirements for the light source used in the process. Ordinary light lacks this coherence property, and sources of coherent light have been 'commercially' available only in the recent years.

For holography, the first requirement of the light sources is of monochromaticity. Monochromaticity is required because the fringe pattern generated by the interference process is a function of the wavelengths of the illumination. If the spectrum of the light is broad, each wavelength component produces its own separate pattern, and the result of many wavelength components is to average out the fringes to a smooth distribution.

The other requirement is of spatial coherence (see Chapter 5). If the source lacks spatial coherence (i.e. if it is broad), then each element of the source produces interference fringes that are displaced from those of other elements; the sum of many such sets of fringes averages to some very nearly uniform value and the fringe pattern is absent.

It is possible to meet both coherence requirements using traditional sources, such as a mercury-arc lamp. Monochromaticity is obtained by passing the light through an optical device, such as a monochromator or a narrow-band colour filter. This process discards all spectral components except those in a narrow band. Spatial coherence is obtained by focusing the light onto a pinhole. Since only a small fraction of the total light output of the lamp can be focused onto the pinhole, the traditional source is quite inefficient, and only an extremely small fraction of the total light emission is available for illumination of the object.

The light produced by a laser, on the other hand, is highly mono-chromatic and has extraordinary spatial coherence, thus making the wasteful processes described above unnecessary. The available light is several orders of magnitude greater than the monochromatic, spatially coherent light available from other sources. Hence the laser is greatly superior to all other known sources for wavefront reconstruction photography and is certainly in large part responsible for the interesting results that have been achieved.

Aside from the stringent coherence requirements on the source, the most important precaution in making a hologram is the stability of the platform on which the hologram is made. Specifically, it is essential that the film, the scattering object and any mirrors used to direct the reference beam to the film be motionless with respect to one another during the exposure.

As laser sources improve, wavefront reconstruction photography may emerge from the laboratory and become, *through its remarkable three-dimensional imaging properties, an important photographic method for simulation and training devices and for applications in which a highly exact reproduction of the object is required.*

Pennington* has described some of the progress made in holography. With pulsed laser sources, instant holograms have been made of three-dimensional volumes that contain, for example, a fog like suspension of small particles. Once the image has been reconstructed, a microscope can be focused on an individual particle, and particles can be studied and counted. Multicolour holograms, which utilize the fact that photographic emulsion has thickness, have also been made. Just as the reflection of X-rays in a crystal can reveal the atomic spacing within the crystal (see Chapter 6), successive layers in the emlusion can reflect a particular colour strongly if their spacing is right for that colour and angle of illumination.

Example

Conceptually, the simplest form of a hologram is one for which the object is just a single, infinitely distant point, so that the object wave at the recording medium is a plane wave [Fig. 7-5(a)]. If the reference wave is also plane and incident on the recording medium at an angle to the object wave, show that the hologram will consist of a series of Young's interference fringes.† Show that when the hologram is illuminated with a plane wave [Fig. 7-5(b)], the transmitted light consists of a zero-order wave travelling in the direction of the illuminating wave and two first order waves.

* K. S. Pennington, 'Advances in Holography', *Scientific American*, February 1968.
† This will be similar to a grating but the intensity distribution will be $\cos^2$ type (see Fig. 4-12).

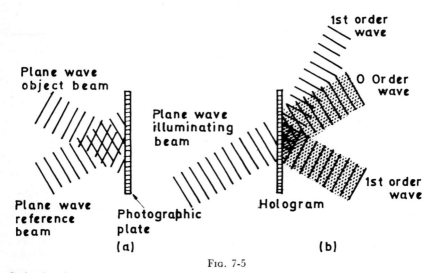

FIG. 7-5

Solution*

It is left as an exercise to show that when the two beams (Fig. 7-5) super-pose, the intensity distribution is of the Cos² type.

In order to calculate the intensity of the diffraction pattern, we first note that for a single slit (Fig. 6-3) one can calculate the intensity distribution by carrying out a simple integration. We consider a small element ds along the width of the slit at a distance s from the centre (Fig. 6-3). The amplitude at an angle θ (i.e. at the point P in Fig. 6-3), due to this element would be proportional to $[\mathrm{Sin}\,(\omega t - kr - ks\,\mathrm{Sin}\theta)]ds$ where $k = 2\pi/\lambda$.

Thus the resultant electric field will be of the form†

$$E = C \int_{-b/2}^{+b/2} \mathrm{Sin}\,[\omega t - kr - ks\,\mathrm{Sin}\theta]\,ds$$

where C is a constant. Splitting the integral, we get

$$E = C\,[\mathrm{Sin}\,(\omega t - kr) \int_{-b/2}^{+b/2} \mathrm{Cos}\,(ks\,\mathrm{Sin}\theta)\,ds$$

$$- \mathrm{Cos}\,(\omega t - kr) \int_{-b/2}^{+b/2} \mathrm{Sin}\,(ks\,\mathrm{Sin}\theta)\,ds]$$

$$= 2C\,\mathrm{Sin}\,(\omega t - kr)\,\frac{\mathrm{Sin}\left(\dfrac{kb}{2}\,\mathrm{Sin}\theta\right)}{\dfrac{kb}{2}\,\mathrm{Sin}\theta}$$

* The solution is a bit mathematical, hence it may be omitted in the first reading.

† For further details, see F. A. Jenkins and H. E. White, *Fundamentals of Optics*, Chapter 15, Third Edition, McGraw-Hill (1957).

where the second integral is zero because the integrand is an odd function

of s. Thus the amplitude is of the form $\dfrac{\text{Sin}\left[\dfrac{\pi b \text{ Sin}\theta}{\lambda}\right]}{\dfrac{\pi b \text{ Sin}\theta}{\lambda}}$ which is precisely

of the same form as Eq. 6-14.

Now in the given problem

$$E = C \int_{-b/2}^{+b/2} \text{Cos}^2 \alpha s \text{ Sin} \left[\omega t - kr - ks (\text{Sin}\theta - \text{Sin } i)\right] ds$$

where i is the angle of incidence (see Problem 9, Chapter 6), the term $\text{Cos}^2 \alpha s$ represents the intensity distribution of the hologram due to the superposition of the two plane waves, and b, the width of the hologram. Now we can write the above integral as

$$E = \tfrac{1}{2}C \int_{-b/2}^{+b/2} (1 + \text{Cos} 2\alpha s) \Big[[\text{Sin}(\omega t - kr) \text{ Cos} \{ks (\text{Sin}\theta - \text{Sin } i)\}]$$
$$- \text{Cos } (\omega t - kr) \text{ Sin} \{ks (\text{Sin}\theta - \text{Sin } i)\} \Big] ds$$

or

$$E = C \text{ Sin } (\omega t - kr) \Big[\int_0^{b/2} \text{Cos} \{ks (\text{Sin}\theta - \text{Sin } i)\} ds$$
$$+ \tfrac{1}{2}\int_0^{b/2} \text{Cos} [\{k (\text{Sin}\theta - \text{Sin } i) + 2\alpha\}s] ds$$
$$+ \tfrac{1}{2}\int_0^{b/2} \text{Cos} [\{k (\text{Sin}\theta - \text{Sin } i) - 2\alpha\}s] ds \Big]$$

If we carry out the integration, then for a large value of b the first term will be appreciable only in directions given by $\text{Sin}\theta - \text{Sin } i = 0$, which gives the zero order wave; whereas the second and third terms will be appreciable only when $k (\text{Sin}\theta - \text{Sin } i) \pm 2\alpha = 0$, which give the two first order waves. For $b \to \infty$ these will be the only directions where we will have any intensity.

7-7 Holography with Sound

We conclude this chapter by mentioning that high frequency sound, like coherent light, can be used to construct a hologram. A laser beam is then employed to reconstruct the acoustical hologram into a recognizable pictorial image. Since sound waves can penetrate opaque objects ranging from living tissues to metal structures, the new imaging technique has promising applications in many areas of medicine and technology.

In order to produce an acoustical hologram the scene to be recorded is 'illuminated' with a pure tone of sound waves instead of a laser beam.

The objects in the scene disturb the sound waves and produce interference patterns analogous to those produced by light waves, for example, acoustical holograms can be produced in the form of a ripple pattern on a water surface (Fig. 7-6) when sound waves 'ensonified' from below interfere with those from a reference source. A laser and telescope are used to view the 'real time' image. For a detailed discussion on acoustical holography one may look up some of the references given at the end of this chapter.

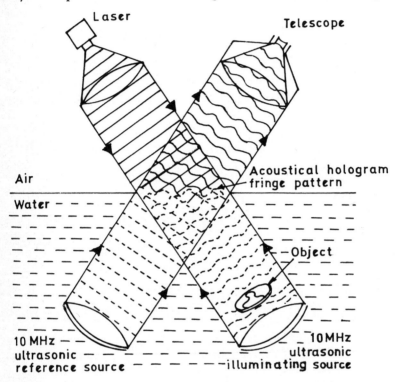

FIG. 7-6: Acoustical holography [After A. F. Metherell, 'Holography with Sound', *Science Journal* Vol. 4 p. 57 November (1968)].

SUGGESTED READING

LEITH, E. N. and J. UPATNIEKS, 'Photography by Laser', *Scientific American*, Vol. 212 p. 24-35, (June 1965). In this article the authors have qualitatively described the theory and the method of holography, illustrating the excellent reconstructions that can be obtained by it.

METHERELL, A. F., 'Acoustical Holography', *Scientific American*, Vol. 221, p. 36 (October, 1969); 'Holography with Sound', *Science Journal* Vol. 4, p. 57, (November 1968). Both the popular articles discuss the construction of holograms using high frequency sound waves.

PENNINGTON, K. S. 'Advances in Holography', *Scientific American*, Vol. 218, p. 40, (February 1968). Recent progresses Holography have been discussed.

SMITH, H. M. *Principles of Holography*, Wiley Interscience (1969). A thorough mathematical treatment of holography has been given in this book. (The mathematical level is quite high.)

VERBIEST, R. *'Holographie'*: An Optics Technology Report. This report, in French, discusses the elementary theory of holography.

Particle Nature of Radiations: Photons

Two voices are there: one is of the sea, one of the mountains; each a mighty voice.

— William Wordsworth

8-1 Introduction

We have studied the propagation, diffraction, interference and polarization of light. All these optical phenomena can be quantitatively explained on the basis of the wave nature of light. However, there are many other optical phenomena which exhibit a different aspect of the nature of light and suggest that it should be regarded not as a wave but as a stream of particles. In this chapter we will discuss several phenomena which exhibit the particle nature of light and will interpret them in terms of the so-called *photon theory.* In Chapter 9 we will show that the wave and particle aspects of radiation are not necessarily mutually inconsistent but that they can be regarded as two aspects of a single fundamental phenomenon.

8-2 The Photoelectric Effect

Towards the end of the 19th Century a number of experiments revealed the emission of electrons from a metal surface when light (particularly ultraviolet light) is incident on it. This phenomenon is known as the photoelectric effect. Fig. 8-1 shows an apparatus to study the photoelectric effect. Monochromatic light, falling on metal plate A, will liberate

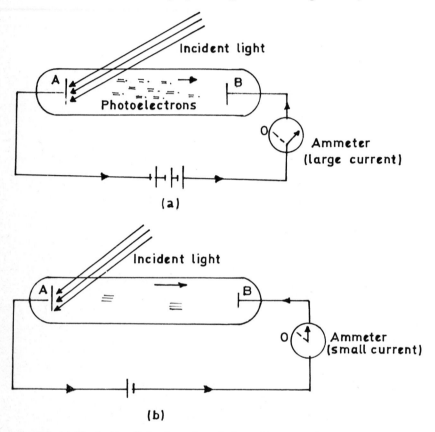

(a)

(b)

FIG. 8-1: (a) The incident light ejects photoelectrons from the metal plate A. These electrons are collected by a plate which is kept at a higher potential with respect to A by means of a battery. The arrows show the normal direction of the current (which is opposite to the direction of flow of electrons).

(b) A method for detecting the photoelectrons of maximum kinetic energy. As the electrode B is made more negative, the slower photoelectrons are repelled before they can reach it. Finally, for a particular value of the applied nagative voltage, the current becomes zero indicating that this voltage corresponds to the maximum photoelectron energy.

photoelectrons, which can be detected as a current if they are attracted to the metal plate B, by means of a potential difference V applied between A and B. The ammeter measures this *photoelectric current*. Fig. 8-2 is a plot of the photoelectric current, developed in an apparatus like that of Fig. 8-1, as a function of the potential difference V. If the value of V is made large enough, the photoelectric current reaches a certain limiting value at which all photoelectrons ejected from plate A are collected by B.

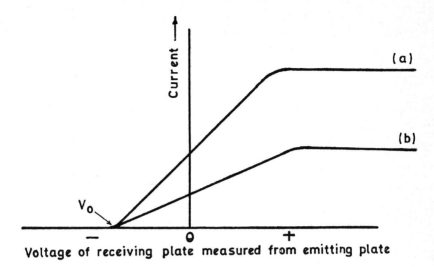

FIG. 8-2: Plots of photoelectric current versus the potential difference, V, between the two electrodes A and B shown in Fig. 8-1. V is positive when the plate B in Fig. 8-1 is positive with respect to the emitting surface A. Note that the current becomes zero only when the receiving plate is a volt or so negative. This value is the cut-off potential, V_0. [In curve (*b*) the incident light intensity has been reduced to one half that of curve (*a*); the frequency is the same.]

If V is reversed in sign, the photoelectric current does not immediately drop to zero, which proves that the electrons are emitted from A with a finite velocity. Some will reach the plate B inspite of the fact that the electric field opposes the motion [Fig 8-1(*b*)]. However, if this reversed potential difference is made large enough, a value V_0 (the *stopping potential*) is reached at which the photoelectric current does drop to zero (Fig. 8-2). This potential difference V_0, multiplied by the electron charge e, measures the kinetic energy K_{max} of the fastest ejected photoelectron. In other words,

$$K_{max} = eV_0 \tag{8-1}$$

For V_0 equal to 2 Volts (which is a typical value) the value of K_{max} is given by

$$K_{max} = (1 \cdot 6 \times 10^{-19} \text{ coulomb}) \times (2 \text{ Volts})*$$
$$= 3 \cdot 2 \times 10^{-19} \text{ joules}$$
$$= 3 \cdot 2 \times 10^{-12} \text{ ergs}$$
$$= 2 \text{ eV}$$

The above described phenomenon of photoelectric effect, at first glance

* Since $e = 1 \cdot 6 \times 10^{-19}$ Coulombs (in MKS units)
  $= 4 \cdot 8 \times 10^{-10}$ cgs units

does not surprise us, for light waves carry energy, and some of the energy absorbed by the metal may somehow concentrate on individual electrons and reappear as kinetic energy. Upon closer inspection of the data, however, we find that the photoelectric effect can hardly be explained in so straight forward a manner.

The first peculiarity of the photoelectric effect is that a bright light yields more electrons than a dim light, but their average energy remains the same.* Thus, K_{max} turns out to be independent of the intensity of the light as shown by curve (b) in Fig. 8-2, in which the light intensity has been reduced to one-half. That V_0 is independent of the light intensity has been tested over intensities range of 10^7.

A second peculiarity of the photoelectric effect is the fact that even when the metal surface is faintly illuminated, the photoelectrons leave the surface immediately. This behaviour contradicts the wave theory of light, which predicts that if the beam has a very low intensity, a certain period of time must elapse on the average before any individual electrons accumulate enough energy to leave the metal.†

An example will illustrate why the instantaneous emission of photoelectrons is so remarkable. A detectable photoelectric current results when sodium is irradiated with 10^{-10} watts/sq.cm of violet light.‡ There are about 10^{16} atoms per square cm‡‡ in the upper ten layers of sodium so that, if we arbitrarily assume that the incident light is absorbed in these layers, each atom receives energy at the rate of 10^{-26} watt, somewhat less than 10^{-7} eV/second. Energy of the order of 1 eV is needed by an electron in order to make it emerge from a metal surface, so that, on the average, 10^7 seconds ($\approx$ few months) would have to elapse before any individual electron would accumulate enough energy to become a photoelectron. Actually as we said above, there is no detectable time lag.

Finally, another major feature of the photoelectric effect that cannot be explained in terms of the wave theory of light is the fact that the photoelectron energy depends on the frequency of the light employed. At frequencies below a certain critical frequency (which is characteristic of the particular metal), no electrons are emitted. Above this threshold frequency, the photoelectrons have a range of energies from zero to a certain maximum value, and this maximum energy increases with increasing frequency. High frequencies result in high maximum

---

* Wave theory suggests that the kinetic energy of the photoelectrons must increase as the light beam is made more intense.

† One experiment that was designed to reveal the time lag if it were longer than 3×10^{-9} sec, gave a negative result [E. O. Lawrence and J. W. Beams, *Physical Review*, Vol. 32, p. 478 (1928)].

‡ 1 watt $= 1$ joule/sec $= 10^7$ ergs/sec
$= 6 \cdot 24 \times 10^{18}$ electron Volts/sec
(1 eV $= 1 \cdot 6 \times 10^{-12}$ erg)

‡‡ Show this.

photoelectron energies, low frequencies in low photoelectron energies. Thus, a faint blue light produces electrons with more energy than those produced by a bright yellow light, although the latter yields a greater number of them.

Fig. 8-3 is a plot of maximum photoelectron energy against the

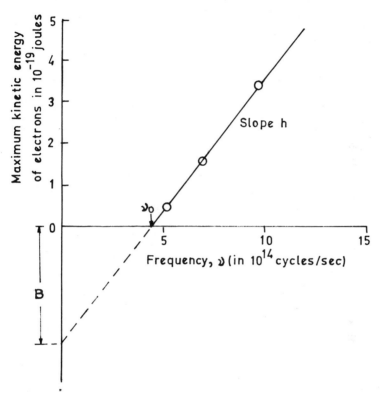

FIG. 8-3: Results of typical experiments with the apparatus shown in Fig. 8-1 and using various light sources and filters of different colours. The plot shows the variation of the maximum kinetic energy of the emitted photoelectrons with frequency for sodium. The cut-off frequency is 4.39×10^{14} cycles/sec. (The quantity measured is the stopping potential, V_0.)

frequency v, of the incident light in a particular experiment. It is seen that for any particular substance, the curve of maximum kinetic energy versus frequency is a straight line with a slope h. And this slope h is the same for all materials. Only the intercept, B, changes from one material to another (Fig. 8-3). Consequently, the plots for all substances can be described by the relation

$$K_{max} = -B + hv \qquad (8\text{-}2)$$

where h is always the same, B a constant in any particular experiment and ν, the frequency of light.

8-3 Einstein's Photon Theory

Einstein* succeeded in explaining the photoelectric effect by making a remarkable assumption, namely, that the energy in a light beam travels through space in concentrated bundles, called *photons*. Each photon carries the energy $h\nu$ (or perhaps the photon is the energy $h\nu$). When the photon hits the surface it disappears, and the energy $h\nu$ is transferred. Some of this energy is needed to bring the photoelectron out of the surface that becomes the threshold value B ($= h\nu_0$). Photons with less energy than that value cannot eject any electron. But if a photon has more energy than the threshold value, the surplus appears as kinetic energy of the emitted electron. However, some electrons may come out by indirect paths having lost some energy in atomic collisions on the way out.

Einstein's photon hypothesis meets all the objections against the wave theory interpretation of the photoelectric effect. For example, doubling the light intensity merely doubles the number of photons and thus doubles the photo electric current;† it does not change the energy ($= h\nu$) of the individual photons or the nature of the individual photoelectric processes.

Further, the fact that there is no measurable time lag between the impinging of the light on the surface and the ejection of the photoelectron follows immediately from the photon theory because the required energy is supplied in a concentrated bundle. It is not spread uniformly over a large area, as in the wave theory.

Thus, according to Einstein's photon theory, in any beam of light of frequency ν, each photon carries an energy

$$E = h\nu \qquad (8\text{-}3)$$

This is the famous Einstein-Planck relation. The constant h is known as the Planck's constant and is directly measurable from the experiment discussed above. The value of h is 6.62×10^{-27} ergs second. From the

---

* A. Einstein (1879-1955). Formerly director of the Kaiser Wilhelm Institute in Berlin, Einstein in 1935 came to the Institute for Advanced Study at Princeton, U.S.A. His work on photoelectric effect appeared in *Annals der. Physik*, Vol. 17, p. 132 (1905). It is one of the marvels of the history of physics that in a single year, 1905, Einstein produced this theory of photoelectric effect, the theory of Brownian motion and the Special Theory of Relativity. In Einstein's own words, "It strikes me as unfair, and even in bad taste, to select a few...for boundless admiration, attributing superhuman powers of mind and character to them. This has been my fate".

† It is the phenomenon of photoelectric effect which is used to measure light intensity in a photomultiplier tube. A special light-sensitive surface (like a thin layer of antimony-cesium) emits photoelectrons which are collected by a dynode (kept at a higher potential). The dynode emits secondary electrons which are collected by the second dynode. There are usually quite a few dynodes and the current builds up.

value of h, we can easily find the energy of photons. Visible light has a wavelength of about 5×10^{-5} cm . Its frequency is

$$\nu = \frac{c}{\lambda} = \frac{3 \times 10^{10} \text{ cm/sec}}{5 \times 10^{-5}} = 6 \times 10^{+14} \text{ sec}^{-1} \tag{8-4}$$

Consequently, an average visible photon carries the energy $E = h\nu$ equal to

$$(6 \cdot 62 \times 10^{-27} \text{ ergs sec}) \times (6 \times 10^{+14} \text{ sec}^{-1})$$
$$\approx 4 \times 10^{-12} \text{ ergs} = 2 \cdot 5 \text{ eV}$$

Thus, the whole range of visible light photons extends from an energy of around 2 eV for the red to somewhat above 3 eV for the blue. (These energies, of a few eV per photon, are exactly in the lowest energy range which can cause important chemical changes presumably because they are energies high enough to disturb the electrons within an atom or even eject them. For these energies, then, but not for lower values, photographic chemicals and photoelectric surfaces are sensitive.)

From the value for the energy of a single photon, we can find the number of photons streaming by in any radiation beam whose total energy rate we know. Then it is interesting to compare the total number of photons striking a photoelectric surface with the number of photoelectrons ejected. Even for a very good surface, only a fraction of the photons eject photoelectrons. (The rest send electrons deeper into the surface or are wasted in some other way.)

8-4   Photon Momentum and Compton Effect

For the sake of completeness, we would like to mention that according to the electromagnetic theory of light if an object completely absorbs an energy U, from a parallel light beam that falls on it then the light beam will transfer to the object a linear momentum given by U/c (see Chapter 3, Section 3-8 where experiments which show that electromagnetic waves transport momentum have been discussed). On the photon picture we imagine this momentum to be carried along by the individual photons, each photon transporting linear momentum in amount $p = h\nu/c$, where $h\nu$ represents the photon energy. Thus, we can write*

$$p = \frac{E}{c} = \frac{h\nu}{c} = \frac{h}{\lambda} \tag{8-5}$$

---

* The relationship $p = E/c$ is stated here as a simple fact. It is required by Maxwell's electromagnetic theory of light as waves and by the description (in special theory of relativity) as particles. Certainly the equation $p = E/c$ is a simple one. How does it compare with the relation between kinetic energy K, and momentum p ($= mv$) for a particle described by Newtonian dynamics? In this case we have $p = mv$, $K = \frac{1}{2}mv^2$, $p = \dfrac{2K}{v}$. Just put $v = c$ but now there is a factor of 2. It represents an essential departure from Newton's mechanics for high speeds.

We can verify the photon momentum directly when a single photon collides with an electron. If we use energetic photons, those of X-rays ($h\nu \approx 1000$ eV) or of nuclear gamma rays ($h\nu \approx 10^6$ eV), the atomic binding energy of electrons ($\approx$ few eV) becomes small in comparison. Thus, for such collisions, the electron can be considered to be free and it is knocked out with (nearly) all the energy and all the momentum lost by the photon. This new process, which is schematically shown in Fig. 8-4, is called the *Compton effect*, after its discoverer Arthur H. Compton.

In a Compton collision the photon is not absorbed. After the collision, it appears with reduced energy and momentum, moving in some new direction (Fig. 8-4). The electron recoils from the collision, also carrying

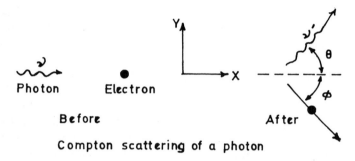

Compton scattering of a photon

Fig. 8-4: A photon of frequency ν is incident on an electron at rest. On collision, the photon is scattered at an angle θ, with decreased frequency ν', while the electron moves off with a velocity v, in direction ϕ. The photon energy ($= h\nu$) is so large compared to the electron binding energy that, before the collision, the electron may be assumed to be at rest.

off some energy and some momentum. In each collision of this kind, the energy and momentum are simultaneously conserved. Conservation of energy, leads to

$$h\nu = h\nu' + E_k \tag{8-6}$$

where ν' is the frequency of the scattered photon and E_k, the final kinetic energy of the electron. The conservation of x-and y-components of momentum leads to the following equations:

$$\frac{h\nu'}{c} \cos\theta + p \cos\phi = \frac{h\nu}{c} \tag{8-7}$$

and

$$\frac{h\nu'}{c} \sin\theta - p\sin\phi = 0 \tag{8-8}$$

where p is the momentum of the electron after collision . In Newtonian mechanics the kinetic energy E_k will be related to p by the relation:

$$E_k = \frac{p^2}{2m} \qquad (8-9)*$$

Eqs. 8-7 and 8-8 can be rewritten in the form:

$$p \cos \phi = \frac{h\nu}{c} - \frac{h\nu'}{c} \cos\theta \qquad (8-10)$$

$$p \sin\phi = \frac{h\nu'}{c} \sin\theta \qquad (8-11)$$

If we square and add, we obtain

$$p^2 = \frac{h^2\nu^2}{c^2} + \frac{h^2\nu'^2}{c^2} - \frac{2h^2\nu\nu'}{c^2} \cos\theta \qquad (8-12)$$

If we use Eqs. 8-6 and 8-12, we obtain

$$h\nu = h\nu' + \frac{p^2}{2m}$$

$$= h\nu' + \frac{h^2}{2mc^2} [\nu^2 + \nu'^2 - 2\nu\nu' \cos\theta]$$

or

$$(h\nu - h\nu') = \frac{h^2}{2mc^2} (\nu - \nu')^2 + \frac{h^2 2\nu\nu'}{2mc^2} (1 - \cos\theta)$$

or

$$(h\nu - h\nu')\left[1 - \frac{h\nu - h\nu'}{2 mc^2} \right] = \frac{h^2}{mc^2} \nu\nu' (1 - \cos\theta)$$

If we multiply the above equation by $\dfrac{c}{h\nu\nu'}$ we would obtain

$$\lambda' - \lambda = \left[1 - \frac{h\nu - h\nu'}{2mc^2} \right]^{-1} \frac{h}{mc} (1 - \cos\theta) \qquad (8-13)$$

Eq. 8-13 is valid only in the non-relativistic approximation, i.e. when the speed of the electron (after the collision) is small compared to that of c. Now according to the theory of relativity the mass of a particle, which is moving with speed v, is given by

$$m = \frac{m_0}{\sqrt{1 - \left(\dfrac{v}{c}\right)^2}} \qquad (8-14)$$

where m_0 is referred to as the *rest mass* and c is the speed of light in vacuum. Further, the total energy, E, and the momentum, p, of a particle which is moving with velocity v are given by

* $p = mv$ and $E_k = \frac{1}{2} mv^2$. Therefore $E_k = \dfrac{1}{2m} p^2$.

$$E = mc^2 = \frac{m_0 c^2}{\sqrt{1 - \left(\dfrac{v}{c}\right)^2}} \tag{8-15}$$

and

$$p = m\boldsymbol{v} = \frac{m_0 v}{\sqrt{1 - \left(\dfrac{v}{c}\right)^2}} \tag{8-16}$$

Using Eqs. 8-15 and 8-16 we can obtain

$$E^2 = p^2 c^2 + m_0{}^2 c^4 \tag{8-17}$$

The total energy conservation condition can be written as

$$h\nu + m_0 c^2 = h\nu' + mc^2$$

or

$$\frac{hc}{\lambda} - \frac{hc}{\lambda'} + m_0 c^2 = mc^2 \tag{8-18}$$

Further, the condition for the conservation of the x- and y-components of momentum gives us

$$\frac{h}{\lambda} = \frac{h}{\lambda'} \; \mathrm{Cos}\theta + mv \; \mathrm{Cos}\phi \tag{8-19}$$

and

$$0 = \frac{h}{\lambda'} \; \mathrm{Sin}\theta - mv \; \mathrm{Sin}\phi \tag{8-20}$$

Using Eqs. 8-14, 8-18, 8-19 and 8-20, if we carry out simple algebraic manipulations (Problem 5) we would obtain

$$\Delta\lambda \; (= \lambda' - \lambda) = \frac{h}{m_0 c} \; (1 - \mathrm{Cos}\theta) \tag{8-21}$$

Thus, the 'Compton shift' $\Delta\lambda$ depends only on the scattering angle θ and not on the initial wavelength. If we compare Eq. 8-13 with Eq. 8-21, we find that the 'non relativistic' expression will be valid only when

$$\frac{h\nu - h\nu'}{2mc^2} \ll 1$$

i.e. when the energy transferred to the electron is small compared to $2 \, mc^2 \; (\approx 1 \text{ MeV})$.

The scattering phenomenon which we have just discussed was observed experimentally by A.H. Compton in 1922. The experimental arrangement is shown in Fig 8-5. Monochromatic X-rays (of known wavelength) are directed at a target, and the wavelengths of the scattered X-rays are determined at various angles θ. *The results, shown in Fig. 8-6, exhibit the*

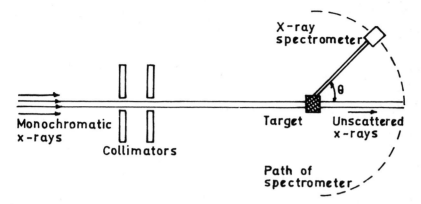

FIG. 8-5: Experimental set-up for observing Compton effect. The spectrometer measures the wavelength of X-rays using the principle of Bragg reflection.

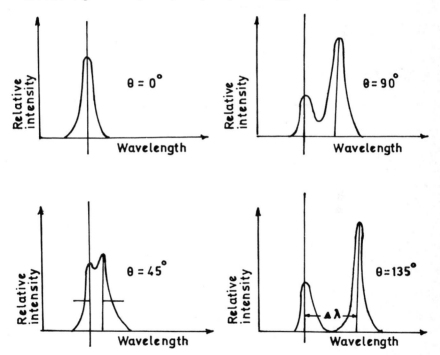

FIG. 8-6: A typical set of measurements for observing Compton effect. The solid vertical line on the left corresponds to the wavelength of incident X-rays, λ, and that of the right to λ'. Note that the shift $\Delta\lambda$ increases with the angle of scattering.

wavelength shift predicted by Eq. 8-21, but at each angle the scattered X-rays included a substantial portion having the initial wavelength. This can be explained in the following manner: In the derivation we assumed that the scattering particles are free to move, a reasonable assumption since many of the electrons in matter are only loosely bound to their parent atoms. Other electrons, however, are very tightly bound and, when struck by a photon, the entire atom recoils instead of the single electron. In this event the value of m_0, which should be used in Eq. 8-21 is that of the entire atom, usually tens of thousands of times greater than that of an electron, with the resulting Compton shift almost undetectable. For example, for a graphite target, m_0 has to be replaced by approximately $12 \times 1840 \times m_e$, where m_e is the electron rest mass.

Example
Light of wavelength 2.5×10^{-5} cm falls on an aluminium surface. In aluminium 4.2 eV are required to remove an electron. Calculate (a) the kinetic energy of the fastest electron, (b) the stopping potential and (c) the cut-off wavelength.

Solution
It can easily be shown (see Problem 1 at the end of this chapter) that the wavelength and energy of a photon are related by the following equation:

$$\lambda E = 12397 \text{ eV\AA}$$

Therefore, for $\lambda = 2500$ Å, $E = 4.96$ eV. Thus the kinetic energy of the fastest electron will be 0.76 eV and the stopping potential will be 0.76 Volts. Further, $\lambda_{\text{cut-off}} = \dfrac{12397}{4.2} \approx 3000$ Å

Example
A 600 watt sodium vapour lamp radiates uniformly in all directions. Find the number of photons (per second per sq. cm) striking a small area of a table top which is at a distance of about 3 metres from the lamp. (The surface area of the table top may be assumed to be at right angles to the direction of propagation.)*

Solution
The intensity will be given by

$$I = \frac{600}{4\pi (3)^2} \approx 5 \text{ watt/m}^2 \approx 3.12 \times 10^{19} \text{ eV/sec m}^2$$

(This would correspond to an amplitude of 61.3 V/m.)
Now, since $\lambda \approx 5890$ Å the photon energy will be given by (see Problem 1)

$$E = \frac{12397}{5890} \text{ eV} = 2.1 \text{ eV}$$

* What will happen if the top is not at right angles?

The number of photons, $\mathcal{N}$, striking 1 sq. cm ($= 10^{-4}$ m²) of the table top will be given by

$$I = \mathcal{N} \times E$$

or

$$\mathcal{N} = \frac{3 \cdot 12 \times 10^{19} \text{ eV/sec m}^2}{2 \cdot 1 \text{ eV}} \approx 1 \cdot 5 \times 10^{19} \text{ photons/sq. m/sec}$$

$$= 1 \cdot 5 \times 10^{15} \text{ photons/sq cm/sec.}$$

8-5 Rest Mass of the Photon

From the special theory of relativity, the total energy E, is related to the rest mass m_0 and momentum p, by the relation (see Eq. 8-17):

$$E^2 = m_0{}^2 c^4 + c^2 p^2$$

Now, for a photon $E = pc$
Thus,

$$m_{\text{ph}} = 0, \text{ (the subscript 'ph' stands for photon)}$$

i.e. *the rest mass of the photon is zero.*

At first sight this result might appear slightly peculiar because the photon has some particle properties and it ought to have a mass when observed in its rest frame. However, *three is no inertial frame in which the photon is at rest* — electromagnetic radiation propagates with velocity c, in every inertial frame. A photon at rest is therefore a meaningless concept.

One might say that an object which can never be at rest should not be called a 'particle'. However, it has become customary to talk about 'massless particles' of which the 'photon' and the 'neutrino' are examples. Finally, it is purely a matter of choice how we define the word 'particle'. A photon is no billiard ball — it merely has some properties in common with billiard balls.

8-6 Splitting of Photon

Is it possible to split a photon of frequency ν into two parts, such that each part carries some fraction of the energy $h\nu$, but still have the frequency ν? In order to answer the above question, let us consider again the photoelectric effect. In the apparatus, described by Fig. 8-1, we set the retarding potential in the cell at such a value such that the photo-sensitive surface will detect a wavetrain, provided the energy carried by the wavetrain exceeds a certain minimum value E_{min}, such that

$$h\nu > E_{\text{min}} > \tfrac{3}{4} h\nu$$

where ν is the frequency of light. (We arbitarily pick 3/4, as a number larger than $\tfrac{1}{2}$ but smaller than 1.) Therefore if we concentrate the entire energy of the incident wavetrain into the photocell,* then the detector will click. However, if only half of this energy reaches the cell, the detector will not click because then the energy imparted to the emitted electron could not possibly overcome the retarding potential.

According to the clasical picture, an experimental arrangement shown in Fig. 8-7 should split a wavetrain. Light from a source† of low inten-

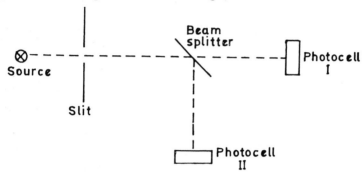

FIG. 8-7: An experimental arrangement which 'splits' a light beam.

sity is allowed to fall on a beam splitter, such as a half-silvered mirror (similar to the one used in a Michelson interferometer; see Chapter 4). The experiment can be so arranged that the intensity of the transmitted beam equals the intensity of the reflected beam, so that the intensity of each one of these beams is half the intensity of the beam emerging from the source through the slit. This is indeed a possible realistic experiment and we do find that the intensities of the transmitted and reflected beams are equal.

Let us now consider what will happen when a single wavetrain arrives at the mirror. Classically, we expect that the beam is split into two parts, and in such a way that the energy carried by the transmitted part of the wavetrain is half of the energy of the incident wavetrain. Therefore, neither the photocell I nor the photocell II should ever click. However, the results of experiment show that the light which goes through is still blue, of frequency v and the detector of photocell I does click as long as $hv > E_{min}$, which shows that the energy of the transmitted beam comes in packets of hv. What does happen when the partially silvered mirror is inserted, is that the counting rate is only half of its value in the absence of the mirror. Thus, we may conclude that a photon cannot be split into two photons of the same frequency which carry only a fraction of the energy of the originl photon, i.e. photons do not behave like classical wavetrains in this respect. This conclusion

---

* The photocell is connected to a loudspeaker, which produces a 'click' when a wavetrain is detected.

† The emission of light may be from a mercury atom which has been excited in a collision. The emitted light is blue, of frequency v.

is further supported by the experimental results on Compton effect, emission of X-rays, and other experimental results. In the theoretical analysis of these phenomona, we make the assumption that the relation, $E = h\nu$, always holds and that 'fractional photons' do not exist.

In Chapter 9, we will consider some other experimental results which also have a bearing on the question of whether photons can be 'split'.

8-7 Photons and Electromagnetic Waves

In what has been discussed above where is the wave picture of light? It may seem to have no relationship to what we have just been discussing, but we cannot forget all the evidence for light waves we discussed in the earlier chapters. Somehow photons and electromagnetic waves must fit together and this leads to *quantum mechanics*, an elementary introduction to which will be given in Chapter 9. Here we will look again at the relation between waves and photons.

For radio waves with frequency of 1 mega cycle per second the photons will have the energy

$$= (6{\cdot}62 \times 10^{-27} \text{ erg sec}) \ (10^6 \text{ sec}^{-1})$$
$$= 6{\cdot}6 \times 10^{-21} \text{ ergs}$$
$$= 4 \times 10^{-9} \text{ eV}$$

Such small photon energies mean that a radio signal cannot be detected unless it contains very many photons. The lower limit of a good communications receiver is about 10^{10} photons/second. Among so many photons the average behaviour alone is detectable. Thus, the large number of photons guarantee that the graininess will not show up unless we try very hard to detect it, and hence the result is the smooth transfer of energy to matter which we expect on the basis of wave theory. We might suspect that there are no photons in light of long wavelength. But we now have experimental evidence of the absorption and emission of 'radio-frequency' photons by atoms and molecules. So we know that both the photon and the wave description apply.

As we go to higher frequencies, we pass to the region of visible light. Here the number of photons is large in any light beam that appears bright to our eye. Sunlight, for example, represents a rain of some 10^{17} photons per sq. cm per second. In such a burst the individual arrival of photons are not easily noticed. At the other extreme, however, we can still make out the glow of a faint luminous source which sends to our eye a hundred photons on the area of the pupil each second. So for visible light, both the wave and the photon sort of observations are possible.

The photographic plate and the photoelectric cell are, like the eye, capable of responding to the smooth flow of intense, moderate beams no less than to the fluctuating and chancy arrival of a few photons per second. It is for this reason that it was with the study of visible light and not in the study of the long wavelength portions of the electromagnetic wave spectrum, that the photon idea was first conceived and found staisfactory.

As the frequency of light increases and the photon energy rises, the photon description appears more and more dominant. The photon energy in the X-ray region is large enough so that single photons are easily detectable. The arrival of a single gamma-ray photon, associated with still higher frequency and correspondingly shorter wavelength, is energetically a sizable event. It can be 'counted' with good probability by a varity of 'counters'. Gamma-ray photons, which have energies as great as 10^{14} eV, have been found in the upper atmosphere. Such a photon produces tens of millions of recoil electrons.

On the other hand, as the frequency rises and the wavelength correspondingly diminishes, and wave effects are less obvious. For example, interference and diffraction become more difficult to demonstrate. As we have seen before, the wavelength of X-rays are of the same order as the interatomic spacings in crystals ($\approx 10^{-8}$ cm). Thus, for X-ray, we can hardly make slits narrow enough to exhibit a detectable interference pattern. Although X-ray interference is widely used, the essential diffracting apparatus is rarely man made. Instead we employ the regular atomic patterns of crystals. For gamma-rays, not even crystals will work as diffraction obstacles.

We recapitulate by mentioning that within the vast range of wavelengths, there are three or more regions of approximation which are of interest. In one of these, a condition exists in which the wavelengths involved are very small compared with the dimensions of the equipment available for their study; furthermore the photon energies are small compared to the energy sensitivity of the equipment. Under these conditions we can make a rough first approximation by a method called geometrical optics. If, on the other hand, the wavelengths are comparable to the dimensions of the equipment, which is difficult to arrange with visible light but easier with radio waves, and if the photon energies are still negligibly small, then a very useful approximation can be made by studying the behaviour of the waves (disregarding the photons). This method is based on the classical theory of electromagnetic radiation. Next, if we go to very short wavelengths, where we can disregard the wave character but the photons have a large energy compared with the sensitivity of our equipment, things get simple again. This is the simple photon picture which we have discussed above. It should, however, be noted that Newton's corpuscular picture of light is by no means restored by the photon. It is no more true than the pure electromagnetic picture with its waves of continuously distributed energy. *The true nature of light is more subtle.* The complete picture which unifies the whole into one model, is the so called 'quantum mechanics' or, to be more precise, quantum electrodynamics.

8-8   Matter Waves: Wave Particle Duality for Other Atomic Objects

In 1923, the French physicist Louis de Broglie asked the following question: Since light, which we had thought was a continuous wave, has also a photon nature, can it be that the particles of matter have wave like properties associated with them? He started with the formula for the momentum of a photon

$$\text{momentum} = \frac{h\nu}{c} = \frac{h}{\lambda} \tag{8-22}$$

Hence, for a photon

$$\lambda = \frac{h}{\text{momentum}} \tag{8-23}$$

de Broglie suggested that Eq. 8-23 for wavelength is a general one, applying to material particles as well as to photons. The most direct evidence for matter waves is the observation that beams of particles are diffracted. Indeed, Davisson and Germer in the U.S.A. and G.P. Thomson in the U.K. independently demonstrated that streams of electrons are diffracted when they are scattered from crystals whose atoms have appropriate spacing. The diffraction patterns they observed were in complete conformity with the electron wavelength predicted by Eq. 8-23. Furthermore by changing the momentum of electron the wavelength could also be changed.

For example, an electron ($m = 9 \cdot 10908 \times 10^{-28}$ gm)* moving with 90 eV of kinetic energy moves at a speed of $5 \cdot 6 \times 10^8$ cm/sec. (Show this.) Its momentum is therefore,

$$(0 \cdot 91 \times 10^{-27} \text{ gm}) \times 5 \cdot 6 \times 10^8 \text{ (cm per sec)}$$
$$= 5 \cdot 0 \times 10^{19} \text{ gm cm/sec}$$

Therefore,

$$\lambda = \frac{6 \cdot 6 \times 10^{-27} \text{ erg sec}}{5 \times 10^{-19} \text{ gm cm/sec}}$$
$$\approx 1 \cdot 3 \text{ Å}$$

which is in the range of atomic dimensions; hence such electrons can easily be diffracted by crystals.

The de Broglie relation holds good for every thing; what sort of particles are used is irrelevant. For example, for a microbe whose mass is about 10^{-12} gm having a speed appropriate to a lethargic snail, say about 10 cm per day, the de Broglie wavelength would be given by

---

* Notice the accuracy with which the mass (and similarly the charge) of an electron has been measured. Yet, it can undergo diffraction as if it were a wave.

$$\lambda = \frac{6 \cdot 62 \times 10^{-27} \text{ erg sec}}{10^{-12} \text{ gm} \times 10^{-4} \text{ cm/sec}}$$

or

$$\lambda \approx 6 \times 10^{-11} \text{ cm}$$

The particle involved in this sort of motion, which is about as small a mechanical system as we might expect to find outside of atomic physics, has a wavelength as short as that of a photon of several million electron volts of energy. Such a photon is a nuclear-gamma ray and it is almost impossible to find its wave properties through diffraction. Similarly, the moving microbe of the same wavelength shows us only normal particle properties. We cannot detect its wave nature. Its physical behaviour is quite perfectly described by Newtonian mechanics. An ordinary body like a tennis ball has a mass about 10^{14} times as much. There is no hope of seeing its wave properties at all, and we can understand why the wave nature of matter took so long to discover.

In recent years the intense neutron beams* (available from nuclear reactors) have been used to perform diffraction experiments. The results are so good that one now uses neutron diffraction patterns to find out more about the positions of atoms in crystal gratings.† In a recent paper,‡ Shull has reported a set of observations on the diffraction of monochromatic slow neutrons (wavelength 4·43 Å) upon passage through fine slits of width between 21 and 4·1 microns (1 micron = 10^{-6} metres). The layout of the experimental arrangement is shown in Fig. 8-8. A white neutron beam is Bragg reflected (by a particular set of planes) from a suitably oriented silicon single crystal. The monochromatic beam is now diffracted by a test slit and the intensity of neutrons in a particular direction was obtained by Bragg reflecting the neutrons from another silicon single crystal. The neutrons were then counted by a detector.

A typical intensity distribution is shown in Fig. 8-9. The distribution can be shown to obey the relation (cf. Eq. 6-16):

$$I(\theta) = I_0 \frac{\text{Sin}^2 \left(\frac{\pi b \text{ Sin}\theta}{\lambda} \right)}{\left[\frac{\pi b \text{ Sin}\theta}{\lambda} \right]^2}$$

where b is the width of the slit. According to the above relation the half intensity should occur at an angle of diffraction given by

---

* For a neutron ($m = 1\cdot67482 \times 10^{-24}$ gm) whose kinetic energy is about 0·01 eV, the corresponding wavelength is about 3Å; of the same order as interatomic spacing.

† See, for example, G. E. Bacon, *Neutron Diffraction*, Second Edition, Oxford University Press (1962).

‡ C. G. Shull, 'Single Slit Diffraction of Neutrons', *Physical Review*, Vol. 179, p. 752 (1969).

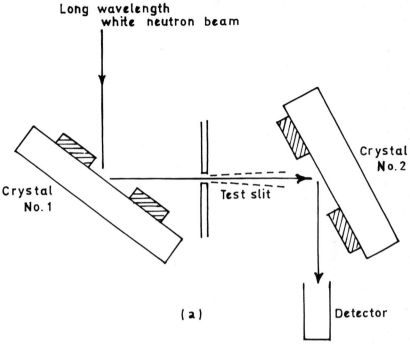

(a)

Test slit (detail)

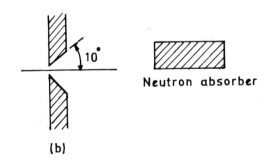

Neutron absorber

(b)

FIG. 8-8: (a) The layout of the experimental arrangement of Shull for observing single slit diffraction of neutrons. Crystal No. 1 is used as a monochromator. Crystal No. 2 'Bragg reflects' the diffracted neutrons. The shaded portions are strong absorbers of neutrons. (b) The detail of the slit construction. (After C.G. Shull *Phys. Rev.*, Vol. 179, p. 752, 1969.)

$$\mathrm{Sin}\,\theta_{1/2} = 0.444\,\frac{\lambda}{b}$$

For $\qquad \lambda = 4.43\ \mathring{\mathrm{A}}$ and $b = 4.1$ microns

$$\theta_{1/2} = \mathrm{Sin}^{-1}\frac{0.444 \times 4.43}{4.1 \times 10^4}$$

$$\approx 4{\cdot}8 \times 10^{-5} \text{ radian}$$

$$\approx 10 \text{ sec}$$

Thus, the half width should be around 20 seconds which agrees very well with the experimental findings (Fig. 8-9).

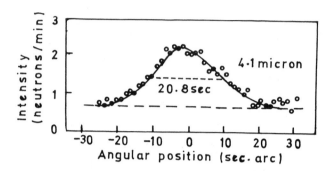

FIG. 8-9: The intensity versus angular position of the crystal No. 2 (Fig. 8-8) corresponding to a slit width of 4.1 microns. The base line was established from intensity measurements well removed from the central position. The low intensity precluded the observation of subsidiary maxima beyond the central peak (After C. G. Shull *Phys. Rev.*, Vol. 179, p. 752, 1969).

The above Fraunhofer diffraction observations indicate that the neutron wavefront as it approaches the slit must be coherent over a transverse width at least that of the largest slit studied by Shull, namely 21 microns.

SUGGESTED READING

BORN, M. *The Restless Universe*, Dover (1951). This is one of the most readable accounts of modern physics ever written. In it, a Nobel Laureate takes the reader step by step through modern developments in our understanding of molecules, atoms, subatomic particles, and nuclear physics providing his own remarkable insights.

HOFFMANN, B. *The Strange Story of the Quantum*, Second Edition, Dover (1959). A completely non-mathematical account of the growth of ideas leading to our present knowledge of the atom.

WEISSKOPF, V. W. 'How Light Interacts with Matter', *Scientific American*, September 1968. The article gives a readable and extensive discussion on the ineraction of light with matter.

WICHMANN, E. H. *Quantum Physics: Berkeley Physics Course Volume 4*, McGraw-Hill Preliminary Edition (1967). Although of a slightly higher level, Chapters 1, 4, 5 and 9 can be read with great profit.

PROBLEMS

1. Show that for a photon the wavelength and the energy are related by the following equation
$$\lambda E = 12397 \text{ eV Å}$$

2. A helium-neon laser continuously emits 1 milli watt of red light ($\lambda \approx 6300\text{A}$) in a beam with a diameter of 1·4 mm. What would be the number of photons crossing an unit area per unit time? ($2\cdot 1 \times 10^{17}$ photons/sq. cm/sec)

3. Green light ejects protoelectrons from a certain surface. Do yo expect electrons to be ejected when the surface is illuminated with (a) red light or (b) violet light? Give reasons.

4. Solar radiation falls on the earth at a rate of 2 cal/cm² minute. Assuming an average wavelength of 5500 Å, how many photons cross 1 sq. cm in one second?

5. Using Eqs. 8-14, 8-18, 8-19, and 8-20 derive Eq. 8-21.

6. Show, by analyzing a collision between a photon and a free electron (using relativistic mechanics) that it is impossible to conserve momentum and energy simultaneously if a photon gives all its energy to the free electron. In other words, photoelectric effect cannot occur for completely free electrons; the electrons must he bound in a solid or in an atom (which takes up the recoil energy).

7. Estimate the number of photons emitted per second by a 100 watt light bulb emitting 10% of its power in the visible region (Assume $\lambda \approx 5000$ Å). What is the light pressure exerted by these photons when they strike (normally) (i) a black body or (ii) a plane mirror 5 metres away?

8. A yellow lamp ($\lambda \approx 6000$ Å) is radiating at a power of 1 watt. Calculate the number of photons entering the pupil of the eye at a distance of 1000 metres (Assume the pupil diameter to be 2 mm).

9. The work function* for potassium is 2·0 eV. When ultraviolet light of wavelength 3·5 $\times 10^{-5}$ cm falls on a potassium surface, show that the maximum energy of the photoelectrons is 1·6 eV.

10. For tungsten, the threshold wavelength for photoelectric emission is 2300 Å. What wavelength of light must be used to eject electrons with a maximum energy of 1·5 eV?
$$(\lambda = 1800 \text{ Å})$$

11. If light transfers energy by means of separate quanta, why do we not perceive a faint light as a series of tiny flashes?

12. An X-ray photon of wavelength 0·20 Å is scattered by 90° after colliding with an electron initially at rest. Show that the shift in wavelength is 0·024 Å (Use Eq. 8-14).

13. A 250 keV photon undergoes a Compton scattering. The kinetic energy of the recoil electron is 200 keV. Calculate the wavelength of the scattered photon.

$$(0\cdot248\text{Å})$$

---

* The work function is the minimum energy needed to dislodge an electron from the illuminated metal surface.

Quantum Behaviour of Photons and Uncertainty Principle

> The completest knowledge of the laws of nature does not carry with it
> the power of prediction, nor of mastery over Nature. *If* the universe is a
> machine its levers and wheels are too fine for our hands to manipulate.
> We can learn and guide its large scale motions only. Beneath our veiled
> sight it quivers its eternal quest.
>
> — Max Born

9-1 Introduction

Isaac Newton (in his OPTICKS, first printed in 1704) described light
as a stream of particles. Then, early in the 19th Century the notion that
light consists of waves, a view already expressed by Christiaan Huygens
in the 17th Century, came into ascendance As we have seen in the earlier
chapters, number of experimental observations were explained quantitatively on the wave theory. In the later part of the 19th Century the wave
nature of light was shown to fit into the electromagnetic theory propounded by James Clerk Maxwell. Later, however it was found that
light did indeed sometimes behaves like a particle (see Chapter 8).
Actually light may be said to behave in a more complex way than either
waves or particles, but these are useful approximations. Arthur
Schawlow* has given an interesting analogy: "It (i.e. the complex
behaviour of light) might be likened to the elephant, as perceived by the
blind men in the fable: to the one who touched the tail, the elephant

---

* Arthur L. Schawlow in the 'Introduction' to *Lasers and Light*, A. H. Freeman and Co.(1969).

was like a rope; to another, who touched a leg, it was like a tree; to others it was like a wall or a hose".

There is one interesting factor to note — electrons also behave like light. Historically, the electron was thought to behave like a particle, and then it was found that in many respects it behaves like a wave. In fact all atomic objects (electrons, photons, neutrons, protons and so on) are 'particle-waves' or whatever one may like to call them, i.e. in some of the experiments they behave like waves and in some other experiments they behave like particles (actually they are like neither). So, in this chapter, whatever we learn about the properties of photons (viz. their quantum behaviour) will apply also to all 'particles'. While understanding the quantum behaviour of photons we will try to resolve the wave-particle controversy. We will first examine a phenomenon which is impossible to explain in any classical way, and has in it the 'basic quantum physics'.

It should be mentioned that because atomic behaviour is so unlike ordinary experience it is very difficult to get used to, and it appears quite mysterious to everyone, both to the beginner and to the experienced physicist. According to Feynman: "Even the experts do not understand it the way they would like to, and it is perfectly reasonable that they should not, because all of direct, human experience and of human intuition applies to large objects. We know how large objects will act, but things on a small scale just do not act that way. So we have to learn about them in a sort of abstract or imaginative fashion and not by connection with our direct experience."

9-2 Quantum Behaviour of Photons

In order to understand the quantum behaviour of photons, we will compare and contrast their behaviour, in a particular experimental* set-up, with the more familiar behaviour of particles like a stream of bullets, and with the behaviour of waves like water waves. These experiments are discussed in great detail by Feynman; we will only give a brief account of them.

9-3 Experiment No. 1: Interference Experiment with Bullets

Let us consider an experimental set-up as shown in Fig. 9-1. The apparatus consists of a gun which randomly shoots (in all directions) a stream of identical and indestructible† bullets. In front of the gun

---

* Some of the experiments discussed in this chapter are adapted from *The Feynman Lectures on Physics, Vol. III*, by R. P. Feynman, R. B. Leighton and Matthew Sands, Addison-Wesley (1965) and from *Quantum Physics: Berkley Physics Course No. 4*, by Eybind H. Wichmann, Preliminary Edition, McGraw-Hill (1967).

† By indestructible bullets we imply that the bullets on striking the surface of the wall do not split into pieces.

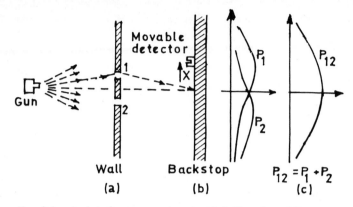

Fig. 9-1:   An interference experiment with bullets. P_1 and P_2 are the probabilities of the arrival of the bullets with hole 1 and with hole 2 open respectively. P_{12} is the probability when both holes are open.

there is a wall with two small holes symmetrically situated with respect to the gun as shown in Fig. 9-1. The holes are just about large enough to let a bullet through. We shall assume that when a bullet happens to hit one of the holes, it bounces off the edges of the hole, and gets scattered to a different direction in a random way. (Thus, after bouncing off, it may end up anywhere on the wall behind.) Beyond the two holes there is a backstop (e.g. a thick wall of wood) which 'absorbs' the bullets when they strike it. On this backstop there is a movable detector (like a box containing sand) which can detect and count the bullets that arrive at a particular point in a given time. (Any bullet which enters the 'box of sand' gets stopped and accumulated. When we wish we can empty the box and count the number of bullets that have been caught.) The position of the detector can be changed at will. With this apparatus we can find the probability* of a bullet, which passes through the holes in the wall, to arrive at the backstop at a distance x from the centre.

Now since the bullets are assumed to be indestructible, when we find something in the detector it is always one bullet or an integral number of bullets. If the rate at which the gun shoots out the bullets is considerably reduced, we find that at any given moment either nothing arrives, or one and only one bullet arrives at the backstop. Thus the 'bullets always arrive in identical lumps'.

For a given rate at which the bullets are fired from the gun we make three measurements:

(i)   We cover up hole 2 and measure the rate† at which the bullets

* By probability we imply the following: If we assume that the gun always shoots at the same rate during the measurements, then the probability is just proportional to the number that reach the detector in some standard time interval.

† It should be noted that since the bullets are fired at random the detector should be allowed to count for a fairly long time, so as to minimize statistical errors.

arrive at the detector for various positions of the detector. This counting rate gives us the probability that a bullet which passes through hole 1 arrives at a certain distance x from the centre. This probability, P_1, is plotted in Fig. 9-1(b). We call the probability P_1 because the bullets come through hole 1. As expected, the maximum of the curve P_1 occurs at the value of x, which is on a straight line with the gun and hole 1.

(ii)   Similarly, if we repeat the experiment with hole 1 closed and hole 2 open, the spatial dependence of the count rate will be similar to that shown by P_2; because in this case we will count only those bullets which pass through hole 2.

(iii)   We carry out the same measurements with both the holes open. The corresponding probability distribution, P_{12}, is plotted in Fig. 9-1(c). The results of experiment will show that

$$P_{12} = P_1 + P_2 \qquad (9\text{-}1)$$

because in the last case the bullets which pass through either of the holes are detected and hence the probabilities just add up. The effect with both holes open is the sum of the effects with each hole open alone.

The curve P_{12} is shown to have a maximum at $x = 0$; it may have a minimum at $x = 0$ if the holes (which are symmetrically situated with respect to $x = 0$) are pushed far away.

The main conclusion that we can draw from the this experiment is that there is no 'interference' between the bullets which arrive from hole 1 with those which arrive from hole 2. Further, the bullets always arrive in identical lumps and the detector measures the probability of arrival of a lump.

9-4   Experiment No. 2: Interference Experiment with Water Waves*

Let us consider an experiment with water waves. The apparatus, shown in Fig. 9-2, consists of a shallow trough of water in which we have a small object which is moved up and down by a motor. The periodic movement of this object gives rise to circular waves. As in the previous experiment, we have to the right of the source a wall with two holes and beyond that is a second wall, which is an 'absorber' so that there is no reflection of the waves that arrive there (something like a sand beach). On this absorber we have a detector which measures the 'intensity' of the wave motion and can be moved back and forth in the x-direction. The 'detector' can be thought of as measuring the height of the wave motion but whose scale is calibrated in proportion to the square of the actual height, so that the reading is proportional to the intensity of the wave.

---

* This experiment has been discussed in Chapter 4 in greater detail. We point out here some of the salient features of the experiment.

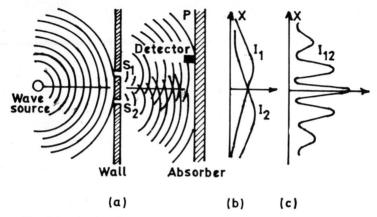

FIG. 9-2: An interference experiment with waves. I_1 and I_2 are the intensity distributions with hole S_1 and with hole S_2 open respectively. I_{12} is the intensity distribution when both holes are open.

Thus, the reading of the detector is proportional to the rate at which energy is carried to the detector.

When we carry out measurements on the above apparatus, we would find that the intensity of the wave, as measured by the detector, can have any value. Thus, when the source moves a very short distance, the intensity recorded by the detector is very small, and on the other hand if there is more motion at the source, the intensity recorded will be larger. Thus, there would not be any 'lumpiness' in the wave intensity.

As in Experiment 1, if we close hole S_2 and if hole S_1 is kept open then we would obtain an intensity pattern something like I_1 shown in Fig. 9-2(b). In this case the waves originating at the source are diffracted at hole S_1, and circular waves spread out from the hole.* The measured intensity pattern is essentially proportional to the square of the amplitude of the diffracted wave. A similar intensity pattern, I_2, will be obtained if hole S_1 is closed and hole S_2 is open.

Now if both the holes are open we get a curve, I_{12}, as shown in Fig. 9-2(c). This observed intensity I_{12} is certainly not the sum of I_1 and I_2 (cf. Eq. 9-1). We are indeed observing the interference of the two waves. Wherever the waves (from the two holes) arrive in phase the wave peaks add together to give a large amplitude and the detector records a large intensity. This gives rise to a maximum in the curve I_{12} and we have 'constructive interference.' This will happen at all points P, for which

$$S_1 P \sim S_2 P = n\lambda, \quad n = 0, 1, 2, \ldots.$$

(See Chapter 4.) On the other hand, when the two waves arrive at the detector out of phase (i.e. having a phase difference of π) we will have

---

* The diameter of the hole is assumed to be small compared to the wavelength.

destructive interference and will get a low value for the intensity, I_{12}.

The relation between I_1, I_2 and I_{12} can be expressed in the following way:

The instantaneous height of the water wave at the detector from hole S_1 can be written as

$$a_1 \, \mathrm{Cos} \, (\omega t + \phi_1)$$

where the amplitude a_1 is, in general, a function of the position of the detector. The intensity, I_1, recorded by the detector is proportional to $a_1{}^2$. Similarly, the instantaneous height of the water wave at the detector from hole S_2 can be written as

$$a_2 \, \mathrm{Cos} \, (\omega t + \phi_2)$$

where again a_2 may be a function of the position of the detector. The intensity, I_2, is proportional to $a_2{}^2$. Finally, when both holes are open, the wave heights add (according to the superposition principle) to give the resultant height as

$$a_1 \, \mathrm{Cos} \, (\omega t + \phi_1) + a_2 \, \mathrm{Cos} \, (\omega t + \phi_2)$$
$$= a \, \mathrm{Cos} \, (\omega t + \phi) \tag{9-2}$$

where

$$a = [a_1{}^2 + a_2{}^2 + 2 \, a_1 a_2 \, \mathrm{Cos} \, \delta]$$

$$\phi = \tan^{-1} \left[\frac{a_1 \, \mathrm{Sin} \phi_1 + a_2 \, \mathrm{Sin} \, \phi_2}{a_1 \, \mathrm{Cos} \phi_1 + a_2 \, \mathrm{Cos} \, \phi_2} \right]$$

and $\delta \, (= \phi_1 - \phi_2)$ is the phase difference between the two disturbances. The measured intensity I_{12} will be proportional to a^2. Thus, omitting the proportionality factor, we may write

$$I_1 = a_1{}^2, \; I_2 = a_2{}^2, \; I_{12} = a_1{}^2 + a_2{}^2 + 2a_1 a_2 \, \mathrm{Cos} \delta \tag{9-3}$$

or

$$I_{12} = I_1 + I_2 + 2 \sqrt{I_1 I_2} \, \mathrm{Cos} \, \delta \tag{9-4}$$

The last term in Eq. 9-4 is the interference term. From Eq. 9-4 we can see that I_{12} can vary between $(\sqrt{I_1} - \sqrt{I_2})^2$ (for $\delta = \pi$) and $(\sqrt{I_1} + \sqrt{I_2})^2$ (for $\delta = 0$). This gives rise to a wavy-like intensity pattern as shown in Fig. 9-2(c). We conclude by noting that for water waves the intensity can have any value and it exhibits interference.

9-5 Experiment No. 3: Interference Experiment with Photons*

Next we imagine a similar experiment with photons with an arrangement shown in Fig. 9-3. The opaque screen has two small holes S_1 and S_2. The light source illuminates the two slits with light (photons) of a well defined frequency ω. For simplicity, we assume that the holes

---

* The corresponding experiment with electrons has been discussed by Feynman. The main conclusions are, however, the same.

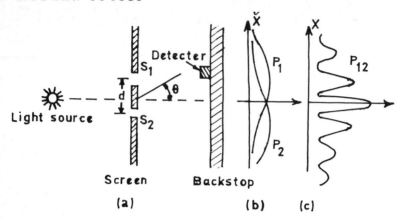

FIG. 9-3: Interference experiment with photons.

are identical in size which is very small compared to the wavelength, We further assume that the separation, d, between the holes is comparable to the wavelength, and that the distance of the source from the screen is large compared to d.

We measure the intensity of the diffracted light by means of a photocell*; the intensity is then proportional to the counting rate observed with the cell. Let this photocell be connected to a 'loudspeaker' which would produce a 'click' whenever a photon hits the photocell.

The first thing that we notice with our experiment is that we hear sharp clicks from the detector (i.e. from the loudspeaker) and all 'clicks' are of the same intensity. If we move the detector around, the rate at which the clicks appear is faster or slower but the size (i.e. the loudness) of each click is always same.† If we lower the intensity of the light source, the rate of clicking slows down, but still the intensity of each click is the same. Further, if we put two detectors at the backstop, one or the other would click, but both will never click at the same time. (However, if there were two clicks close together in time, our ear might not sense the separation.) We may, therefore, conclude that whatever arrives at the backstop arrives in lumps. Let us now see what the classical wave theory says about the intensity distribution to the right of the screen. We have already assumed that the diameter of the hole is small compared to the wavelength of light used. This would imply that if either hole is covered, the angular distribution of the diffracted radiation is a smooth function of the angle θ. However, if both the holes are open we would obtain the interference pattern‡ with intensity varying as:

* We can also use a fluoroscent screen; the intensity distribution can then be directly recorded.

† The loundness of the click is assumed to depend on the energy given to the photocell by the photon.

‡ See the discussion of Young's double slit experiment in Chapter 4.

$$I = I_1 + I_2 + 2\sqrt{I_1 I_2}\, \text{Cos}\,\delta \qquad (9\text{-}5)$$

where I_1 and I_2 are the intensities produced by hole S_1 and hole S_2 respectively and δ is the phase difference between the two waves (see Eq. 9-4). For the experimental arrangement shown in Fig. 9-3, δ would be given by

$$\delta \simeq \frac{2\pi}{\lambda}\, d\, \sin\theta \qquad (9\text{-}6)$$

If the measurements are carried out at a large distance compared to d, then I_1 and I_2 would be very nearly equal and we will obtain (cf. Eq. 4-21)

$$I \simeq 4I_1\, \text{Cos}^2\, \delta/2 \qquad (9\text{-}7)$$

The intensity distribution, I, when both the holes are open is thus equal to the product of the intensity, I_1, when one of the holes is open and the factor $4\,\text{Cos}^2\,\delta/2$, which describes the effect of interference of waves emerging from the two holes. We note that because of this interference we will observe zero intensity in certain directions provided $4d/\lambda > 1$. In certain other directions the intensity will be four times as large as in the experiment with one hole open.

In Chapter 8 we had discussed the particle nature of light and had also learnt about the 'impossibility' of splitting the photons. In view of this fact that the photons cannot be split we might be tempted to conclude that the prediction contained in Eq. 9-5 must be wrong. For, we would say that since the photons cannot be split, each photon has come either through hole S_1 or through hole S_2. Let us write this in the form of a 'proposition':

Proposition A: Each photon *either* goes through hole S_1 or it goes through hole S_2.

Assuming the above proposition, all photons that would arrive at the detector can be divided into two classes: (1) Those that come through hole S_1, and (2) those that come through hole S_2. (This is then similar to the experiment with bullets.) Thus our observed curve (where both the holes are open) must be the sum of the effects of the photons which come through hole S_1 and the photons which come through hole S_2. This idea can be checked by experiment. First, we cover hole S_2 and make a measurement for the photons that come through hole S_1. From the count rate, we get the curve P_1 [Fig. 9-3(b)]. The result seems quite reasonable. In a similar way we measure P_2, the probability distribution for the photons that come through hole S_2. When both the holes are open and if Proposition A is correct then we would expect an intensity distribution P_{12} which should be the sum of P_1 and P_2:

$$P_{12} = P_1 + P_2 \qquad (9\text{-}8)$$

The distribution P_{12} experimentally *obtained* [Fig. 9-3(c)] with both holes open is clearly not the sum of P_1 and P_2 and indeed *the experimental evidence is conclusively in favour of the wave theory prediction** (Eq. 9-7).

Before we reject the prediction given by Eq. 9-8 let us consider the following proposition:

Proposition B: The interference phenomenon described by Eq. 9-7 arises because of some kind of 'interaction' between several photons.

The above proposition implies that with a sufficiently intense light source we will have several photons in transit at the same time, i.e. several photons passing through the holes simultaneously and we might wonder whether the interference effects are possibly a '*many photon phenomena*'. This might lead us to aruge that the prediction expressed by Eq. 9-8 is correct for extremely feeble sources, such that only one photon is effectively in transit at a time, whereas the prediction expressed by Eq. 9-7 is correct for sufficiently intense sources. Thus, can it be that the wave properties demonstrated so clearly in the interference experiments apply only to strong light, where somehow the many individual photons interact one with another to produce the wave like results?

The answer to this question is *no*; and an experiment directly bearing on this question was carried out in 1909 by Geoffrey Taylor, at the University of Cambridge.† The experimental set up of Taylor (shown in Fig. 9-4) consisted of a light-tight box with a small lamp which casts

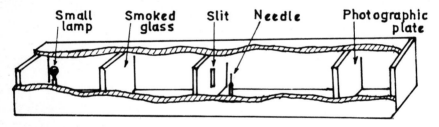

FIG. 9-4:   The layout of G. I. Taylor's experiment which proved that the diffraction pattern was correctly given even if only one photon was present in the apparatus at a time. The entire apparatus was contained in a light-tight box.

the shadow of a needle on a photographic plate. The dimensions were so chosen that the diffraction bands around the shadow of the needle were plainly visible. The intensity of light was greatly reduced and conse-

---

* We can regard the present experiment as the prototype of a large class of interference experiments among which we note measurements with diffraction gratings and X-ray diffraction experiments with crystals.

† G. I. Taylor, 'Interference Fringes with Feeble Light', *Proceedings of the Cambridge Philosophical Society*, Vol. 15, p. 114 (1909).

quently longer exposures were needed to get a well exposed plate. The intensity was made so low that there was only one (and often no) photon inside the box at a given time. (One can find the number of photons in any given length of light beam from the intensity of the light source and from the energy of one photon — see Problem 8 of Chapter 8.) Taylor found the photons were spaced farther apart on the average than the length of the box. So usually there was only one (and often no) photon in the box. For such small intensity, to get a well exposed plate, the exposure had to be made over a very long time, and, Taylor in his experiment made an exposure lasting for several months. Even on this plate the diffraction bands were perfectly clear. Interference therefore takes place even for every single photon, thus ruling out Proposition B.

Alternatively, one may argue that perhaps some of the photons go through hole S_1, and then go around through hole S_2 and then around a few more times, or by some other complicated path....then by closing hole S_2, we changed the chance that a photon that *started out* through hole S_1 would finally get to the detector. However, this type of argument would fail if we notice that there are some points at which very few photons arrive when both holes are open, but which receive many photons if we close one hole, so closing one hole increased the number from the other. Many ideas have been put forward to explain the curve for P_{12} in terms of individual photons going around in complicated paths, but none of them has succeeded in giving a satisfactory explanation. However, as noted before, the mathematics for relating P_1, P_2 and P_{12} is the same as we had for water waves (cf. Eqs. 9-4 and 9-7).

We therefore conclude that the photons arrive in lumps, like the indestructible bullets, but the probability of arrival of these lumps is distributed like the distribution of a wave. It is in this sense that a photon behaves 'sometimes like a particle and sometimes like a wave'.

9-6 Watching the Photons

Let us conduct an experiment with the help of some ingenious device so that we know through which slit a photon has come through. To our apparatus we add a light source* (of a much smaller frequency) placed behind the wall and between the two holes as shown in Fig. 9-5. Suppose the photons emitted by this light source could be scattered by the incoming photons so that when a photon passes, however does it pass, it will scatter some light to our eye, and we can see where the photon goes.

* We are assuming that the photocell is insensitive to the light produced by the second light source. This is possible because the second light source has a much smaller frequency. However, it should be pointed out that this experiment has never been done and is essentially a "thought" experiment. We know the results that would be obtained because there are many experiments that have been done, in which the scale and proportions have been chosen to show the effects we describe.

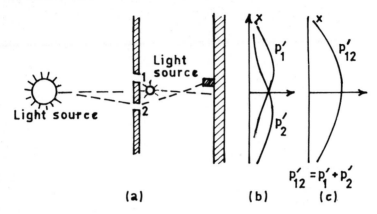

FIG. 9-5: Watching the photons. A strong light source placed behind the
wall and between the holes finds out whether the photon went through
hole 1 or hole 2.

The result of the experiment would be, that every time we hear a click
from the photocell on the backstop; we also see a flash of light either near
hole 1 or near hole 2, but never at both holes at the same time. (Thus
Proposition A would appear to be true.) Next, let us track the photons
and find out what they are doing. For each position of the photocell we
will count the photons that arrive and also keep track of which hole they
passed through by watching for the flashes. We can make an observation
table which would consist of two columns. Whenever we see a flash near
hole 1 we will put a count in column 1 implying that a photon has passed
through hole 1; and similarly when we see a flash near hole 2 we will
put a count in column 2. From the number recorded in column 1 we get
the probability P'_1 that a photon will arrive at the detector via hole 1;
and from the number recorded in column 2 we get P'_2. Such measurements can be carried out for various positions of the detector and we get
the curves for P'_1 and P'_2 shown in Fig. 9-5(b).

To find the total probability of a photon to arrive at the photocell by
any route we must just add P'_1 and P'_2, because for every photon
received by the photocell it has been detected to come either from hole
1 or hole 2. Thus we obtain

$$P'_{12} = P'_1 + P'_2 \qquad (9\text{-}9)$$

Eq. 9-9 implies that although we have succeeded in watching from
which hole our photons come through, we no longer get the interference
curve P_{12}, but a new curve, P'_{12}, showing no interference. If we do not
try to detect the photons the interference pattern would be restored.
Thus, we conclude that *when we look at the photons the distribution of them
on the screen is different than when we do not look.* Further, if our device is
such that only some of the photons are detected then those 'seen at hole

1' have a distribution like P'_1; those 'seen at hole 2' have a distribution like P'_2 (so that those 'seen at either hole 1 or hole 2' have a distribution like P'_{12}); and those 'not seen at all' have a 'wavy' distribution just like P_{12} of Fig. 9-3(c). Thus, if the photons are not seen, we have interference. We can understand this by noting that when we do not see the photons, they are not disturbed and an interference pattern is observed. On the other hand, when we do see the photons they are disturbed and the interference pattern is destroyed.

Is there not some way we can see the photons without disturbing them? Possibly the photon (emitted by the light source near the holes) gives a big jolt to the photons and therefore disturbs them. So let us try to reduce this jolt; we know (from Chapter 8) that the momentum carried by a photon is inversely proportional to its wavelength ($p = h/\lambda$). So, in order to reduce the jolt we may use a light source (which is placed near the holes) of large wavelength. If we go on increasing the wavelength, at first the result does not seem to change. Then a new phenomena occurs. In Appendix F it is shown that in a microscope, due to the wave nature of light, there is a limitation on how close two spots can be seen as two separate spots. This distance is of the order of the wavelengths of light. Thus, when we make the wavelength longer than the distance between the holes we see a fuzzy flash and we can no longer tell through which hole the photon passed through. We just know it went somewhere. So, when the 'jolts' given to the photon are small enough P'_{12} begins to look like P_{12} i.e. we begin to see interference effects.

We, therefore, find that it is impossible to devise a set-up in such a way that one can tell which hole the photon passed through and at the same time not disturb the interference pattern. Then the question arises: Is proposition A true or false? The answer is, 'if one has a piece of apparatus which is capable of determining whether the photons go through hole 1 or hole 2, then one *can* say that it goes through hole 1 or hole 2. But, when one does not try to tell which way the photon goes, when there is nothing in the experiment to disturb the photons, then one may say that the photon came through *both* holes: *partly through hole 1 and partly through hole 2.*' This answer is in the spirit of quantum physics.

Thus, in an experiment in which we observe the two hole interference pattern we cannot tell through which hole any particular photon came. The interference pattern can arise only if the photons go partly through both slits, and it is then meaningless to ask through which one of the holes the photon came through.

As mentioned earlier, the quantum behaviour of atomic objects (electrons, protons, neutrons, photons, and so on) is the same for all, they are all 'particle-waves' or whatever we want to call them. So what we have learnt about the properties of photons will also apply to all 'particles.'

(a)   (b)

FIG. 9-6: Interference pattern with bullets, (a) actual and (b) observed.

Now if the motion of all particles like photons and electrons must be described in terms of waves, then why did we not see an interference pattern with the bullets? It turns out that for the bullets the wavelengths were so small ($= h/mv$) that the interference patterns become very fine. They would be so fine that with any detector of finite size one could not distinguish the separate maxima and minima. Fig. 9-6(a) shows the probability distribution one might predict for bullets using quantum mechanics. The rapid waggles represent the pattern one gets for waves of very short wavelength. Any physical detector measures the smooth curve shown in Fig. 9-6(b).

9-7   The Probabilistic Interpretation of Matter Waves

In the preceding sections we have established a series of facts which seem to indicate unequivocally that light (and similarly electrons, protons and other atomic objects), behaves in some cases like a wave and in other cases like corpuscles. How are these contradictory aspects to be reconciled? The interpretation generally accepted at present was put forward by Max Born* and runs as follows:

First, we note that once the existence of matter waves as either an inevitable companion or inherent property of atomic objects has been established, it is obvious that this must somehow influence the final localization of the particles themselves. Thus, if we let a monochromatic light beam fall upon a grating, the intensity of the diffracted wave is, in every point of space, proportional to the square of the wave amplitude at that point. But the intensity is nothing but an expression for the energy carried by the diffracted beam. This energy, however, as is shown by the photoelectric and Compton effects (see Chapter 8) is not evenly spread over the entire wavefront but is somehow condensed in point like entities called photons. Therefore, it follows that the wave intensity for each volume of space is proportional to the average number of photons contained in that volume. In other words, the carriers of energy associated with a 'matter-wave' are the 'particles' whose distribution in space we study.

So far we have considered only the meaning of the wave associated with a multi-particle beam. Yet, experimentally, the phenomenon of diffraction can be observed by individual particles (see the discussion

---

* See M. Born, *Physics in My Generation: A Selection of Papers*, pp. 183, Pergamon (1956); *Atomic Physics*, Chapter IV, Blackie (1957).

Max Born (1882-1970) was awarded the Nobel Prize in 1954 for his work in quantum mechanics.

on Taylor's experiment). So an interpretation has to be found also for the limiting case in which the beam is reduced to a single particle. Thus, we have to find what information the wave supplies about each single particle. But this is immediately achieved if we divide the function which gives the average number of particles found in each elementary volume of space by the total number belonging to the original beam. Hence we have the probabilistic interpretation — the square of the wave amplitude determines, at every point of space, the probability for the occurence of its associated particles at that point.

Many physicists have felt that there is something paradoxical about the reasonings given above. They argue as follows:

Consider the diffraction of a light beam by a small hole. The screen on which we measure the intensity of the diffracted beam* is at a very large distance (say one light year)† away from the hole. After the photon has been emitted it spreads out like a spherical shell. By the time the wave arrives at the detector the energy carried by the wave has spread out over a very large region in space, say within a spherical shell of radius one light year. How is it then possible that the entire amount of this energy can suddenly become concentrated within the photocell in case the cell does register? It should take more than a year for the energy within the 'far-side' of the shell to reach the photocell, otherwise we violate the principle that no signal can propagate faster than light.

The fallacy in the above argument lies in the fact that the energy does not spread out in the way described by the classical theory. The classical expression for energy density refers to the average energy density which we will observe for a large number of photons, but it does not describe the energy density associated with a single photon. For the latter case one may only talk of the probability of finding it in a particular direction.

9-8 Explanation of the Interference Pattern produced by an Apparatus like Michelson Interferometer

Let us next consider the interference experiment in which we have a beam of light which is passed through an interferometer so that it splits up into two components and the two components are subsequently made to interfere. Let us suppose the incident beam consists of only a single photon and inquire what will happen to it as it goes through the apparatus. As in the earlier case, the photon can be described going partly into each of the two components into which the incident beam is split. According to Dirac‡, the photon is then in a state given by the superposition of the

* The intensity is being measured by a photocell.

† Light travels with a speed of 1,86,000 miles per second. So, a distance of one light year would correspond to $186000 \times 3 \cdot 1 \times 10^7 \approx 5 \cdot 5 \times 10^{12}$ miles (1 year $\approx 3 \cdot 1 \times 10^7$ seconds).

‡ P. A. M. Dirac, *The Principles of Quantum Mechanics*, Chapter I, Oxford (1958).

P. A. M. Dirac (1902-   ) was awarded the Nobel Prize in 1933 for his contribtuions in quantum mechanics.

two states associated with the two components. For a photon to be in a definite state it need not be associated with one single beam of light but may be associated with two or more beams of light which are the components into which the original beam has been split. *If, however, we try to determine the energy in one of the components then the result is either the whole photon or nothing at all.* Thus, the photon must change suddenly from being partly in one beam and partly in the other to being entirely in one of the beams. As before, this sudden change is due to the disturbance in the state of the photon which the observation necessarily makes. It is impossible to predict in which of the two beams the photon will be found, only the probability of either result can be calculated. Once again, since each photon goes partly into each of the beams, in general, each photon interferes only with itself (see also the discussion on interference of two photons).

One would like to emphasize here an important difference between classical and quantum mechanics. We have been trying to measure the probability of a photon to arrive on the screen at a given circumstance. To quote Feynman:

> We have implied that in our experimental arrangement (or even in the best possible one) it would be *impossible* to predict exactly what would happen. We can only predict the odds! This would mean, if it were true, that physics has given up on the problem of trying to predict exactly what will happen in a definite circumstance. Yes! physics *has* given up. *We do not know how to predict what would happen in a given circumstance,* and we believe now that it is impossible — that the only thing that can be predicted is the probability of different events. It must be recognized that this is a retrenchment in our earlier ideas of understanding nature. It may be a backward step, but no one has seen a way to avoid it.

We conclude this section by noting that whereas the wave amplitude associated with a photon can be discussed as in classical wave theory (or, more precisely, as in classical electromagnetic theory), all quantities which depend quadratically on the amplitude must be interpreted in terms of probabilities. A photon can be 'split' in the sense that the wave can be divided into two, or several parts by half-silvered mirrors or other devices, just as in classical electromagnetic theory. However, a photon cannot be 'split' in the sense that we can detect, say with a photocell, a 'fractional photon' which carries only some fraction of the energy $\hbar\omega$, where ω is the frequency of the photon. These ideas constitute a clear departure from the ideas of classical electromagnetic theory. However, it would not be proper to say that the classical theory has been completely overturned—we have merely found limitations of the classical theory.

9-9   Interference of Two Different Photons

We digress here for a moment and ask ourselves whether two different photons can interfere with each other? Let us modify the arrangement of Fig. 9-3, such that each slit is illuminated by a separate light source as shown in Fig. 9-7. Then is it possible for a photon from source 1 to

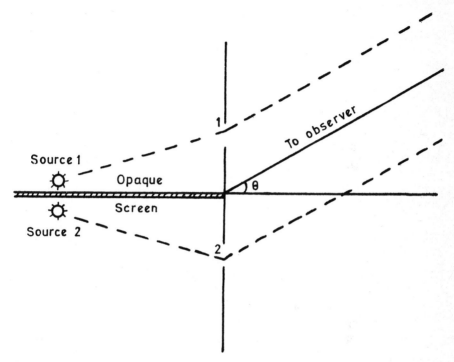

Fig. 9-7:   Interference of two photons.

interfere with a photon from source 2? The answer is yes and two photons do interfere with each other, but that we do not *readily* see any interference pattern with the set-up of Fig. 9-7. It's sometimes mentioned that wavetrains of light from two different sources cannot interfere, but such statements are wrong. The fact is that the superposition of two light waves of almost the same frequencies gives rise to beats in precisely the same way as do radiowaves from two different antennae (after all radio waves also consist of photons). It is true, however, that these beats are difficult to observe with light. The first demonstration of beats resulting from the superposition of incoherent beams was given by Forrester* *et al.* in 1955. Later with the development of the laser beams such experiments on beats

---

* A. T. Forrester, R. A. Gudmundsen and P. O. Johnson, 'Photoelectric Mixing of Incoherent Light', *Physical Review*, Vol. 99, p. 1691 (1955).

became much easier to perform. Javan *et al.** and Lipsett and Mandel†
determined the coherence times of the He-Ne laser and the ruby laser
respectively, by superposition of beams from two independent sources.
(Some of these experiments and the expected intensity distribution are
briefly discussed in Chapter 5.) These experiments tend to indicate that
the interference between two independent light beams does produce an
interference pattern. The detailed quantum theory of such phenomena
is much beyond the scope of this book. It involves a lot more physics than
we have discussed.‡

9-10 Looking at Pictures

We will briefly discuss another kind of circumstantial evidence for the
probabilistic interpretation of matter waves. Rose‡‡ has presented evidence that the image formed on the retina and interpreted in the brain
may reveal a randomness due to the arrival of individual photons. More
relevant to the present discussion is the impersonal evidence (as cited by
Rose) that is provided by cameras or photomultipliers.

Now, in a photographic emulsion, the development of all 10^{10} silver
atoms in a grain of the emulsion can be triggered if just a few silver atoms
are activated by the original photographic exposure. Plate 12, which has
been reproduced from Rose's paper, shows how the quality of a photograph improves as the amount of light used to produce the image is
increased by a factor of 10^4 in five (roughly equal) steps. The last picture
has acceptable detail, and actually corresponds to the arrival of about
3×10^7 photons. The photographs suggest that light is not smoothly
spread, but that the picture is built up by the impact of very localized
concentrations of radiant energy. The point at which any individual
photon arrives appears to be a matter of chance. The early stages of
building up the picture will have a featureless character$ as shown in the
first photograph of Plate 12. Nevertheless, as more and more photons
arrive and make their contributions to the pattern, we see that the total
intensity distribution, the integrated distribution of energy, approaches

* A. Javan, E. A. Ballik and W. L. Bond, 'Frequency Characteristics of a Continuous—
Wave He-Ne Optical Maser', *Journal of the Optical Society of America*, Vol. 52, p. 96 (1962).

† M. S. Lipsett and L. Mandel, 'Coherence Time Measurements of Light from Ruby
Optical Masers', *Nature*, Vol. 199, p. 553 (1963).

‡ The complete theory of such phenomena has been worked out by L. Mandel and E. Wolf
in their paper 'Coherence Properties of Optical Fields', *Reviews of Modern Physics*, Vol. 37,
p. 231 (1965).

‡‡ A. Rose, 'Quantum Effects in Human Vision', *Advances in Biology and Medical Physics*, 5,
p. 211 (1957). See also *Physics—A New Introductory Course, Parts I & II*, Chapters 8 and 9, by
A. P. French, A. M. Hudson and N. H. Frank, Prepared at Massachusetts Institute of Technology Science Teaching Centre, Revised Preliminary Edition (1965).

$ Similarly, the early stages of building up an interference pattern with very weak light
have the same chaotic and featureless character as the first photograph in Plate 12.

more and more close to what a wave theory will predict.*

Example

In Taylor's experiment, it was found that an energy of 5×10^{-13} Joule per second reaches the photographic plate. (This value was found by comparing the average blackening of the plate with that produced in 10 seconds by a candle 6 feet away, without any absorbing screen of smoked glass.) Assuming $\lambda = 5 \times 10^{-5}$ cm and the distance between the slit and the photographic plate to be about 100 cm, calculate the average distance between the photons.

Solution

The energy of each photon, E, will be given by:

$$E = h\nu = \frac{hc}{\lambda}$$

$$= \frac{6 \cdot 625 \times 10^{-27} \text{ erg sec} \times 3 \times 10^{10} \text{ cm/sec}}{5 \times 10^{-5} \text{ cm}}$$

$$\approx 4 \times 10^{-12} \text{ ergs}$$

$$\approx 4 \times 10^{-19} \text{ Joules}$$

Therefore, the number of photons hitting the photographic plate per second, N, will be given by

$$N = \frac{5 \times 10^{-13}}{4 \times 10^{-19}} = 1 \cdot 25 \times 10^6 \text{ photons/sec}$$

Thus, the average time that elapses between the arrival of one photon and the next will be about 8×10^{-5} second. Hence, if L, represents the average distance between one photon and the next in the beam of light, then

$$L \approx (3 \times 10^{10} \text{ cm/sec}) \times (8 \times 10^{-5} \text{ sec})$$

$$\approx 2 \cdot 4 \times 10^6 \text{ cm}$$

Since the box is about 100 cm long, and if we were asked 'how many photons are in the box at any chosen instant' then the answer would be 'none' most of the time.

9-11 The Uncertainty Principle†

Perhaps the best-known consequence of the association of waves with particles in motion is contained in the celebrated uncertainty relations

* Thus, if in an interference pattern the intensity at a minima is not zero, then there is a small (but finite) probability that the first 10 photons will land up on the minima. However, for a very large number of photons, the photons will arrive according to the predictions of the wave theory.

† Heisenberg's views on this subject are described in the book *The Physical Principles of the Quantum Theory*, by W. Heisenberg, (Dover Publications, New York).

of Heisenberg.[†] The uncertainty principle can be stated as follows:

If a measurement is made on an object and the x-component of its momentum is measured with an uncertainty Δp, then one cannot, at the same time, know its x-position more accurately than $\dfrac{h}{\Delta p}$, where h is a definite fixed number. It is the Planck's constant and its value is approximately $6\cdot627 \times 10^{-27}$ ergs seconds. Thus

$$\Delta x\,\Delta p > h \qquad\qquad (9\text{-}10)$$

i.e. the uncertainties in the position and momentum of a particle at any instant must have their product greater than the Planck's constant.

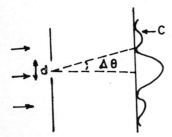

FIG. 9-8: Diffraction of photons through a slit.

In order to conceive an ideal experiment that brings out the uncertainty inherent in the simultaneous measurement of the position and the momentum of a particle let us consider the diffraction of photons passing through a slit. Suppose a light beam, with a certain energy, is coming from far away source, so that the light rays coming are essentially horizontal (Fig. 9-8). We will concentrate on the vertical component of the momentum. All of these photons have a certain horizontal component p_x, say, in a classical sense. So the vertical component of the momentum p_y, before the photon goes through the hole, is definitely known. Since the photons have come from a source which is very far away, their vertical component of momentum is zero. Let us suppose it goes through a slit of width d. Then after it has come out through the slit the vertical position is known with considerable accuracy, namely within $\pm\,d/2$. Thus, the uncertainty in the y- coordinate, Δy, is of order[‡] d. Now we might also want to say, that $\Delta p_y = 0$ because the momentum is horizontal. However, that would be wrong since *we knew the momentum to be horizontal before it entered the slit, but afterwards we have no knowledge about the momentum.* Before the photons passed through the slit, we did not know their vertical positions. Now that we have found the vertical position by having the photon come through the slit, we have lost our information on the vertical momentum. This is because there would be diffraction of the waves after they go through the slit. Therefore there is a certain probability that photons coming out of the slit are not coming out exactly in straight lines. The pattern is spread out by the diffraction

[†] W. Heisenberg (1901-   ) introduced the uncertainty relations and was awarded the Nobel Prize in 1932 for the creation of quantum mechanics.

[‡] Here we need not concern about numerical factors of 2.

effect and the angle of spread, which we can define as the angle of the first minimum (shown as $\Delta\theta$ in Fig. 9-8), is a measure of the uncertainty in the final angle.*

The vertical component p_y has a spread which is approximately equal to $p_0 \Delta\theta$, i.e.

$$\Delta p_y \approx p_0 \Delta\theta \qquad (9\text{-}11)$$

where p_0 is the horizontal momentum. For the single slit diffraction pattern, the first minimum occurs at an angle such that (see Chapter 6)

$$\Delta\theta \approx \frac{\lambda}{d} \qquad (9\text{-}12)$$

$$\therefore \quad \Delta p_y \approx \frac{p_0 \lambda}{d} \qquad (9\text{-}13)$$

Thus, if we make the value of d smaller and make a more accurate measurement of the position of the photon, the diffraction pattern gets wider. Hence, the narrower we make the slit, the wider the pattern gets, and the more likelihood that we would find the photon has vertical momentum. Since $\Delta y \approx d$ we find

$$\Delta p_y \Delta y \approx p_0 \lambda \qquad (9\text{-}14)$$

But the wavelength times the momentum is Planck's constant h. (This is true not only for photons but for every particle.) Thus, we obtain

$$\Delta p_y \; \Delta y \approx h$$

Many a time people argue (incorrectly) in the following way: When the particle arrived from the left, its vertical component was zero. And now that it has gone through the slit, its position is known. Both position and momentum seem to be known with arbitrary accuracy. This argument is wrong, because the uncertainty relation does not refer to this. It is not correct to say, 'I knew what the momentum was before it went through the slit, and now I know the position,' because now the momentum knowledge is lost.

Example
A 10 gm lead ball appears to be at rest near the edge of a table. An experimenter makes repeated measurements of its position and finds the results are all within $\pm 10^{-6}$ cm of each other. Calculate the uncertainty in momentum. How long would it take the block to move a detectable distance?

Solution

$$\Delta p \sim \frac{h}{\Delta x} = \frac{6 \cdot 6 \times 10^{-27} \text{ erg sec}}{10^{-6} \text{ cm}}$$

* Majority of the photons which reach the screen fall in the vicinity of the central maximum, represented in Fig. 9-8 by a large peak.

or

$$\Delta p \sim 6\cdot6 \times 10^{-21} \text{ gm cm/sec}$$

or

$$\Delta v \sim 6\cdot6 \times 10^{-22} \text{ cm/sec}$$

For the particle to move a distance of 10^{-6} cm, it will take about

$$\frac{10^{-6}}{6\cdot6 \times 10^{-22}} \approx 1\cdot5 \times 10^{15} \text{ sec} \approx \tfrac{1}{2} \times 10^8 \text{ years} = 50 \text{ million years.}$$

The above type of consideration may appear quite mysterious, for it is not the type of thing that we expect from traditional thinking. In fact during the development of quantum physics, it was the physicists themselves who were most disturbed, since the experimental evidence was adding up to a model of the physical world which violated fundamental assumptions about the very nature of fundamental thinking. Indeed, in 1926, physicists gathered in Copenhagen to interpret the new findings of quantum physics, and later Heisenberg had remarked about the conference 'I remember discussions with Bohr* which went through many hours till very late at night and ended almost in despair; and when at the end of the discussion I went alone for a walk in the neighbouring park I repeated to myself again and again the question: Can nature possibly be as absurd as it seemed to us in these atomic experiments?'†

Let us look deeper into the connection between the laws of classical mechanics and the quantum conditions. In classical theory, the fundamental concept concerning motion is that any moving particle occupies, at any given moment, a certain position in space and possesses a definite velocity characterizing the time changes of its position on the trajectory. If someone is asked why he believes that any moving particle occupies at any given moment a certain position and the particle describes in a course of time a definite curve (called the trajectory) he will most probably answer 'Because I see it this way, when I observe the motion.' Let us analyze how this classical motion of the trajectory is formed and see if it really will lead to a definite result. Let us imagine a physicist, supplied with any kind of the most sensitive apparatus, trying to pursue the motion of a material body thrown from the wall of his laboratory. The physicist decides to make his observation by 'seeing' how the body moves and for this purpose he uses a small light source (Fig. 9-9). Of course to see the moving body he must illuminate it and, knowing that light in general produces a pressure on the body and might disturb its motion, he decides to use short-flash illumination, only at the moments

* Niels Bohr (1885-1962) had introduced the idea that the electron moved about the nucleus in well defined orbits and was awarded the Nobel Prize in 1922 for the study of structure and radiation of atoms.

† From W. Heisenberg, *Physics and Philosophy*, Harper and Row, New York (1958).

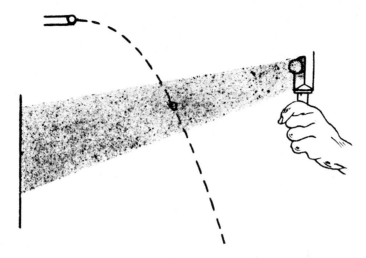

Fɪɢ. 9-9:   Attempt to persue the motion of a little material body.

when he makes the observation. For his first trial he wants to observe only 10 points on the trajectory and thus he chooses his flashlight source so weak that the total effect of light pressure during 10 successive liluminations should be within the accuracy he neesd. Thus, flashing his light 10 times during the fall of the body, he obtains, with the desired accuracy, 10 points on the trajectory.

Now he wants to repeat the experiment and to get 100 points. He knows that 100 successive illuminations will disturb the motion too much and therefore, preparing for the second set of observations, chooses his flashlight 10 times less intense. For the third set of observations, desiring to have 1000 points, he makes the flashlight 100 times fainter than it was originally.

Proceeding in this way, and constantly decreasing the intensity of his illumination, he can obtain as many points on the trajectory as he wishes to, without increasing the possible error above the limit he had chosen at the beginning. This highly idealized, but in principle quite possible, procedure represents the logical way to construct the motion of a trajectory by 'looking at the moving body' and in the frame of classical physics, it is quite possible.

But let us see what happens if we introduce the quantum limitations and take into account the fact that the action of any radiation can be transferred only in the form of light quanta (i.e. photons). We have seen that our observer was constantly reducing the amount of light illuminating the moving body and we must now expect that he will find it impossible to continue to do so as soon as he comes down to one photon. Either all or none of the total photons will be reflected from the moving body, and in the latter case the observations cannot be made. Of course we have seen that the effect of collision with a photon decreases with increasing wavelength, and our observer, knowing it too, will certainly try to use for his observations light of increasing wavelength to compensate for the number of observations. But here he will meet with another difficulty.

It is well known that when using light of certain wavelengths one cannot see details smaller than the wavelength used. Thus, by using longer and longer waves, he will spoil the estimate of each single point and soon will come to the stage where each estimate will be uncertain by an amount comparable to the size of all his laboratory and more. Thus, he will be forced finally to a compromise between the large number of observed points and uncertainty of each estimate and will never be able to arrive at an exact trajectory as a mathematical line such as that obtained by his classical colleagues. His best result will be a rather broad washed-out band and, if he bases his notion of the trajectory on the result of his experience, it will be rather different from a classical one.

The collision with a photon will, because of the law of conservation of momentum, introduce an uncertainty in the momentum of the particle comparable with the momentum of the photon. Thus,

$$\Delta p_{\text{particle}} \approx \frac{h}{\lambda}$$

and since the uncertainty of position of the particle Δq, is nearly equal to the wavelengths, λ we obtain:

$$\Delta p \, \Delta q \approx h$$

A more sophisticated version of the above experiment consists in the determination of the position of an electron by means of the γ-ray microscope in the following way.

If we wish to determine the position of an electron by optical means by illuminating it and observing the scattered light then for a greater resolution we must employ light of the shortest possible wavelength (say γ-rays). However, the employment of such short wave radiation implies the occurrence of a Compton scattering process when the electron is irradiated, so that the electron experiences a recoil, which to a certain extent is indeterminate. Let us analyse this more carefully:

Let the electron under the microscope be irradiated in any direction with light of wavelength λ (Fig. 9-10). Then using the expression for the

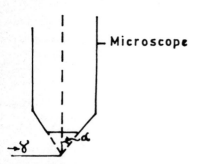

FIG. 9-10: Determination of the position of an electron by means of the γ-ray microscope.

resolving power of the microscope, its position can only be determined subject to the possible error

$$\Delta x \sim \frac{\lambda}{\text{Sin } \alpha}$$

where α is the angular aperture. According to the corpuscular view, the electron suffers a Compton recoil of the order of magnitude $h\nu/c$, the direction of which is undetermined to the same extent as is the direction in which the photon flies off after the process. Since the photon is actually observed in the microscope, this indeterminateness of direction is given by the angular aperture α. The component momentum of the electron perpendicular to the axis of the microscope after the process is therefore, indeterminate to the extent Δp, where approximately

$$\Delta p \sim \frac{h\nu}{c} \text{ Sin} \alpha$$

Thus, the order of magnitude relation

$$\Delta p \, \Delta x \simeq h$$

holds good here also.

What the above equation says is the fact that with the best of all possible microscopes, the product of the two uncertainties is (of the order of) the Planck's constant; practical limitations which we have not considered here, like the problem of casting a perfect lens, for example, will increase the uncertainty product. But h is the absolute minimum which can ever be achieved.

Till now we have considered a few specific examples but it turns out that a similar restriction appears in all such problems. Whether one is trying to design the best of microscopes or undertaking any other experiment which requires measuring simultaneously the position and momentum of the object, the minimum uncertainty product always makes its appearance to limit the available knowledge.

We should mention that all bodies in nature are subject to the above uncertainty principle, but the Planck's constant is such a small number that for bodies which we usually see, the uncertainties are ridiculously small. For example, for a small dust particle with mass 0.000,0001 gm both position and velocity can be measured with an accuracy of 0.000,000,01%. However, for an electron (whose mass is roughly 10^{-29} gm) the product $\Delta v \Delta q \approx 100$. In fact one can estimate the size of the atom from the uncertainty principle.[*]

Thus, as pointed out by Marganeau[†] 'the physicist, while still fond

---

[*] This estimate is given in *The Feynman Lectures on Physics*, Vol. III by R. P. Feynman, R. B. Leighton and M. Sands, Chapter 2 (see also Problem 5).

[†] H. Marganeau in *Quantum Theory* Vol. I, (Edited by D. R. Bates), Academic Press, New York and London (1961).

of mechanical models wherever they are available and useful, no longer regards them as the ultimate goal of all scientific description; he recognizes situations where the assignment of a simple model, especially a mechanical one, no longer works and where he feels called upon to proceed directly under the guidance of logical and mathematical considerations and at times with the renunciation of the visual aspects which classical physics would carry into the problem.'

We conclude this chapter by quoting Max Born.* 'Physicists of today have learned that not every question about the motion of an electron or a photon can be answered, but only those questions which are compatible with the uncertainty principle.'

SUGGESTED READING

BAKER, ADOLF, *Modern Physics and Antiphysics*, Addison-Wesley (1970). This is an undergraduate book on the 20th century physics written primarily for non-scientists. The symmetries, relativity and quantum physics are treated in a non-mathematical manner with emphasis on the processes whicn led to the discoveries. The last six chapters 'Waves and Particles', 'Light is a Wave', 'A Wave is a Particle', 'A Particle is a Wave', 'The Heisenberg Uncertainty Principle' and 'The World of Quantum Mechanics' are directly relevant to the concepts developed in this chapter.

BORN, M., *Atomic Physics*, Eigth Edition, Blackie (1969). The fourth chapter 'Wave-Corpuscles' gives an excellent account of the wave-particle duality.

ENRICO, CANTORE, *Atomic Order: An Introduction to the Philosophy of Microphysics*, The MIT Press (1969). Chapter V 'Interacting Properties of Microparticle' discusses wave-like interactions, statistical aspects of atomic events and uncertainty principle. Chapter VII 'Epistemological Implications of Atomic Physics, discusses the philosophical implications.

DIRAC, P. A. M., 'The Evolution of the Physicists Picture of Nature', this article appears in *Project Physics Reader* 5, *Models of the Atom*, Prepared by Harvard Project Physics, (1963). It gives an account of how physical theory has developed in the past and how it might be expected to develop in the future. One of the dramatic statements made by Prof. Dirac in this article is, '*I think one can make a safe guess that uncertainty relations in their present form will not survive in the physics of future*'.

FEINBERG, G., 'Light', *Scientific American*, Vol. 219, p. 50, September 1968. The author discusses the dual nature of light, summarizes the historical evidence for both the wave and the particle properties, and shows how these aspects are reconciled in the quantum mechanical picture.

FEYNMAN R. P., R. B. LEIGHTON and MATHEW SANDS, *The Feynman Lectures on Physics*, Vol. III, Addison-Wesley (1965). The first two chapters give a beautiful introduction to quantum physics.

GAMOW, G., *Mr. Tompkins in the Wonderland*, Cambridge University Press (1940). In the second and the fourth dream of Mr. Tompkins, Professor Gamow has speculated in a delightful manner what our world would be like if the Planck's constant had a value of 1 erg sec, so that non-classical ideas would be apparent to our sense perception.

HEISENBERG, W., *The Physical Principles of the Quantum Theory*, Dover Publication (First published by the University of Chicago Press, 1930). The book (although at places mathematical) gives us the unique opportunity of learning the significance of the uncertainty principle from the one who is responsible for its formulation.

* Max Born, *Atomic Physics*, Chapter IX, Blackie (1962).

OPPENHEIMER, J. ROBERT, *The Flying Trapeze: Three Crises for Physicists*, Harper & Row (1969).
The second chapter 'Atom and Field' is relevant to the discussion in this chapter.

WICHMANN, EYVIND H., *Quantum Physics; Berkeley Physics Course*, Vol. 4 (Preliminary edition)
McGraw-Hill (1967). Chapter 4 'Photons' and Chapter 5 'Material Particles' give a
lucid description of the quantum behaviour of photons and other particles. In this book
reference is made to a volume of selected reprints of original papers *Quantum and Statistical
Aspects of Light* published by American Institute of Physics. Several of the papers in this
volume (especially the paper by G. I. Taylor) are interesting and relevant to the discussion
of this chapter.

PROBLEMS

1. (a) An electron of energy 100 eV is passed through a circular hole of radius 5×10^{-4} cm.
    What is the uncertainty introduced in the angle of emergence?
    (b) What would be the corresponding uncertainty for a 200 gm lead ball thrown with a
    velocity 2×10^3 cm/sec through a hole 25 cm in radius?   [(a) 2·6 sec (b) $6·8 \times 10^{-29}$ sec]

2. A linearly polarized monochromatic beam of light is incident on a polaroid whose axis for
    transmitting light is tilted at an angle θ to the direction of polarization of the incident beam.
    Classically, what is the ratio of the transmitted intensity to the incident intensity? What
    does the polaroid do for the case of a single incident photon?

3. In Young's experiment for producing interference pattern (see Fig. 4-8), the circular holes
    S, S_1 and S_2 are each 1 mm in diameter. A 100 watt sodium vapour lamp (which radiates
    uniformly in all directions) is placed at a distance of 500 cm from the hole S. The distance
    $SS_1 = SS_2 = 100$ cm. The screen is also at a distance of 100 cm from the holes S_1 and S_2.
    Assuming the light to be monochromatic with wavelength 5890 Å, calculate the average
    time that elapses between the arrival of one photon and the next. Also find the distance
    between one photon and the next in the beam of light. On an average at any chosen instant,
    how many photons are in the space between the holes and the screen?

4. A microscope of numerical aperture (see Appendix G) 1·25 is focussed on a particle of
    mass 10^{-4} gm. If the wavelength of the illuminating light be 5×10^{-5} cm calculate the
    values of Δp_x and Δx as predicted by Heisenberg's principle.

5. If we measure the position of the electron in a hydrogen atom and if the spread in position
    be of the order a (that is, the distance of the electron from the nucleus is usually about a),
    then the momentum of the electron (which is of the same order as the spread in momentum)
    is roughly $\hbar/a.$ $\left(\hbar = \dfrac{h}{2\pi} \right)$ (a) In the CGS system of units, show that the total energy is

    $$E = \frac{\hbar^2}{2ma^2} - \frac{e^2}{a}$$

    where e is the charge of the electron.*
    (b) Suppose the atom is going to arrange itself to make some kind of compromise so that
    the energy is as little as possible. Show that for a minimum value of E, a has the value

    $$a_0 = \frac{\hbar^2}{me^2} \approx \tfrac{1}{2}\text{Å}$$

    (The result is amazing, because we can get an estimate of the size of the atoms from the
    uncertainty relations. Atoms are completely impossible from the classical point of view,
    since the electrons would spiral into the nucleus.)†

6. The origin of spectral lines are mostly from the decay of an excited state of an atom via

---

* In the MKS system of units (Appendix A) e^2 is the charge of an electron squared divided by
$4\pi\varepsilon_0$.

† See *The Feynman Lectures on Physics*, Vol. III, Addison-Wesley (1965), for a good discussion
on this point.

photon emission to the ground state.* The mean life time, τ, of a state can be assumed to be a measure of the duration of the emission process. (The time, τ is essentially the coherence time defined in Chapter 5, Secs. 5-2 and 5-3.) Associated with a finite value of τ there is always a frequency spread $\Delta\nu$ given by (see Chapter 5)

$$\tau\Delta\nu \sim 1$$

The processes which lead to the above equation are of a special nature and therefore the above relation between τ and $\Delta\nu$ has the form of an equality. Show that for a general oscillatory process

$$\tau\Delta\nu \geqslant 1$$

From above derive the uncertainty relation $\Delta E\,\Delta t \geqslant h$, which can be physically described as follows: If we consider any time dependent process, then the uncertainty ΔE in the energy of the process, and Δt in the 'time at which the process takes place' must be restricted through the inequality $\Delta E\,\Delta t \geqslant h$.

7. Consider a modification of the apparatus shown in Fig. 9-3. Let the wall with the holes consists of a plate mounted on rollers so that it can move freely up and bown (in the x-direction). By watching the motion of the plate carefully we can tell which hole the photon went through.

Discuss qualitatively, that, if we determine the momentum of the plate with sufficient accuracy so that from the recoil measurement we know which hole the photon had gone tnrough then the uncertainty in the X-position of the plate will, according to the uncertaintly principle, be enough to smear out the interference pattern. (This point is discussed in section 1-8 of *Feynman lectures in Physics*, Volume III.)

---

* For definiteness we are discussing atoms but our considerations are completely general and apply equally well to nuclei and molecules.

Some Mathematical Considerations of Wave Motion

A-1   The Wave Equation

We have seen in Chapter 1 that the function $f(x + vt)$ describes a disturbance propagating in the positive direction of x with speed v and $f(x - vt)$ describes a disturbance propagating in the negative direction of x with the same speed v. Although $f(x + vt)$ describes the wave, there is no 'physics' in the description. By this we mean that the behaviour of the medium has not been dictated by any general law of motion or by any conservation law. Such laws enter through the *wave equation*, which governs the system and for which $f(x \pm vt)$ is a solution. The situation is similar to that in elementary mechanics: Newton's laws govern the behaviour of material objects, but the equations of motion describe how they actually move in a particular situation. The wave equation, like Newton's second law, is a differential equation. That is, it states a relationship between one or more derivatives of some function and certain external constants. Thus, the equation

$$F = m \frac{d^2 r}{dt^2} \tag{A-1}$$

specifies how the state of motion of a material object changes in response to external forces, and the (one dimensional) wave equation

$$\frac{\partial^2 f}{\partial x^2} = \frac{1}{v^2} \frac{\partial^2 f}{\partial t^2} \tag{A-2}$$

specifies how the function f changes in the medium as a wave propagates through. The usefulness of the wave equation is that its solutions are always of the form $f(x \pm vt)$; hence whenever we encounter such an equation, we know that waves will result.

That the solution of Eq. A-2 is indeed of the form $f(x \pm vt)$ can easily be shown by making the following transformations:

$$\left.\begin{array}{l} \eta = x - vt \\ \xi = x + vt \end{array}\right\} \tag{A-3}$$

Thus, in terms of the new independent variables ξ and η,

$$\frac{\partial f}{\partial x} = \frac{\partial f}{\partial \xi}\frac{\partial \xi}{\partial x} + \frac{\partial f}{\partial \eta}\frac{\partial \eta}{\partial x}$$

$$= \frac{\partial f}{\partial \xi} + \frac{\partial f}{\partial \eta}$$

$$\frac{\partial^2 f}{\partial x^2} = \frac{\partial}{\partial \xi}\left(\frac{\partial f}{\partial \xi} + \frac{\partial f}{\partial \eta}\right)\frac{\partial \xi}{\partial x} + \frac{\partial}{\partial \eta}\left(\frac{\partial f}{\partial \xi} + \frac{\partial f}{\partial \eta}\right)\frac{\partial \eta}{\partial x}$$

$$= \frac{\partial^2 f}{\partial \xi^2} + 2\frac{\partial^2 f}{\partial \xi \partial \eta} + \frac{\partial^2 f}{\partial \eta^2}$$

Similarly

$$\frac{\partial^2 f}{\partial t^2} = v^2\left[\frac{\partial^2 f}{\partial \xi^2} - 2\frac{\partial^2 f}{\partial \xi \partial \eta} + \frac{\partial^2 f}{\partial \eta^2}\right]$$

Substituting the above expressions for $\dfrac{\partial^2 f}{\partial x^2}$ and $\dfrac{\partial^2 f}{\partial t^2}$ in Eq. A-2, we obtain

$$\frac{\partial^2 f}{\partial \xi \partial \eta} = 0 \tag{A-4}$$

or

$$\frac{\partial F}{\partial \xi} = 0$$

where

$$F(\xi, \eta) = \frac{\partial f}{\partial \eta} \tag{A-5}$$

Obviously, F should be independent of ξ and may, therefore, be an *arbitrary* function of η. Thus,

$$F\left(= \frac{\partial f}{\partial \eta}\right) = F_1(\eta) \tag{A-6}$$

If we integrate Eq. A-6 with respect to η we would obtain

$$f = \int F_1(\eta)\,d\eta + \text{constant}$$

The constant of integration can obviously be an arbitrary function of ξ. Further, the integral of an arbitrary function of η can also be an arbitrary function. Thus, the general solution of the wave equation is

$$f = f_1(\eta) + f_2(\xi)$$

or

$$f(x,t) = f_1(x - vt) + f_2(x + vt) \qquad \text{(A-7)}$$

The former representing a disturbance propagating in the $+x$ direction with speed v and the latter representing a disturbance propagating in the $-x$ direction.

[In three dimensions the differential equation is

$$\frac{\partial^2 f}{\partial x^2} + \frac{\partial^2 f}{\partial y^2} + \frac{\partial^2 f}{\partial z^2} = \frac{1}{v^2}\frac{\partial^2 f}{\partial t^2} \qquad \text{(A-8)}$$

and the most general solution involving plane waves is

$$f = f_1(lx + my + nz - vt) + f_2(lx + my + nz + vt) \qquad \text{(A-9)}$$

in which

$$l^2 + m^2 + n^2 = 1 \qquad \text{(A-10)}$$

One may write other solutions of the wave equation. For example, if we transform Eq. A-8 to spherical co-ordinates r, θ, ϕ then the equation of wave motion becomes

$$\frac{\partial^2 f}{\partial r^2} + \frac{2}{r}\frac{\partial f}{\partial r} + \frac{1}{r^2 \, \text{Sin}\,\theta}\frac{\partial}{\partial \theta}\left(\text{Sin}\,\theta\, \frac{\partial f}{\partial \theta}\right)$$
$$+ \frac{1}{r^2 \, \text{Sin}^2\theta}\frac{\partial^2 f}{\partial \phi^2} = \frac{1}{v^2}\frac{\partial^2 f}{\partial t^2} \qquad \text{(A-11)}$$

If we are interested in solutions possessing spherical symmetry (i.e. independent of θ and ϕ), the above equation simplifies to

$$\frac{\partial^2 f}{\partial r^2} + \frac{2}{r}\frac{\partial f}{\partial r} = \frac{1}{v^2}\frac{\partial^2 f}{\partial t^2} \qquad \text{(A-12)}$$

This may be written as

$$\frac{\partial^2}{\partial r^2}(rf) = \frac{1}{v^2}\frac{\partial^2}{\partial t^2}(rf) \qquad \text{(A-13)}$$

showing (cf. Eqs. A-2 and A-7) that it has solutions

$$rf = f_1(r - vt) + f_2(r + vt) \qquad \text{(A-14)}$$

where f_1 and f_2 again being arbitrary functions. Thus, the solutions are of the form

$$f = \frac{1}{r} f_1(r - vt) + \frac{1}{r} f_2(r + vt) \qquad \text{(A-15)}$$

which represent outgoing and incoming spherical waves; cf. Eq. 1-34.]

The obvious question that now comes to our mind is how did we derive Eq. A-2? The answer is, that for the propagation of a wave under a particular situation, the wave equation has to be derived from physical considerations and from fundamental laws of physics. For example, if we pluck a stretched string, the transverse displacement y, of the particles of the wire can be shown to satisfy the following differential equation:

$$\frac{\partial^2 y}{\partial x^2} = \frac{\rho}{T}\frac{\partial^2 y}{\partial t^2} \qquad (A\text{-}16)$$

where ρ is the mass per unit length of the string and T, the tension in the string. Thus, in this case the speed of propagation of a transverse wave is given by

$$v = \sqrt{\frac{T}{\rho}} \qquad (A\text{-}17)$$

The proof of Eq. A-17 runs as follows:

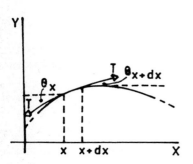

FIG. A-1: Section of a long string under tension T.

Consider a long string which is under tension T (Fig. A-1). The string has been pulled transversely in the y- direction so that a displacement wave travels along the string in the x-direction. We consider a small element dx and apply Newton's second law of motion to it in order to find how the wave moves along the string.

Since ρ denotes the mass per unit length of the string, the mass of the element dx is $\rho\, dx$. The net force in the y- direction acting on this element is (see Fig. A-1):

$$T \operatorname{Sin} \theta_{x+dx} - T \operatorname{Sin} \theta_x \qquad (A\text{-}18)$$

We will consider only small transverse displacements of the string, so that the restoring force will vary linearly with displacement. This implies that θ in Fig. A-1 will be small, so that we may substitute Sin θ with Tanθ. Now Tanθ is simply the slope of the string, i.e. it equals $\partial y/\partial x$. (We must use partial derivatives because the transverse displacement y, depends not only on x but also on t.) The net force in the y-direction is then

$$T\left(\frac{\partial y}{\partial x}\right)_{x+dx} - T\left(\frac{\partial y}{\partial x}\right)_x$$

$$= T\frac{\partial}{\partial x}\left(\frac{\partial y}{\partial x}\right)dx$$

$$= T\frac{\partial^2 y}{\partial x^2}dx \qquad (A\text{-}19)$$

The mass of the element dx of the string is ρdx and its transverse acceleration is $\partial^2 y/\partial t^2$. Hence if we apply Newton's second law to the transverse motion of the string, we would obtain

$$\rho dx \, \frac{\partial^2 y}{\partial t^2} = T \, \frac{\partial^2 y}{\partial x^2} \, dx$$

or

$$\frac{\partial^2 y}{\partial x^2} = \frac{\rho}{T} \, \frac{\partial^2 y}{\partial t^2} \tag{A-20}$$

which is the wave equation, the general solution of which is $y \, (x \pm vt)$.

Another pertinent example is the propagation of sound waves in a gas. For a sound wave propagating in the x- direction, it can be shown that*

$$\frac{\partial^2 \xi}{\partial x^2} = \frac{1}{dp/d\rho} \, \frac{\partial^2 \xi}{\partial t^2} \tag{A-21}$$

where ξ is the displacement of a layer of the gas (along the direction of propagation of the wave) and p and ρ represent the pressure and density respectively. For a perfect gas

$$\frac{dp}{d\rho} = \frac{\gamma p}{\rho} \tag{A-22}$$

where $\gamma \, (= c_p/c_v)$ is the ratio of the two specific heats. Eq. A-21 tells us that the speed of propagation of sound waves in a perfect gas would be given by

$$v = \sqrt{\frac{\gamma p}{\rho}} \tag{A-23}$$

Next, we would like to consider the propagation of electromagnetic waves. However, before we discuss electromagnetic waves, it is necessary to introduce the MKS system of units and Maxwell equations.

A-2 The MKS System of Units

We first note that in the MKS system of units length is measured in metres, mass in kilograms and time in seconds. The units of force and energy are Newtons (abbreviated as nt) and Joules respectively and it can be easily shown that

$$\left.\begin{array}{l} 1 \text{ nt} = 1 \text{ kg m/sec}^2 = 10^5 \text{ dynes} \\ \\ 1 \text{ Joule} = 1 \text{ kg (m)}^2/\text{sec}^2 = 10^7 \text{ ergs} \end{array}\right\} \tag{A-24}$$

and

where dynes and ergs are the corresponding units in the CGS system.

Next, we define the unit of current. It is based on an experimental fact that when two long parallel wires separated by a distance d carry currents i_a and i_b then the two conductors attract each other by a force, the magnitude of which is given by

* See C. A. Coulson, *Waves*, p. 87-90, Oliver and Boyd (1955).

$$F_b = \frac{\mu_0 \, l \, i_b i_a}{2\pi d} \tag{A-25}$$

where F_b is the force experienced by a length l of the wire which carries the current i_b (Fig. A-2). The constant μ_0 will be discussed later.

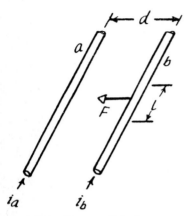

This attraction between long parallel wires is used to define the ampere. Suppose that the wires are 1 metre apart ($d = 1 \cdot 0$ metre) and that the two currents are equal ($i_a = i_b = i$). If this common current is adjusted until, by measurement, the force of attraction per unit length between the wires is 2×10^{-7} nt/metre, the current is defined to be 1 ampere. Thus, if we have

$$i_a = i_b; \quad d = 1 \text{ m}$$

$$\frac{F_b}{l} = 2 \times 10^{-7} \text{ nt/m}$$

FIG. A-2: Two parallel wires carrying currents attract each other.

then

$$i_a = i_b = 1 \text{ amp.}$$

Substituting the above values in Eq. A-25 we get

$$2 \times 10^{-7} \text{ nt/m} = \frac{\mu_0 \, (1 \text{ amp})^2}{2\pi \times (1\text{m})}$$

or

$$\mu_0 = 4\pi \times 10^{-7} \text{ nt/(amp)}^2 \tag{A-26}$$

The quantity μ_0, which is known as the permeability constant in free space, has an assigned value of $4\pi \times 10^{-7}$ nt/(amp)^2.

The MKS unit of charge is the Coulomb (abbreviated as coul). A Coulomb is defined as the amount of charge that flows through a given cross-setion of a wire in 1 second if there is a steady current of 1 ampere in the wire, i.e.

$$q = it \tag{A-27}$$

where q is in Coulombs, i in amperes and t in seconds.

The Coulomb's law in the MKS system of units is then written as

$$F = \frac{1}{4\pi\varepsilon_0} \frac{q_1 q_2}{r^2} \tag{A-28}$$

where F is the magnitude of the force that acts on each of the two charges q_1 and q_2 at a distance r apart. The quantity $\dfrac{1}{4\pi\varepsilon_0}$ is the proportionality constant. In the MKS system we can measure q_1, q_2, r and F in Eq. A-28 in ways that do not depend on Coulomb's law. There is no choice about the so-called *permittivity constant* ε_0; it must have that value which makes the right-hand side of Eq. A-28 equal to the left-hand side. This (measured) value turns out to be

$$\varepsilon_0 = 8 \cdot 85418 \times 10^{-12} \text{ coul}^2/\text{nt}/(\text{m})^2$$

$$\left(\frac{1}{4\pi\varepsilon_0} \approx 9 \times 10^9 \text{ nt } (\text{m})^2/\text{coul}^2 \right)$$

The MKS unit of potential difference* is Joule/coul. This is used so often that a special unit, the volt is used to represent it, i.e.

$$1 \text{ volt} = 1 \text{ Joule/coul}$$

The unit for the electric field is volt/metre which is identical to nt/coul. This is obvious from the fact that in a uniform electric field E the force on a test charge q_0 is $q_0 E$. (Thus, the unit of the electric field can be written as nt/coul.) Further if we assume that the test charge q_0 is moved by an external agent and without acceleration, from A to B along the straight line connecting them (in a direction opposite to that of E) then the work done is

$$W_{AB} = Fd = q_0 Ed$$

or

$$V_B - V_A = \frac{W_{AB}}{q_0} = Ed$$

The above equation shows that the unit for electric field is volt per metre.

Finally, we define the magnetic induction B. We assume that there is no electric field present. Let us fire a positive test charge q_0 with arbitrary velocity v through a point P. If a sideways deflecting force F acts on it, we assert that a magnetic field is present at P and we define the magnetic induction B of this field in terms of F and other measured quantities.

If we vary the direction of v through point P, keeping the magnitude of v unchanged, we find, in general, that although F will always remain at right angles to v, the magnitude of F will change. For a particular orientation of v (and also for opposite direction $-v$) the force F becomes zero. We define this direction as the direction† of B.

Having found the direction of B, we are now able to oreient v so that the test charge moves at right angles to B. We will find that the force F is now a maximum, and we define the magnitude of B from the measured magnitude of this maximum force $F_\perp$, or

---

* The electric potential difference between two points A and B, $V_A - V_B$, is defined

$$V_A - V_B = W_{AB}/q_0$$

where W_{AB} is the work that must be done in moving a test charge q_0 from A to B.

† The direction of B is still not completely specified. Actually the direction of B is such that

$$F = q_0 v \times B$$

v, q_0 and F being the measured quantities. For a more detailed treatment of the definition of magnetic induction see D. Halliday and R. Resnick, *Physics*, Part II Chap. 33, John Wiley (1966).

$$B = \frac{F_\perp}{q_0 v} \tag{A-29}$$

The unit of B that follows from Eq. A-29 is (nt/coul)/(metre/sec). This is given the special name Weber/metre². Recalling that a coul/sec is an ampere,

$$1 \text{ Weber/m}^2 = \frac{1 \text{ nt}}{\text{coul m/sec}} = \frac{1 \text{ nt}}{\text{amp m}}$$

An earlier unit for B, still in common use, is the gauss; the relationship is

$$1 \text{ Weber/m}^2 = 10^4 \text{ gauss}$$

Further, we may write

$$\mu_0 = 4\pi \times 10^{-7} \ nt/(\text{amp})^2 = 4\pi \times 10^{-7} \text{ Weber/amp m}$$

A-3 Maxwell Equations

We would now write down the basic equations of electromagnetism, namely the Maxwell equations. The derivation of these equations will not be given here, since it would involve a rather extensive review of the principles of electricity and magnetism.* Instead we shall merely state the equations in their simplest form, applicable to empty space, and prove that they predict the existence of waves having the properties of light waves.

Maxwell equations may be written as four vector equations, but we shall express them by differential equations. For *vacuum* these become, using a right-handed set of coordinates

$$\frac{\partial E_x}{\partial x} + \frac{\partial E_y}{\partial y} + \frac{\partial E_z}{\partial z} = 0 \tag{A-30}$$

$$\frac{\partial B_x}{\partial x} + \frac{\partial B_y}{\partial y} + \frac{\partial B_z}{\partial z} = 0 \tag{A-31}$$

$$-\frac{\partial B_x}{\partial t} = \frac{\partial E_z}{\partial y} - \frac{\partial E_y}{\partial z} \tag{A-32}$$

$$-\frac{\partial B_y}{\partial t} = \frac{\partial E_x}{\partial z} - \frac{\partial E_z}{\partial x} \tag{A-33}$$

$$-\frac{\partial B_z}{\partial t} = \frac{\partial E_y}{\partial x} - \frac{\partial E_x}{\partial y} \tag{A-34}$$

$$\mu_0 \varepsilon_0 \frac{\partial E_x}{\partial t} = \frac{\partial B_z}{\partial y} - \frac{\partial B_y}{\partial z} \tag{A-35}$$

$$\mu_0 \varepsilon_0 \frac{\partial E_y}{\partial t} = \frac{\partial B_x}{\partial z} - \frac{\partial B_z}{\partial x} \tag{A-36}$$

---

* For the derivation of Maxwell equations, see D. Halliday and R. Resnick, *Physics*, Part II, Supplementary Topic V, John Wiley (1966).

$$\mu_0 \varepsilon_0 \frac{\partial E_z}{\partial t} = \frac{\partial B_y}{\partial x} - \frac{\partial B_x}{\partial y} \tag{A-37}$$

These partial differential equations give the relations in space and time between the vector quantities E and B. Thus E_x, E_y and E_z are the components of E along the three rectangular axes, and B_x, B_y and B_z those of B. The electric and magnetic fields are measured in the MKS system of units.

Eq. A-30 merely expresses the fact that no free electric charges exist in a vacuum. The impossibility of a free magnetic pole gives rise to Eq. A-31. Eqs. A-32 - A-34 express Faraday's law (see Chapter 1) of induced electromotive force (emf). Thus, the quantities occurring on the left side of Eqs. A-32 - A-34 represent the time rate of change of the magnetic field, and the spatial distribution of the resulting electric fields occur on the right side. These equations do not give directly the magnitude of the emf, but only the rates of change of the electric field along the three axes.

The most important contribution of Maxwell in giving these equations was the statement of Eqs. A-35 - A-37. The equations come from an extension of Ampere's law for the magnetic field due to an electric current. The right hand sides of Eqs. A-35 - A-37 give the distribution of magnetic field in space, but the quantities on the left side do not, at first sight, seem to have anything to do with electric current. They represent the time rate of change of the electric field. But Maxwell regarded this as the equivalent of a current, the *displacement current*, which flows as long as the electric field is changing and which produces the same magnetic effects as an ordinary conduction current.

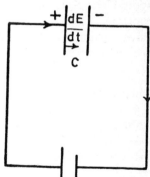

The concept of displacement current can easily be understood if we consider a parallel plate condenser connected to a battery as shown in Fig. A-3. The apparatus is assumed to be in vacuum and with vacuum between the condenser plates. As the condenser gets charged, current flows through the wires, but no current of the ordinary sort flows between the plates of the condenser. By considerations of continuity, Maxwell was led to assume that a changing electric field in

FIG. A-3: Illustrating the concept of displacement current.

the space between the plates is equivalent of a displacement current.

A-4 The Wave Equation for Electromagnetic Waves

Let us consider plane waves propagating in the x-direction, so that the wavefronts are planes parallel to the y-z plane. Thus, at any instant the electric and magnetic fields must be constant over the whole plane, and their partial derivatives with respect to y and z must be zero. Therefore, Eqs. A-30 to A-37 take the form

$$\frac{\partial E_x}{\partial x} = 0 \qquad \text{(A-38)}$$

$$\frac{\partial B_x}{\partial x} = 0 \qquad \text{(A-39)}$$

$$-\frac{\partial B_x}{\partial t} = 0 \qquad \text{(A-40)}$$

$$-\frac{\partial B_y}{\partial t} = -\frac{\partial E_z}{\partial x} \qquad \text{(A-41)}$$

$$-\frac{\partial B_z}{\partial t} = -\frac{\partial E_y}{\partial x} \qquad \text{(A-42)}$$

$$\mu_0\,\varepsilon_0\,\frac{\partial E_x}{\partial t} = 0 \qquad \text{(A-43)}$$

$$\mu_0\,\varepsilon_0\,\frac{\partial E_y}{\partial t} = -\frac{\partial B_z}{\partial x} \qquad \text{(A-44)}$$

$$\mu_0\,\varepsilon_0\,\frac{\partial E_z}{\partial t} = -\frac{\partial B_y}{\partial x} \qquad \text{(A-45)}$$

If we consider Eq. A-38 and Eq. A-43, it appears that the longitudinal component E_x is a constant in both space and time; similarly B_x is also a constant. These components can therefore have nothing to do with the wave motion, but must represent constant fields superimposed on the system of waves. For the waves, we may write

$$E_x = 0, \quad B_x = 0$$

which implies that the waves are transverse.

Further, we notice that Eq. A-41 and Eq. A-45 involve E_z and B_y, while Eq. A-42 and Eq. A-44 involve E_y and B_z. Thus, if we assume that the electric vector has only a y-component (and therefore $E_z = B_y = 0$) we would obtain

$$-\frac{\partial B_z}{\partial t} = \frac{\partial E_y}{\partial x}$$

and

$$\mu_0\,\varepsilon_0\,\frac{\partial E_y}{\partial t} = -\frac{\partial B_z}{\partial x}$$

If we differentiate (partially) the first equation with respect to x and the second equation with respect to t, we would obtain

$$-\frac{\partial^2 B_z}{\partial x\,\partial t} = +\frac{\partial^2 E_y}{\partial x^2}$$

and

$$\mu_0 \, \varepsilon_0 \, \frac{\partial^2 \, E_y}{\partial^2 t} = - \frac{\partial^2 \, B_z}{\partial t \, \partial x}$$

Eliminating the derivatives of B_z, we get

$$\frac{\partial^2 \, E_y}{\partial x^2} = \mu_0 \, \varepsilon_0 \, \frac{\partial^2 \, E_y}{\partial t^2} \tag{A-46}$$

In a similar manner, we find

$$\frac{\partial^2 B_z}{\partial x^2} = \mu_0 \varepsilon_0 \, \frac{\partial^2 B_z}{\partial t^2} \tag{A-47}$$

Both Eqs. A-46 and A-47 have just the form of the wave equation for plane waves (Eq. A-2) with E_y and B_z, respectively playing the part of the displacement f in the two cases. Comparing Eqs. A-46 and A-47 with Eq. A-2 we find that the speed of an electromagnetic wave is given by*

$$c = \frac{1}{\sqrt{\varepsilon_0 \mu_0}} \tag{A-48}$$

and substituting the values of ε_0 and μ_0 we get

$$c = \frac{1}{\sqrt{8 \cdot 8542 \times 10^{-12} \, \dfrac{\text{coul}^2}{\text{nt m}^2} \times 4\pi \times 10^{-7} \, \dfrac{\text{nt}}{(\text{amp})^2}}}$$

Recalling that an ampere is a coul/sec we have $c = 3 \cdot 0 \times 10^8$ m/sec which happens to be the speed of light in free space. Sinusoidal solutions of Eqs. A-46 and A-47 are of the form

$$E_y = E_m \, \text{Sin} \, (kx - \omega t) \tag{A-49}$$

$$B_z = B_m \, \text{Sin} \, (kx - \omega t) \tag{A-50}$$

where

$$\frac{\omega}{k} = c = \frac{1}{\sqrt{\varepsilon_0 \mu_0}}.$$

and it can easily be shown that

$$B_m = \frac{1}{c} \, E_m \tag{A-51}$$

The two waves are interdependent; neither can exist without the other. Both are transverse waves, and are propagated in vacuum with a velocity equal to $1/\sqrt{\varepsilon_0 \mu_0}$.

It is worthwhile pointing out that if we had not made the assumption of plane wavefronts, then a more complicated manipulation of Eqs. A-30 to A-37 would have led to

* If one worked with the Gaussian system of units, one would have obtained the speed of propagation of the electromagnetic waves as the ratio of the electromagnetic unit of current and electrostatic unit of current. (See footnote on page 96)

$$\frac{\partial^2 E_x}{\partial x^2} + \frac{\partial^2 E_x}{\partial y^2} + \frac{\partial^2 E_x}{\partial z^2} = \varepsilon_0\mu_0 \frac{\partial^2 E_x}{\partial t^2} \tag{A-52}$$

$$\frac{\partial^2 E_y}{\partial x^2} + \frac{\partial^2 E_y}{\partial y^2} + \frac{\partial^2 E_y}{\partial z^2} = \varepsilon_0\mu_0 \frac{\partial^2 E_y}{\partial t^2} \tag{A-53}$$

$$\frac{\partial^2 E_z}{\partial x^2} + \frac{\partial^2 E_z}{\partial y^2} + \frac{\partial^2 E_z}{\partial z^2} = \varepsilon_0\mu_0 \frac{\partial^2 E_z}{\partial t^2} \tag{A-54}$$

and three similar equations satisfied by B_x, B_y and B_z which are indeed the general wave equations (cf. Eq. A-8).

A-5   The Superposition Principle

We conclude by noting that the principle of superposition follows directly from the wave equation. Thus, if $f\,(x-vt)$ and $g\,(x-vt)$ are solutions to the wave equation, then their sum is also a solution. Indeed, any linear combination* of solutions is a solution, so that $C_1 f\,(x+vt) + C_2 g\,(x-vt) + C_3\,h(x+vt)$ is a wave too. This statement can easily be proved if we note that the functions $f\,(x+vt)$, $g\,(x-vt)$ and $h\,(x+vt)$ satisfy the following equations:

$$\frac{\partial^2 f}{\partial x^2} = \frac{1}{v^2}\frac{\partial^2 f}{\partial t^2}$$

$$\frac{\partial^2 g}{\partial x^2} = \frac{1}{v^2}\frac{\partial^2 g}{\partial t^2}$$

$$\frac{\partial^2 h}{\partial x^2} = \frac{1}{v^2}\frac{\partial^2 h}{\partial t^2}$$

If we multiply the first equation by C_1, the second by C_2 and the third by C_3 and add we obtain

$$\frac{\partial^2 \Psi}{\partial x^2} = \frac{1}{v^2}\frac{\partial^2 \Psi}{\partial t^2}$$

where
$$\Psi = C_1 f + C_2 g + C_3 h$$

Thus, Ψ also is a solution of the wave equation. This is the basis of the principle of superposition.

SUGGESTED READING

Coulson, C. A., *Waves*, Seventh Edition, Oliver and Boyd, (1955). This book gives a mathematical account of the common types of wave motion. It would be a good text to read for those who want to know more about wave motion.

Halliday, D. and R. Resnick, *Physics, Part II*, John Wiley (1966). This book gives an excellent account of the fundamental principles of electricity and magnetism. Throughout the book MKS system of units are used.

* A linear combination of two functions f and g has to be of the form $c_1 f + c_2 g$ where c_1 and c_2 are constants. Functions like fg or $f^2 + g$ are not linear combinations.

Simple Harmonic Motion

B-1   Periodic Motion

Before we discuss simple harmonic motion, we will define *periodic motion.*
Periodic motion is a motion which repeats itself at regular intervals. A typical
example is the motion of a simple pendulum or of a mass at the end of a
spiral spring (Fig. B-1). The centre of mass in each case makes a to-and
fro-motion along the path *AOB*, taking the same time to describe a

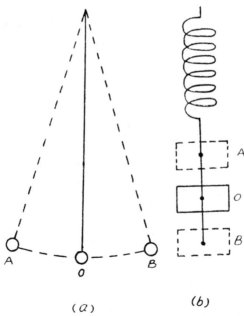

(a)   *(b)*

Fig. B-1:   (a) The periodic motion of a simple
pendulum. The point *O* represents the equilibrium position.
       (b)   The (vertical) periodic motion of
a mass on a spiral spring. The point *O* represents the equilibrium position.

complete cycle from A to B and back again* The *periodic time*, T, is defined as the time taken to execute one vibration. The frequency, ν, of a periodic motion is the number of cycles executed in unit time, i.e.

$$\nu = \frac{1}{T} \qquad \text{(B-1)}$$

B-2 Simple Harmonic Motion

The simplest kind of periodic motion is a *simple harmonic motion*. In order to understand what is simple harmonic motion let us consider a point Q moving around a circle of radius a in the anticlockwise direction with a constant angular velocity ω (Fig. B-2). The angular velocity is expressed in radians per second,† i.e. in one second it rotates through ω radians. It is evident that the angle formed by the radius vector of the moving point with the x-axis will vary with time; in fact in time t the radius vector will rotate through an angle ωt.

After circumscribing the whole circle, the movement will repeat in exactly the same manner. Thus, it is a periodic motion. The period of this motion will be the time, T, required to travel the path of the whole circle, i.e. to rotate through $360°$ ($= 2\pi$ radians). Since the point rotates through ω radians in one second, the time taken for making a complete rotation will be $2\pi/\omega$ seconds. Thus,

$$T = \frac{2\pi}{\omega} \qquad \text{(B-2)}$$

and using Eq. B-1, we obtain

$$\omega = 2\pi\nu \qquad \text{(B-3)}$$

Let P be the foot of the perpendicular drawn from Q, on one of the diameters, say YY', of the circle (Fig. B-2). As Q moves once around the circle in the anticlockwise direction from X, the foot of the perpendicular moves from O to Y, Y to Y' and back to O. The motion will repeat itself after every T seconds.

Let the point Q start from the point X at time $t = 0$. Then in time t the point Q will rotate through an angle ωt and the projection on the y-axis will be given by

$$y = a \, \text{Sin} \, \omega t \qquad \text{(B-4)}$$

where a is the radius of the circle. The angular distance ωt through which

---

* In actual practice, however, we know that due to the presence of frictional forces, the bob of the pendulum (and also the mass on the spring) will eventually come to rest. We are neglecting the presence of such forces.

† The radian is a unit to measure angles, and

$$\pi \text{ radians} = 180°$$

or   1   radian $= 180/\pi = 57 \cdot 2958° = 57° \, 17' \, 44''$

and $1° = 0 \cdot 0174533$ radians.

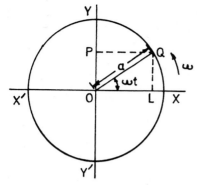

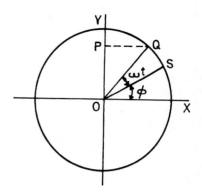

FIG. B-2: The point Q is rotating on the circumference of a circle (of radius a) with constant angular velocity ω. The point P executes simple harmonic motion.

FIG. B-3: Simple harmonic motion with an initial phase ϕ.

the radius vector is displaced from OX is the *phase* of the motion at the instant t. According to Fig. B-2, the phase is zero at $t = 0$.

However, instead of measuring time as above, i.e. $t = 0$ when Q is at X, we may take, without any loss of generality, $t = 0$ when Q is at the point S (Fig. B-3). Thus after t seconds the radius vector will make an angle $\omega t + \phi$ with the x-axis where $\phi = \angle SOX$. The distance OP is now given by

$$y = a \text{ Sin } QOX$$
$$= a \text{ Sin } (\omega t + \phi) \tag{B-5}$$

The phase of the motion at the instant t is now $(\omega t + \phi)$, and the initial phase (i.e. at $t = 0$) is ϕ.

A motion is said to be simple harmonic if the displacement from the equilibrium position varies with time according to Eq. B-5. *Thus, if a point is moving on the circumference of a circle with uniform angular velocity then its projection on any of its diameters executes simple harmonic motion.* The quantity a is called the *amplitude* of the motion.

The velocity and acceleration of the point P are readily obtained by differentiating Eq. B-5. We have

$$y = a \text{ Sin } (\omega t + \phi)$$

Therefore, the velocity, v, and acceleration, f, are given by

$$v = \frac{dy}{dt} = a\omega \text{ Cos } (\omega t + \phi) \tag{B-6}$$

$$f = \frac{d^2y}{dt^2} = -a\omega^2 \text{ Sin } (\omega t + \phi) \tag{B-7}$$

From Eqs. B-5 and B-7 we obtain

$$\frac{d^2y}{dt^2} = -\omega^2 y \qquad (B-8)$$

Table I gives the displacement, velocity and acceleration of a particle, which is executing simple harmonic motion, at various times. The initial phase has been assumed to be zero. A similar table can easily be prepared for $\phi \neq 0$. Table I shows that when the magnitude of the displacement is maximum the velocity is zero and the direction of acceleration is such that it tends to return the particle to its equilibrium position.

TABLE I

DISPLACEMENT, VELOCITY AND ACCELERATION OF A PARTICLE EXECUTING
SIMPLE HARMONIC MOTION WITH ZERO INITIAL PHASE ($\phi = 0$).

Time t	Displacement y	Velocity v	Acceleration f
0	0	$+a\omega$	0
$\frac{1}{4}T$	a	0	$-a\omega^2$
$\frac{1}{2}T$	0	$-a\omega$	0
$\frac{3}{4}T$	$-a$	0	$+a\omega^2$
T	0	$+a\omega$	0

Thus, the negative sign in Eq. B-8 indicates that the acceleration is always in a direction opposite to the displacement and tends to return the system to its equilibrium position. Eq. B-8 therefore leads to another way of defining simple harmonic motion as *the motion of particle in a straight line in which the acceleration of the particle is proportional to its displacement from the equilibrium position and oppositely directed.*

In Fig. B-4(a) is shown the variation of the displacement with time having an initial phase equal to ϕ. The corresponding variation for zero intial phase is shown in Fig. B-4(b). Both the curves show sinusoidal variation of the displacement with time.

B-3 Examples of Simple Harmonic Motion

Example 1

Let us consider a mass m sliding on a frictionless surface. It is connected to rigid walls by means of two identical springs, each of which has zero mass, spring constant* k, and relaxed length l_0 [Fig. B-5(a)]. At the equilibrium position, each spring is stretched to length l, and thus, each spring has tension $k(l - l_0)$ at equilibirum position [Fig. B-5(b)]. Let the mass be displaced from the equilibrium position by a distance x [Fig. B-5(c)]. Then its distance from the left hand wall is $(l + x)$ and

* If a spring has a spring constant k then it implies that if it is stretched by a distance Δx, the tension on the spring equals $k\Delta x$.

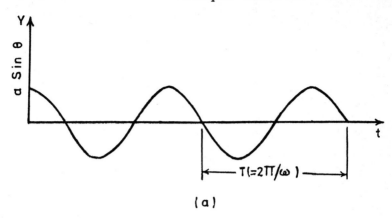

(a)

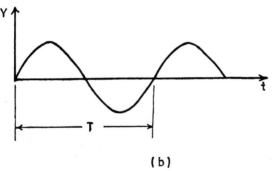

(b)

Fig. B-4: Sinusoidal variation of y with time T, denotes the time period; (a) corresponds to initial phase ϕ and; (b) corresponds to zero initial phase.

from the right hand wall is $(l - x)$ The left hand spring exerts a force, $k (l + x - l_0)$ in the $-x$ direction. The right hand spring exerts a force $k (l - x - l_0)$ in the $+x$ direction. The net force, F_x, in the $+x$ direction will be given by:

$$F_x = k (l - x - l_0) - k (l + x - l_0)$$
$$= - 2 kx \tag{B-9}$$

Newton's second law then gives

$$m \frac{d^2x}{dt^2} = - 2kx$$

or

$$\frac{d^2x}{dt^2} = - \omega^2 x \tag{B-10}$$

where

$$\omega = \sqrt{\frac{2k}{m}}$$

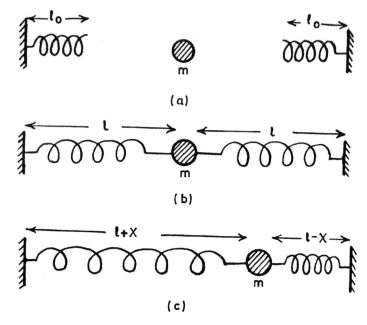

FIG. B-5: Oscillations of a mass m.

The above equation is of the same form as Eq. B-8 and therefore the mass will execute simple harmonic motion with a time period T given by

$$T = 2\pi \sqrt{\frac{m}{2k}} \qquad \text{(B-11)}$$

One can alternatively say that the general solution of Eq. B-10 is of the form

$$x = a \text{ Cos } (\omega t + \phi) \qquad \text{(B-12)}$$

which represents simple harmonic oscillations.

Example 2

A common example of approximate simple harmonic motion is the motion of a simple pendulum, which consists of a bob of mass m on the end of a string of length l (Fig. B-6). We will describe the displacement of the mass from its equilibrium position by vector s.

In order to show that the motion of the mass is simple harmonic, we must show that there is a linear restoring force, a force which is proportional to the negative of the displacement. As we see in Fig. B-6, the weight of the body mg can be broken into two rectangular components. One component is taken along the string and simply keeps the string pulled tight; the other component is perpendicular to the string and supplies a restoring force. The magnitude of this restoring force, F_s, is given by

$$F_s = mg \, \text{Sin}\theta \qquad \text{(B-13)}$$

Now, as long as s is small compared to l, we may approximately write

$$\text{Sin } \theta \approx \frac{s}{l}$$

or

$$F_s \cong \frac{mg}{l} \, s$$

Also the direction of **s** is very nearly parallel to F_s but in the opposite direction. Therefore,

$$F_s = -\frac{mg}{l} \, s \qquad \text{(B-14)}$$

Fig. B-6: The simple pendulum.

Thus, we have a linear restoring force of the form $F_s = -k \, s$ necessary to give simple harmonic motion. Eq. B-14 is of the same form as Eq. B-9 and therefore the time period of oscillation is given by

$$T = 2\pi \sqrt{\frac{l}{g}} \qquad \text{(B-15)}$$

Eq. B-15 shows that to the degree of approximation involved, the period depends only upon the length of the pendulum and the acceleration due to gravity. The period is independent of the mass of the bob and also of the amplitude when the amplitude is small. If the amplitude is large, the period is greater than that for small amplitudes. For a maximum angular displacement of 10° the error is about 0·2%; for 30° about 1·7%.

Example 3
Finally, consider the motion of a stretched string, clamped rigidly at its ends and vibrating in, say, its fundamental mode [see Fig. 1-23 top]. Now at any instant, the string will be in the form shown in Fig. B-7, and

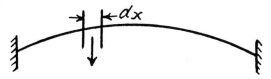

Fig. B-7: A vibrating string.

an element such as dx will experience a resultant force in the direction indicated by the vertical arrow, towards the equilibrium position; the force being proportional to the distance of the element from its equilibrium position. Thus, each element will execute simple harmonic motion.

It should be pointed out that the amplitude and initial phase do not appear in any of the above discussions the values of these will depend on the way in which the body is set in motion by the experimenter. The frequency of the motion is determined entirely by the nature of the oscillating system; but its amplitude and initial phase are decided arbitrarily by the experimenter. This fact is also obvious from the mathematics of the motion, for Eq. B-8 can be solved to give

$$y = a \text{ Sin } (\omega t + \alpha)$$

in which a and α are the arbitrary constants of integration.

Further, in the examples that have been discussed above, there has been an actual body having a definite mass, which has been moving in space, and the displacement has been a distance that the mass has moved from a given position. But a periodic phenomenon may be called simple harmonic even though there is no massive body; the motion of the spot of the light on the scale of a ballistic galvanometer, for example, is approximately simple harmonic. Again, it is not necessary that the quantity, y, which satisfies Eq. B-8 should be a distance. For example, associated with light waves are electric and magnetic fields, the components of which satisfy Eq. B-8 (see Appendix A).

B-4 Phase Difference

We have defined the phase of a simple harmonic motion (at a given instant) as the value of the angle QOX (Fig. B-2). It really denotes at what stage of the cycle the particle has reached a given point. If the phase is zero, the particle is at the position of zero displacement *and moving towards Y*, if it is $\pi/2$ the particle is at the position of maximum positive displacement and moving towards Y' and so on.

The *phase difference* between two simple harmonic motions is an important quantity and is the difference in phase of the two motions at any instant. It will, of course, remain constant if the motions have the same periodic time. If the phase difference ·between two simple harmonic motions is π, one will be at the position of maximum positive displacement while the other ·is at the position of maximum negative displacement or one will be at the point of zero displacement moving in one direction, while the other is at the same place moving in the opposite direction. If the phase difference is $\pi/2$, one is at the position of zero displacement and moving towards Y, while the other is at Y' i.e. the former is always a quarter of a cycle behind the latter. *The phase difference indicates to what extent the two vibrations are out of step with each other;* if it is zero, they are in step; if it is π, they are exactly out of step, if it is $\pi/2$, they are half out of step and so on. The phase difference cannot be more than 2π; for a phase difference of 4π or $9\pi/2$ is identical to saying zero phase difference or $\pi/2$ phase difference respectively.

Let us consider a few examples.

Example 1

Consider a simple pendulum [Fig. B-8(*a*)] long enough so that the bob can be assumed to be moving in a straight line. Thus, the instantaneous displacement of the bob from the equilibrium position can be (approximately) written as

$$x(t) = a \text{ Sin } \omega t \qquad \text{(B-16)}$$

where a is the amplitude and is the maximum displacement of the mass, and $T(=2\pi/\omega)$ the time after which the mass acquires an identical position and velocity.

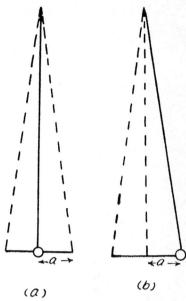

(*a*)   (*b*)

FIG. B-8: Instantaneous positions of two identical simple pendulums vibrating with a phase difference of $\pi/2$.

Eq. B-16 assumes that we have set our clock in such a way that $t = 0$ the particle passes through the equilibrium position. We could have equally well chosen, without any loss of generality, our zero of time in such a way that $x(t)$ would have been given by

$$x(t) = a \text{ Sin } (\omega t + \pi/4) \qquad \text{(B-17)}$$

or

$$x(t) = a \text{ Sin } (\omega t - \pi/2) \qquad \text{(B-18)}$$

Thus, in this case we can have an arbitrary value of ϕ (the initial phase) by properly choosing the 'zero time'.

Next, let us consider two similar pendulums [Figs. B-8 (*a*) and (*b*)] vibrating in such a way that when the first particle passes through the equilibrium position moving towards right then the other particle is at its extreme right position. Thus, if the instantenous position of the first particle is represented by

$$x_1(t) = a \text{ Sin } \omega t \qquad \text{(B-19)}$$

then the position of the other particle, measured from its equilibrium position will be given by

$$x_2 = a \text{ Cos } \omega t$$
$$= a \text{ Sin } (\omega t + \pi/2) \qquad \text{(B-20)}$$

We could have equally well chosen

$$x_1 = a \text{ Sin } (\omega t + \pi/4)$$

but, then x_2 would be given by

$$x_2 = a \, \text{Sin} \, (\omega t + \pi/4 + \pi/2)$$

Thus, the two particles are at a phase difference of $\pi/2$; or, more precisely, the second particle is ahead of phase by $\pi/2$ with respect to the first particle. *This phase difference cannot be removed by any choice of zero time.*

Example 2

Let us consider a circuit containing a resistance through which an alternating current is flowing (Fig. B-9). For this circuit the relation between current and voltage is the same as for direct current. Thus, in Fig. B-9,

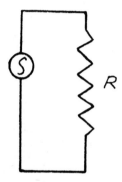

FIG. B-9: Alternating current passing through a resistance, R.

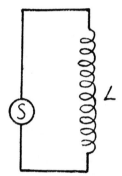

FIG. B-10: Alternating current pasing through an inductance, L.

if I is the instantaneous current in a resistor R when the instantaneous voltage across it is V, we can write, at any instant

$$V = IR$$

and if the voltage is varying sinusoidally, we have

$$\left.\begin{array}{l} V = V_0 \, \text{Sin} \, (\omega t + \phi) \\ I = I_0 \, \text{Sin} \, (\omega t + \phi) \end{array}\right\} \tag{B-21}$$

with

$$V_0/I_0 = R$$

At most places the power supply is 50 cps implying

$$\omega = 2\pi \times 50$$

$$= 100\pi \text{ radians/sec}$$

(The direct current, say from a battery, corresponds to $\omega = 0$.) Eq. B-21 tells us that the voltage and current are in phase.

On the other hand if a coil having a very small resistance is connected, then as the current changes an induced electromotive force is set up in the coil which is in the opposite direction to that driving the current. It can be shown that for the circuit of Fig. B-10, containing a coil of inductance L, the voltage and current are related by

$$\left.\begin{aligned} V &= V_0 \operatorname{Sin} (\omega t + \phi) \\ I &= I_0 \operatorname{Sin} (\omega t + \phi - \pi/2) \end{aligned}\right\} \qquad \text{(B-22)}$$

with

$$\frac{V_0}{I_0} = \omega L$$

This relation between current and voltage is illustrated in Fig. B-11 for the case $\phi = 0$; the current lags on the voltage by a quarter cycle, or a phase difference of $\pi/2$.

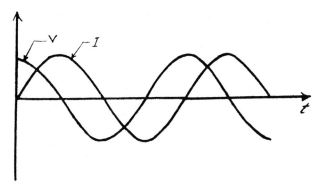

Fig. B-11: The time variation of the voltage and current in the circuit of Fig. B-10.

SUGGESTED READING

CRAWFORD, FRANK S. (Jr.), *Waves: Berkeley Physics Course*, Volume 3, McGraw-Hill (1968). The first chapter discusses the free oscillations of simple systems and contains some interesting examples which have been worked out fully.

RESNICK, R. and D. HALLIDAY, *Physics, Part I*, John Wiley (1966). Chapter 15 'Oscillations' has discussed simple harmonic motion in considerable detail.

Electromagnetic Oscillators

C-1 Introduction

Electromagnetic waves originate in the oscillation of electric charge. The necessary vibratory motion may be produced most simply by means of the Hertzian oscillator described in Chapter 3. Here an oscillatory discharge takes place across the gap (see Fig. 3-2), the charges on the knobs reversing sign every half period. Essentially we have equal positive and negative charges vibrating along the line joining the knobs in simple harmonic motion, the charge of the one sign being out of phase with that of the other by π radians. The mechanism is equivalent to an oscillating electric doublet. We will suppose that the oscillations are maintained at constant amplitude and shall assume that the region surrounding the oscillator is empty space.

C-2 Radiation from an Oscillating Electric Doublet

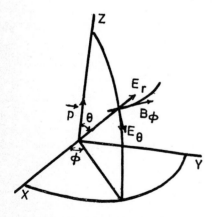

FIG. C-1: Components of electric and magnetic field about an oscillating electric doublet.

Consider a positive charge oscillating with a frequency $\omega/2\pi$ along a section (Fig. C-1) of the z-axis. If a_1 is the amplitude of vibration, the displacement of the charge at any instant is given by

$$z_1 = a_1 \cos \omega t \qquad \text{(C-1)}$$

The displacement of an equal negative charge vibrating with the same frequency but out of phase by π radians, with the first is given by

$$z_2 = a_2 \cos (\omega t + \pi) = - a_2 \cos \omega t$$

where a_2 represents the amplitude of the second charge.

At any instant, then, the distance between the two charges is

$$\mathcal{Z} = \mathcal{Z}_1 - \mathcal{Z}_2 = (a_1 + a_2) \text{ Cos} \omega t \qquad \text{(C-2)}$$

The oscillations of the two charges constitute an oscillating electric doublet. The electric moment of the doublet is

$$p = q\mathcal{Z} = p_0 \text{ Cos } \omega t \qquad \text{(C-3)}$$

where

$$p_0 = q (a_1 + a_2)$$

The electric and magnetic fields due to this doublet can be shown* to be given by

$$E_r = A_1 \text{ Cos} \theta \left\{ \frac{2}{r^3} \text{ Cos } (kr - \omega t) + \frac{2k}{r^2} \text{ Sin } (kr - \omega t) \right\} \qquad \text{(C-4)}$$

$$E_\theta = A_1 \text{ Sin} \theta \left\{ \left(\frac{1}{r^3} - \frac{k^2}{r} \right) \text{ Cos } (kr - \omega t) + \frac{k}{r^2} \text{ Sin } (kr - \omega t) \right\} \qquad \text{(C-5)}$$

$$E_\phi = 0 \qquad \text{(C-6)}$$

$$B_r = B_\theta = 0 \qquad \text{(C-7)}$$

$$B_\phi = A_2 \frac{k}{r} \text{ Sin} \theta \left\{ k \text{ Cos } (kr - \omega t) - \frac{1}{r} \text{ Sin } (kr - \omega t) \right\} \qquad \text{(C-8)}$$

where r, θ and ϕ are the polar co-ordinates of the point at which the electric and magnetic fields are being calculated (Fig. C-1). Further $k = 2\pi/\lambda$, $\omega/k = c$, A_1 and A_2 are constants and in the MKS system of units (see Appendix A)

$$A_2 = \frac{1}{c} A_1 \qquad \text{(C-9)}$$

where c is the speed of light in free space. It should be noted that $\boldsymbol{E}$ is at right angles to $\boldsymbol{B}$.

At great distances from the origin we need retain only those terms which involve $1/r$. Hence E_r has a negligible magnitude compared to E_θ, and we may write

$$E_\theta = - C_1 \frac{k^2 \text{ Sin} \theta}{r} \text{ Cos } (kr - \omega t) \qquad \text{(C-10)}$$

and

$$B_\phi = - C_2 \frac{k^2 \text{ Sin} \theta}{r} \text{ Cos } (kr - \omega t) \qquad \text{(C-11)}$$

At this distance from the doublet a small section of the wavefront is approximately plane and the electric field becomes transverse† to the

---

* See, for example, the references given at the end of this Appendix.

† The transverse property of the wave motion at great distances is common to the fields of all electromagnetic sources, but we must not overlook the fact that in the vicinity of the source there is a longitudinal component in the direction of propagation (Eq. C-4).

direction of propagation. (The magnetic field is always transverse to the direction of propagation.) Eqs. C-10 and C-11 indicate the following important facts:

(i) E_θ is in phase with B_ϕ.

(ii)   Both E_θ and B_ϕ are zero along the axis of the dipole on account of the factor $\mathrm{Sin}\theta$, and are maximum in the equitorial plane. The radiation, therefore, is emitted from the doublet in greatest intensity in direction at right angles to the line of oscillation.

(iii)   Both fields fall off inversely with the distance r. Thus the intensity will obey the inverse square law.

(iv)   Both the fields are inversely proportional to the square of the wavelength, or, directly proportional to the square of the frequency.

(v)   The radiation emitted is polarized with E in the r-z plane and B at right angles thereto.

SUGGESTED READING

PAGE L. and N. I. ADAMS, *Principles of Electricity*, Chapter XVI, D. Van Nostrand and Co. (1955). A beautifully written text but uses Gaussian system of units.

REITZ J. R. and F. J. MILFORD, *Foundations of Electromagnetic Theory*, Chapter 16, Addison-Wesley (1962).

Lasers

The invention of laser* has opened up completely new fields of development in optics. Since the first operation of a laser by Maiman in 1960, the magnitude and quality of scientific and engineering effort which has been devoted to this area has not been equalled since the war-inspired efforts on radar and the atomic bomb. This is due to the fact that the laser beam has certain features which are not present in other light sources. In order to understand this, let us consider how ordinary light beams are produced. All light sources, incandescent lamps, arcs and so on, are essentially hot matter. The atoms are continuously 'pumped' to an excited state. A short time later ($\approx 10^{-8}$ seconds) spontaneous emission takes place and the excited atom becomes normal again by emitting a photon of the same frequency as that which was originally absorbed but in a random direction and with a random phase. Thus, the light that comes from any conventional light source is spatially incoherent (see Chapter 5).

However, the main physical process involved in the emission of a laser beam is the phenomenon of stimulated emission which was predicted by Einstein in 1917. Now, when a photon is absorbed by an atom the energy of the photon is converted to internal energy of the atom. The atom is then raised to an excited quantum state. Later, it may radiate this energy spontaneously, emitting a photon and reverting to the ground state or to some state in between. During the period in which the atom is still excited, it can be stimulated to emit a photon, if it is struck by an outside photon having precisely the energy as the one that would otherwise be emitted spontaneously. *In this stimulated emission (Fig. D-1) the photon forces the excited atom to emit another photon of the same frequency in the*

---

* Laser is an acronym meaning 'Light Amplification through Stimulated Emission of Radiation'. The laser operates at optical frequencies (i.e. corresponding to the visible region). The working of a MASER is also on the same principles but it operates at microwave frequencies (i.e. corresponding to wavelengths of the order of centimetres). The maser principle was discovered by Townes in 1954 and the first operating maser was completed by Gordon in 1955. The possibility of operating devices at light frequencies was pointed out by Townes and Schawlow in 1958; and in 1960, Maiman produced the first operating laser.

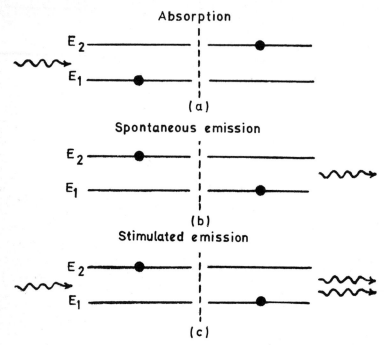

FIG. D-1: (a) When an atom in the ground state absorbs a photon (wavy arrow), it is excited, or raised to a higher energy state.

(b) The excited atom may then radiate energy spontaneously, emitting a photon and reverting to the ground state.

(c) An excited atom can also be stimulated to emit a photon when it is struck by an outside photon. Thus, in addition to the stimulating photon there is now a second photon of the same wavelength and the atom reverts to the ground state.

same direction and in the same phase. The two photons go off together as *coherent* radiation.

Thus, in a stimulated emission one photon will give rise to two photons. Obviously, a chain reaction producing light amplification will be possible only if more than half of these photons go on to stimulate further emission. However, these photons may be absorbed by atoms which happen to be in the ground state. Light amplification will therefore be achieved if there are more atoms in the higher energy state than in the lower energy state. This situation is called population inversion.*

The first laser constructed by Maiman in 1960 was made from a single cylindrical crystal of ruby with its ends ground flat and silvered (Fig. D-2). Ruby consists of aluminum oxide with some of the aluminum atoms replaced by chromium. The relevant energy levels of a chromium

---

* In different types of lasers the problem of obtaining a population inversion is solved in different ways.

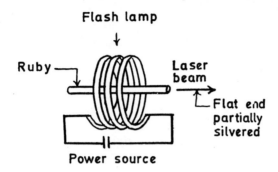

FIG. D-2: Ruby laser is powered by a flash lamp, which
provides optical pumping. Output beam is emitted
through partially silvered end of ruby crystal; other
end is completely silvered.

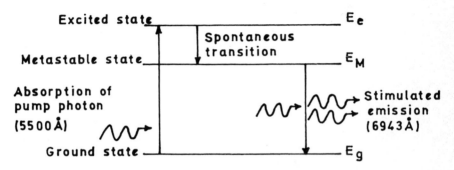

FIG. D-3: Energy levels of a chromium atom involved in a ruby laser.

atom* are shown in Fig. D-3. The half life of normal atomic states is of the
order of 10^{-8} seconds (See Problem 6 of Chapter 9). However the state
designated as E_M in Fig. D-3 has a half life of about 3×10^{-3} seconds
almost a million times greater than the half life of most excited states of
atoms. Such a long-lived state is termed as *metastable state*. In the ruby
laser, atoms in this metastable state are stimulated by red photons (with
$\lambda = 6943$ Å) to make a transition to the ground state and emit another
identical photon.

The population of chromium atoms in the metastable state is made
momentarily much greater than the population in the ground state by
a method known as *optical pumping*. The ruby crystal is illuminated by
a bright flash of light whose wavelength is 5500 Å. These photons are
absorbed by chromium atoms in the ground state which produces transi-

---

* The corresponding energy diagram for a helium-neon laser is discussed in 'Optical Masers'
by A. Schawlow, *Scientific American*, Vol. 204, p. 52 June, 1961.

tion to an excited state shown as E_e in Fig. D-3. After a time lapse of about 10^{-8} seconds, the atoms in the excited state E_e undergo a spontaneous transition to the metastable state E_m. (It just happens that the probability of the excited state E_e making a direct transition to the ground state is very small.)

With a majority of the chromium atoms in their metastable state, a photon with a wavelength 6943 Å, emitted by one atom in a spontaneous transition to its ground state, will stimulate similar transitions in other chromium atoms giving rise to light amplification.

In order to increase the probability of photon interaction and to achieve a unidirectional beam, one exploits reflection from flat parallel ends. The light that is perpendicular to these ends will make repeated traversals of the crystal causing substantial amplification. At the partially silvered end about 1% of the incident photons escape and constitute the emerging laser beam. In the original Maiman's experiment this was a flash of red light lasting 0·3 millisecond with a peak power of 10^4 watts.

Applications

The net effect of all the processes taking place in a laser tube is a beam of radiation that is (1) very intense, (2) almost perfectly parallel, (3) almost monochromatic and (4) spatially coherent. Because of the last two properties interesting interference experiments can be performed with a laser. These have already been discussed in Chapters 5 and 7.

The laser beam being almost perfectly parallel, the spreading is due to the phenomenon of diffraction. This beam can therefore be focussed by a lens in a region whose linear dimension will be of the order of $f\lambda/a$ where f is the focal length of the lens and a the aperture of the beam. Since $f \approx a$, the area of the region where the laser can be focussed is of the order of λ^2. Because a laser beam can be brought to such a sharp focus, it can be used as an exceedingly effective 'drill' to burn through a target* (See Example in Chapter 3, p. 107 bottom).

The highly directional properties of a laser have also been used for determining distances (like that of the moon) with great accuracy.

SUGGESTED READING

SCHAWLOW, ARTHUR L. 'Optical Masers', *Scientific American*, Vol. 204, June 1961. An excellent non-mathematical account of lasers put forward by the person who (along with C. H. Towne) had first described optical masers.

THUMM, W. and D. E. TILLEY, *Physics: A Modern Approach*, Addison-Wesley(1970). Chapters 9 and 10 are relevant for the concepts developed in this chapter.

Lasers and Light (Published by G. H. Freeman and Co.). This book is a collection of 32 articles from *Scientific American*. The last 13 papers are on lasers and their applications. These articles would be an excellent guide to the students who want to know about the properties and applications of laser light.

* Certain types of lasers can produce temperature of about 8000°C at the focal point of a focussed beam.

Appendix E

Optical Activity

If a plane polarised light is incident on a material which consists of molecules which do not have reflection symmetry* then a very interesting effect is observed, viz., as the beam passes through the substance, the direction of polarization rotates about the beam axis. This effect is known as *optical activity*. We will try to give a qualitative explanation of this phenomenon.[†]

Consider an asymmetric molecule in the shape of a spiral as shown in Fig. E-1. (Molecules need not actually be shaped like a corkscrew, in order to exhibit optical activity, but this is a simple shape which we will consider as a typical example of the molecules which do not have reflec-

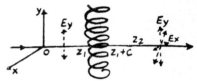

Fig. E-1: A beam of linearly polarized ligh falls on a molecule, which is shaped like a corkscrew, having no reflection symmetry.

tion symmetry.) Let a linearly polarized light beam, with the E-vector oscillating in the y-direction, fall on such a molecule (Fig. E-1). The electric field associated with the wave will drive charges up and down the helix, thereby generating a current in the y-direction and radiating an electric field E_y polarized in the y-direction. However, if the electrons are constrained to move along the spiral, they must also move in the x-direction as they are driven up and down. When a current is flowing up the spiral, it is also flowing into the paper at $z = z_1$ and out of the paper at $(z = z_1 + C)$, where C is the diameter of the spiral. At first sight, it might appear that the current in the x-direction would produce no net radiation, since the currents are in opposite diections on opposite sides of the spiral. However, if we consider the x-components of the electric field arriving at $z = z_2$, we see that the field radiated by the current at $z = (z_1 + C)$ and the field radiated from $z = z_1$ arrive at

---

* Molecule something like a corkscrew, which if viewed through a mirror, would be reversed in the same way that a left-hand glove reflects as a right-hand glove.

† The present discussion is adopted from '*The Feyman Lectures on Physics*', R. P. Feyman, R. B. Leighton and M. Sands, Vol. I, Section 33-5, Addison Wesley (1965).

355

$z = z_2$ separated in time by the amount a/c, and thus separated in phase by $(\pi + \omega C/c)$. Since the phase difference is not exactly π, the two fields do not cancel exactly and one is left with a small (but finite) x-component in the electric field generated by the motion of the electrons in the molecule, whereas the driving electric field has only a y-component. This small x-component, added to the large y-component, produces a resultant field which makes a small angle with respect to the original direction of polarization. As the wave propagates through the medium, the direction of polarization rotates about the beam axis. It can be shown that the existence of optical activity and the sign of rotation are independent of the orientation of the molecules.

Sugar solution is a common example which possesses optical activity. The phenomenon is easily demonstrated with a polariod sheet to produce a linearly polarized beam, a tube containing sugar solution, and a second polariod sheet to detect the rotation of the direction of polarization as the light passes through the sugar solution.

SUGGESTED READING

FEYNMAN, R. P., R. B. LEIGHTON and M. SANDS, *The Feynman Lectures on Physics* Vol. I, Chapter 33, Addison-Wesley (1965)

FOWLES, G. R. *Introduction to Modern Optics*, Chapter 5, Holt, Rinehart and Winston (1968).

Resolving Power of Microscope

If we want to see two spots which are very close to each another, then one might think that all we have to do is to build a microscope (see Chapter 2, Sec. 2-22) with enough magnification. Further, with proper design one can eliminate all the spherical and chromatic aberrations and there is no reason why we can not keep on magnifying the image. However, although we can build a system of lenses with a very high magnification, yet, we still may not be able to see two points that are too close together because of the limitations of geometrical optics.

With a microscope the object is very close to the objective, and the latter subtends a large angle $2i$ at the object plane as shown in Fig. F-1.

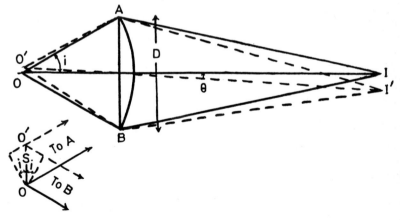

FIG. F-1: Formation of image by a microscope objective.

Suppose we neglect all aberrations, and imagine that for a particular object O (or O') all the rays take exactly the same time to travel from the object to the image I (or I'). Thus, according to geometrical optics, there should be two point images. However, due to the phenomenon of

357

diffraction each image will consist of a disk and a system of rings (see Fig. 6-11) and the two images can be said to be just resolved if the principal maximum of one falls on the first minimum of the other (the Rayleigh's criterion).

Now, while considering the diffraction at a circular aperture (Chapter 6, Sec. 6-16) we had mentioned that the first minimum occurs for an angle of diffraction equal to $1·22\lambda/D$ where D is the diameter of the circular aperture. Thus, for the first minimum the path difference between extreme rays must be* $1·22\lambda$. Hence, for the two points O and O' to be just resolved, the first minimum due to the object at O must fall at I' and, this will happen when the extreme rays OAI' and OBI' differ in path by $1·22$ λ. If O lies on the axis

$$OA = OB$$

and, therefore for

$$OAI' - OBI' = 1·22\lambda \tag{F-1}$$

we must have

$$AI' - BI' = 1·22\lambda \tag{F-2}$$

But I' is the image of the point O', therefore

$$O'A + AI' = O'B + BI'$$

or

$$AI' - BI' = O'B - O'A \tag{F-3}$$

From the inset in Fig. F-1

$$OA - O'A = S \; Sin \; i$$
$$O'B - OB = S \; Sin \; i$$
$$O'B - O'A = 2S \; Sin \; i \tag{F-4}$$

where S is the distance $O'O$. Thus, for the images to be just resolved (using Eqs. F-2, F-3 and F-4) we must have

$$2S \; Sin \; i = 1·22\lambda$$

Or, the smallest distance between two points O and O' (which will produce images I and I' that are just resolved) will be given by

$$S = \frac{1·22\lambda}{2 \, Sin \, i} \tag{F-5}$$

In the above derivation we have assumed that the points O and O' were self luminous objects, such that light given out by each has no constant phase relationship with the other. In actual practice, the objects used in microscopes are illuminated with light from a condenser, and therefore the light scattered by two points are not independent in

---

* On the other hand, for the single slit the first minimum corresponds to a path difference of λ between extreme rays (see Fig. 6-5).

phase. Consequently, the problem becomes considerably complex because the resolving power will then depend also on the mode of illumination of the object. Abbe worked on this problem in considerable detail and concluded that a satisfactory working rule for calculating the resolving power was given by Eq. F-5, omitting the factor 1·22. Further, if the space between the object and the objective is filled with oil (of refractive index μ) then the expression for resolving power would become

$$S = \frac{\lambda}{2\ \mu \, \mathrm{Sin}\ i} \tag{F-6}$$

The product $\mu \mathrm{Sin}\ i$ is characteristic of a particular objective and is called the 'numerical aperture'.* Thus for $\lambda = 5 \times 10^{-5}$ cm, the resolving power would be about $1·5 \times 10^{-5}$ cm. In order to have a better resolution one often uses ultraviolet light, which, because of its shorter wavelength, permits finer details to be examined than would be possible for the same microscope operated with visible light. In an *electron microscope*,† the electron beam may have an effective wavelength of the order of 5×10^{-10} cm which is shorter by almost a factor of 10^5 than visible light ($\lambda \approx 5 \times 10^{-5}$ cm). (An electron beam, under some circumstances, behaves like waves as discussed in Chapter 8.) This permits the detailed examination of tiny objects like viruses.‡ If a virus is examined with an optical microscope, its structure would be concealed by diffraction.

---

* In practice the largest value of $\mu \mathrm{Sin}\ i$ obtainable is about 1·6.

† See, for example, V. K. Zworkyin *et al. Electron Optics and the Electron Microscope*, John Wiley (1945).

‡ The wavelength of the electron waves depend on the voltage through which they have been accelerated (from acceleratng potential from 100 to 10000V, λ varies from about 10^{-8} cm to 10^{-9} cm). Further, it is possible by means of electric and magnetic fields to focus the electron beam and in this way details comparable to the wavelength of the electron can be photographed.

Miller Indices

The position and orientation of a crystal plane is determined by giving the co-ordinates of three non-colinear atoms lying in the plane. However, in the analysis of X-ray photographs and study of crystal structures it turns out to be more useful to specify the orientation of a plane by *Miller indices*, which are determined as follows:

(1) Find the intercepts on the three basis axes in terms of the lattice constants.

(2) Take the reciprocals of these intercepts and reduce to the smallest three integers having the same ratio. The result is enclosed in parentheses (*hkl*). The numbers *h*, *k* and *l* are the Miller indices of the plane and also of any other plane parallel to it.

For the purpose of illustration let us consider the plane shown in Fig. G-1. This cuts the co-ordinate axes with intercepts 3, 2 and 4 respectively

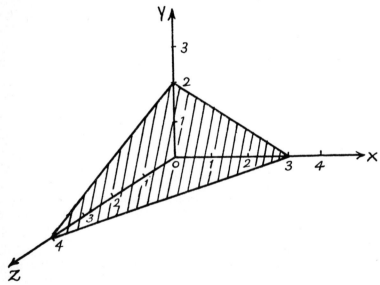

FIG. G-1

in terms of the lattice constants. If we multiply by the common divisor 12, we get the required Miller indices, viz., (463). It can be easily shown by the repitition of the procedure that the plane which has the intercepts 6, 4 and 8 respectively has also (463) as Miller indices. So Miller indices represent a family of parallel planes.

If an intercept is at infinity, the corresponding index is zero. The Miller indices of the planes shown in Fig. 6-26(*i*) and (*ii*) are (100) and (110) respectively.

SUGGESTED READING

AZAROFF, L. V. *Elements of X-ray Crystallography*, McGraw-Hill Book Co (1968).

Author Index

Adams, N.I., 105, 350
Airy, G., 226
Arons, A.B., 60
Azaroff, L.V., 361

Bacon, G.E., 294
Baker, A., 172, 322
Ballard, S.S., 122, 141
Ballik, E.A., 207, 314
Barkla, C.G., 128
Beams, J.W., 280
Beiser, A., 14
Bird, G.R., 120
Bohr, N., **318**
Bond, W.L., 207, 314
Born, M., 52, 65, 202 209, 296, 298, **310**, 322
Borowitz, S., 14
Bragg, L., 253, 260
Bragg, W., 255
Brewster, D., 125

Cantore, E., 322
Cerenkov, P.A., 49
Compton, A.H., 283
Coulson, C.A., 40, 329, 336
Crawford, F.S., 40, 222, 347

Davisson, C., 293
De Broglie, L., 293
Debye, P.J.W., 257
Dibdin, F.J.H., 63
Dirac, P.A.M., 311, 322
Ditchburn, R.W., 31

Einstein, A., **282**, 351

Feinberg, G., 322
Fermat, P., 65
Feynman, R.P., 49, 66, 90, 95, 117, 134, 141,
    211, 299, 312, 321, 323, 324, 355
Fizeau, M., 96
Forrester, A.T., 207, 313
Fowles, G.R., 209, 218, 261

Frank, I.M., 49
Frank, N.H., 314
French, A.P., 314
Fresnel, A., 46, **50**

Gabor, D., 265
Gamow, G., 322
Germer, L.H., 293
Gordon, J.P., 351
Gudmundsen, R.A., 207, 313

Halliday, D., 140, 141, 216, 331, 332, **336, 347**
Heirtzler, J.H., 111, 141
Heisenberg, W., **316**, 318, 322
Henry, G.E., 108
Hertz, H., **98**
Hoffmann, B., 296
Hudson, A.M., 314
Hull, A.W., 257
Huygens, C., **45**, 90, 95, 298

Inglis, S.J., 103

Javan, A., 207, 314
Jenkins, F.A., 61, 141, 188, 214, 260, 273
Johnson, P.O., 207, 313

Kerr, J., 134

Laue, Max von, 257
Lawrence, E.O., 280
Leighton, R.B., 49, 66, 90, 134, 141, 299,
    321, 323, 324, 355
Leith, E.N., 266, 276
Lipsett, M.S., 207, 208, 314
Lipson, H., 209
Lipson, S.G., 209
Lloyd, H., 171
Longhurst, R.S., 186, 188, 246

Magyar, G., 155, 198
Maiman, H., 351
Malus, E., 122

Mandel, L., 155, 198, 207, 208, 209, 314
Marganeau, H., 321
Maxwell, J.C., **17**, 95, 298
Metherell, A.F., 276
Michelson, A.A., **184,** 204
Milford, F.J., 350
Miller, D.C., 360

Newton, Isaac, 44, 90, **182,** 295

Oppenheimer, J.R., 323

PSSC (Physical Science Study Committee), 40, 44, **90**
Page, L., 105, 350
Pantell, R.H., 209
Parrish, M., 120
Pennington, K.S., 273, 276
Planck, M., 282
Prutkoff, K., 1

Rayleigh, Lord, 233, 252
Reitz, J.R., 350
Resnick, R., 140, 141, 216, 331, 332, **333,** 347
Rose, A., 314

Sands, M., 49, 66, 90, 134, 141, 299, 321, 323, 324, 355
Schawlow, A.L., 209, 298, 351, 353; 354
Scherrer, P., 257
Schrödinger, E., 90
Schuster, A., 53
Shamos, M.R., 103

Shull, C.G., 294, 295, 296
Shurcliff, W.A., 122, 141
Sladkova, J., 188
Smith, F.D., 76, 90
Smith, H.M., 276
Snell, W., 59
Sommerfeld, A., 100

Tamn, I.Y., 49
Taylor, G.I., 306, 315
Thomson, G.P., 293
Thumm, W., 354
Tilley, D.E., 354
Townes, C.H., 351
Towne, D.H., 127, 209, 230

Upatnieks, J., 266, 276

Verbiest, R., 276

Waldron, R.A., 40
Webb, R., 260
Weisskopf, V.F., 90, 296
White, H.E., 61, 141, 214, 260, 273
Wichmann, E.H., 296, 299, 323
Wolf, E., 52, 65, 202, 209, 314
Wood, R.W., 51, 52, 61

Young, H.D., 24, 132
Young, T., **24,** 46, 95, 156

Zworkyin, V.K., 359

Subject Index

Aberrations, 75
  chromatic, 76, 88
  spherical, 76
Acoustical hologram, 275
Ampere, definition of, 330
Airy disk, 229
Angular resolution of human eye, 235
Anomalous refraction, 135
Antinodes, 28, 102

Beats, 39
  optical, 205
Birefringence, 128
  applications of, 131
Body centred cubic lattice, 253
  X-ray diffraction from, 256, 259, 263
Bow waves, 48
Bragg condition, 255
Bragg reflection, 255, 287
Brewster's law, 125
  interpretation of, 126

Cerenkov radiation, 49
Chromatic aberration, 76, 88
Circularly polarized wave, 117, 133, 140
Coherence, 193
  spatial, 199, 272
  temporal, 194
Coherence length, 195
Coherent sources, 154
Compton effect, 283
Compton scattering, 283, 320
Conical wavefront, 48
Corpuscular theory, 44
Coulomb's law, 330

De Broglie wavelength, 293
Debye-Scherrer rings, 258, 260
Diffraction, 50
  by circular sperture, 54, 226
    table of intensities, 218
  by circular obstacle, 50, 54

  by double slit, 235
  Fraunhofer, 212
  Fresnel, 213
  by N slits, 243
  by single slit, 213
    table of intensities, 218
    quantum theory of, 316
  of neutrons, 294
  of X-rays, 253
Diffraction grating, 247
  dispersion of, 248
  resolving power of, 251
Diffraction limited beam, 223
  angular width of, 222
Diffraction limited laser beam, 229
Diffraction limited optics, 229
Diffuse reflection, 62
Dipole, oscillating, 21, 348
Dispersion of light, 89
Double refraction, 135
Double slit diffraction pattern, 21

Einstein's photon theory, 282
Electromagnetic oscillators (see oscillating dipole)
Electromagnetic spectrum, 111
Electromagnetic theory, basic consequences of, 97
  and Maxwell equations, 332
Electromagnetic unit of current, 96
Electromagnetic waves, 17, 95
  circularly polarized, 117
  describing the, 112
  detection of, 98
  elliptically polarized, 117
  energy of, 105
  generation of, 21, 98
  intensity of, 105
  momentum of, 108
  reflection of, 100, 103
  refraction of, 103
  representation of, 97
  standing, 31, 100

superposition of, 113
wave equation for, 333
Electrostatic unit of current, 96
Elliptically polarized wave, 117, 133, 140
Energy transport in wave motion, 15, 105
Equation of wave motion, 4, 325
from wave equation, 326
Exit pupil, 83
Extraordinary ray, 136
Eye, angular resolution of, 235

Face centred cubic lattice, 253
X-ray diffraction from, 259, 263
Fermat's principle, 65, 94
and laws of reflection, 65
and laws of refraction, 66
applications of, 67
Fraunhofer diffraction, 212
Frequency, 7, 338
Fresnel's biprism, 163
Fresnel diffraction, 213
Fresnel's mirror arrangement, 170
Fringewidth, 152, 159
Full wave plate, 131

Geometrical optics, 64
applications of, 69
limitations of, 232, 292
Grating (see diffraction grating)

H-sheet, 122
Half period zones, 51
Half wave plate, 131
Heisenberg's uncertainty principle, 316
Hertz's experiments, 98
Hologram, 265, 271, 273
acoustical, 275
recording of, 268
reproduction of, 267
three dimensional, 268
two dimensional, 268
Holography, 264
principle of, 266
requirements of, 272
theory of, 268
with sound, 275
Huygens-Fresnel principle, 50
Huygens' principle, 47
and law of reflection, 57
and law of refraction, 58
Huygens' theory, 46

Intensity, of an electromagnetic wave, 105
of a laser beam, 107

of a wave, 16
Intensity distribution,
of circular aperture diffraction pattern, 228
of double slit diffraction pattern, 237
of N-slit diffraction pattern, 244
by rectangular aperture, 225
of single slit diffraction pattern, 216
quantum theory of, 316
of interference patterns, 161, 299, 301, 303
Interference,
constructive and destructive, 149, 154, 161, 162
in thin films, 174
of two photons, 313
white light, 164
Interference experiment,
with bullets, 299
with electrons, 303
with photons, 303
with water waves, 301
Interference pattern, production of, 147, 156, 163, 170, 184, 299
Interference with white light, 164
Ives' experiment, 32

Kerr effect, 134

Lasers, 351
applications of, 354
principle of, 351
ruby, 353
Laser beam, diffraction limited, 223
intensity of, 107
photography with, 267
spatial coherence of, 204
temporal coherence of, 198
Laue pattern, 257, 258
Lenses, compound (see also thin lenses), 74
Lloyd's mirror arrangement, 171

Magnifying glass, 79
magnification of, 79
Malus' law, 121, 122
Maser, 351
Matter waves, 293, 309
probabilistic interpretation of, 310
Maxwell equations, 332
Metre, standardisation of, 186
Michelson interferometer, 184, 195, 290
explanation of the interference pattern by photon hypothesis, 311
uses of, 186
Michelson's stellar interferometer, 202
Microscope, compound, 80

electron, 359
gamma ray, 320
magnification of, 81
resolving power of, 356
Miller indices, 256, 360
Mirage, 67, 93
MKS system of units, 329

Neutron diffraction, 294
Newton's rings, 182
applications of, 183
Nodal lines, 149
Nodes, 28, 102, 146
Non-reflecting films, 180
Normal spectrum, 249
Numerical aperture, 359

Optic axis, 129, 136
Optical activity, 355
Optical beats, 205
Optical instruments, 78
Ordinary ray, 136
Oscillating dipole, 21
radiation from, 348

Particle waves, 293, 309
Permeability constant, 105, 330, 335
Permittivity constant, 105, 331, 335
Phase, 12, 152, 339
Phase change on reflection, 172
Phase difference, 153, 344
Photoelasticity, 134
Photoelectric effect, 277
explanation by photon theory, 282
Photoelectrons, 278
Photography, by coherent light, 264
by laser, 267
ordinary, 265
by wave front reconstruction, 268
Photons, 282
and electromagnetic waves, 291
energy of, 282
intensity of, 288
interference of, 305, 313
momentum of, 283
rest mass of, 289
splitting of, 289, 312
watching the, 307
Planck's constant, 282
Plane polarized waves, 23, 98, 117
Polarized light,
analysis of, 140
production of, 119, 124, 127
transverse character of, 121

Polaroid, 122
Powder method, 253
Principal maxima, 245
width of, 251
Probabilistic interpretation of matter waves,
310
Pulse, electromagnetic, 22
Gaussian, 7
semicircular, 6, 25
triangular, 40, 41

Quarter wave plate, 131

Radiation, particle nature of, 277
Radiation pressure, 110
Ray treatment of light, 64
Rayleigh's criterion, 233, 252
Rectilinear propagation of light, 50, 55
Reflection at a broken surface, 61
Reflection by a plane mirror, 62
Refraction, by a spherical surface, 63, 70
Resolution, limit of, 231
for optical instruments, 232
Resolving power,
of a grating, 251
of a microscope, 357
of optical instruments, 232
of a telescope, 232
Ruby laser, 353

Secondary maxima, 245
Shock wavefront, 49
Simple cubic lattice, 253
Simple harmonic motion, 337
examples of, 340
Simple pendulum, 342
Single slit diffraction pattern, 213, 223
quantum theory of, 316
Sinusoidal waves, superposition of, 32
Snell's law, 59
Sound waves, 13, 16
diffraction of, 225
Spatial coherence, 199, 272
Spectral line, purity of, 196
Specular reflection, 57
Spherical aberration, 76
Splitting of photons, 282, 312
Spontaneous emission, 352
Standing waves, 27, 31, 100
Stationary light waves, 31
Stationary waves (see standing waves)
Stellar interferometer, 202
Stimulated emission, 352
Stopping potential, 279

Superposition, principle of, 23, 336
  of two sinusoidal waves, 32
  of many simple harmonic vibrations, 35, 38
  of nearly equal frequencies, 38

Taylor's experiment, 306, 315
Telescopes, astronomical, 81
  exit pupil of, 83
  magnifying power of, 82
  normal magnification of, 84
  reflecting, 84
Temporal coherence, 194
  determination of, 195
Thin lens equation, 73
Time period, 9

Uncertainty principle, 316
Uncertainty relation,
  position and momentum, 316, 318, 320
  time-energy, 324
  time-frequency, 324
Unpolarized light, 118

Vibrating string, 27, 327, 343
Volt, definition of, 331

Wave, composition of, 23
  electromagnetic, 17
  in one dimension, 2
  intensity of, 16
  longitudinal, 13
  on a rope, 2, 145
  plane polarized, 23, 98
  propagation of, 20

reflection of, 24
spherical, 16
standing, 27
sinusoidal, 7
transverse, 13
Wave equation, 325
  for electromagnetic waves, 333
  general solution of, 326
  for sound waves, 329
  spherically symmetric solution of, 327
  and superposition principle, 336
  for a vibrating stretched string, 328
Wave-particle duality, 293
Wavefront reconstruction photography, 268
Weber, definition of, 332
Work-function, 297
Water waves, 13
  interference of, 147, 301
Wavefront, 46
Wavelength, 7, 148
  determination of, 150, 163, 186
Wavelets, secondary, 47
Wiener's experiment, 31
Wire grid polarizer, 119

X-rays, diffraction of, 253
  powder method, 257
  electromagnetic character of, 111
  scattering of, 127, 286
X-ray spectrometer, 287

Young's interference experiment, 136, 200

Zone-plate, 55